The physics of active galactic nuclei, the origin of extragalactic jets and the formation of extended extragalactic radio sources are among the most interesting challenges of modern astrophysics and are at the beginning of a new flowering. This book contains the proceedings of the 7th meeting of the Institut d'Astrophysique de Paris which drew together both observers and theorists in this exciting field. In order to obtain a global picture ranging from the reservoir of jet material around the central engine to the extended lobes, observational specialists of extragalactic radio sources and specialists of plasma physics and high-energy physics present the new developments in their fields. Recent observational data at X-ray, optical and radio wavelengths are discussed and new theoretical developments concerning beam and jet formation models are considered. Special treatment is given to plasma physics problems relating to particle acceleration, magnetic reconnection, beam–plasma interaction and coherent emission.

Extragalactic Radio Sources – From Beams to Jets

Extragalactic Radio Sources – From Beams to Jets

7th IAP Meeting
Paris, 2–5 July 1991

Extragalactic Radio Sources – From Beams to Jets

Proceedings of the 7th IAP Meeting
held at the
Institut d'Astrophysique de Paris
Paris, France
2–5 July 1991

Edited by
J. ROLAND
Institut d'Astrophysique de Paris
H. SOL
Observatoire de Paris–Meudon
and
G. PELLETIER
Observatoire de Grenoble

CAMBRIDGE
UNIVERSITY PRESS

Published by the Press Syndicate of the University of Cambridge
The Pitt Building, Trumpington Street, Cambridge CB2 1RP
40 West 20th Street, New York, NY 10011-4211, USA
10 Stamford Road, Oakleigh, Victoria 3166, Australia

First published 1992

Printed in Great Britain at the University Press, Cambridge

A catalogue record for this book is available from the British Library

Library of Congress cataloguing in publication data available

ISBN 0 521 41602 7 hardback

CONTENTS

I. PHYSICAL CONDITIONS AROUND THE CENTRAL ENGINE

Preface

The physics of active galactic nuclei, the origin of extragalactic jets and the formation of extended extragalactic radio sources are among the most interesting challenges of modern astrophysics and are at the beginning of a new flowering. After the improvements in data reduction techniques in the early eighties, spectacular radio maps of extended and compact radio sources were obtained using linked interferometers and VLBI arrays. New fields of crucial investigations such as polarimetric and millimetric VLBI observations have been opened. Together with high spatial, spectral and temporal resolution in the X-ray, UV and optical ranges, they provide a wealth of new information on the physical conditions in the cores of extragalactic radio sources and in the inner regions of the jets. Moreover, the observations of the extended radio lobes and kiloparsec scale jets provide important constraints on the plasma ejected by active nuclei of galaxies. In order to obtain a global picture ranging from the reservoir of jet material around the central engine to the extended lobes, it seemed opportune to organize a meeting with observational specialists of extragalactic radio sources and specialists of plasma physics and high energy physics.

The 7th meeting of the Institut d'Astrophysique de Paris gathered an assembly of about eighty participants in an informal atmosphere, maintaining a balance between observers and theorists in order to favour a comparison of different points of view. Recent observational data at X-ray, optical and radio wavelengths were presented and new theoretical developments concerning beam and jet–formation models were discussed. Special consideration was given to plasma physics problems related to particle acceleration, magnetic reconnection, beam-plasma interaction and coherent emission.

We thank all the participants and speakers who contributed towards making this meeting an important vehicle for the communication and exchange of ideas in these fields.

Many thanks are also due to Y. Bousquet, V. Demailly, C. Douillet, D. Lopes, B. Raban, M. Aubier and C. Aureau for their help before and during the conference and for their help in preparing the proceedings.

Jacques Roland
Hélène Sol
Guy Pelletier

Paris, Décembre 1991

Remerciements

Ce colloque a été organisé par l'Institut d'Astrophysique du CNRS (Paris)

Les organisateurs remercient vivement pour leur aide les organismes suivants:

INSTITUT D'ASTROPHYSIQUE DE PARIS

MINISTERE DE LA RECHERCHE ET DE LA TECHNOLOGIE

INSTITUT NATIONAL DES SCIENCES DE L'UNIVERS

CENTRE D'ETUDE DE SACLAY: SERVICE D'ASTROPHYSIQUE

OBSERVATOIRE DE PARIS-MEUDON

UNIVERSITE PARIS VI

UNIVERSITE PARIS VII

THOMSON CSF

LIST OF PARTICIPANTS

M. Abramowicz	NITF, Nordita, Copenhagen (Danemark)
U. Achatz	MPIfR, Bonn (Germany)
C. Akujor	NRAL, Jodrell Bank (UK)
H. Aller	University of Michigan, Ann Arbor (US)
M. Aller	University of Michigan, Ann Arbor (US)
S. Appl	Landessternwarte, Heidelberg (Germany)
L. Bååth	Onsala Space Observatory, Onsala (Sweden)
D. Batuski	University of Maine, Orono (US)
G. Benford	University of California, Irvine (US)
V. Beskin	Lebedev Institut, Moscow (SU)
G. Bicknell	Mt Stromlo Observatory, Canberra (Australia)
A. Black	Cavendish, Cambridge (UK)
K. Blundell	Cavendish, Cambridge (UK)
C. Boisson	Observatoire de Paris, Meudon (France)
W. Brinkmann	MPIfEP, Garching bei Munchen (Germany)
M. Camenzind	Landessternwarte, Heidelberg (Germany)
T. Cawthorne	Harvard College Observatory, Cambridge (US)
P. Charlot	Observatoire de Paris, Paris (France)
M. Cohen	Caltech, Pasadena (US)
R. Conway	NRAL, Jodrell Bank (UK)
C. Fanti	Istituto di Radioastronomia, Bologna (Italy)
R. Fanti	Istituto di Radioastronomia, Bologna (Italy)
V. Fedorenko	Physico-Technical Institut, Leningrad (SU)
J. Ferreira	Observatoire de Grenoble, Grenoble (France)
R. Fosbury	ESO, Garching bei Munchen (Germany)
D. Fraix-Burnet	STSI, Baltimore (US)
K.D. Fritz	MPIfK, Heidelberg (Germany)
D. Gabuzda	JPL, Pasadena (USA)
G. Ghisellini	Osservatorio di Torino, Pino Torinese (Italia)
R. Goodrich	Caltech, Pasadena (US)
L. Gurvits	Lebedev Institut, Moscow (SU)
G. Henri	Observatoire de Grenoble, Grenoble (France)
Ya. Istomin	Lebedev Institut, Moscow (SU)
T. Jones	University of Minnesota, Minneapolis (USA)
T. Kirchner	ZIfA, Potsdam (Germany)
J. Kirk	MPIfK, Heidelberg (Germany)
B. Komberg	Lebedev Institut, Moscow (SU)
A. Königl	University of Chicago, Chicago (US)
J. Leahy	NRAL, Jodrell Bank (UK)
F. Lebrun	CE Saclay, Service d'Astrophysique (France)

R. Lehoucq	CE Saclay, Service d'Astrophysique (France)
H. Lesch	MPIfR, Bonn (Germany)
R. Liu	Cavendish, Cambridge (UK)
R. Lovelace	Cornell University, Ithaca (US)
F. Macchetto	STSI, Baltimore (US)
J.L. Masnou	Observatoire de Paris, Meudon (France)
S. Massaglia	Istituto di Fisica, Torino (Italia)
A. Matthews	Observatoire de Paris, Meudon (France)
P. Miller	Cavendish, Cambridge (UK)
R. Morganti	Istituto di Radioastronomia, Bologna (Italy)
R. Mutel	University of Iowa, Iowa City (US)
G. Pelletier	Observatoire de Grenoble, Grenoble (France)
J. Perry	IOA, Cambridge (UK)
P. Petitjean	IAP, Paris (France)
M. Pohl	MPIfR, Bonn (Germany)
S. Rawlings	Cavendish, Cambridge (UK)
E. Rokaki	IAP, Paris (France)
J. Roland	IAP, Paris (France)
M. Romanova	Lebedev Institut, Moscow (SU)
H.J. Roeser	MPIfA, Heidelberg (Germany)
F. Rosso	Observatoire de Grenoble, Grenoble (France)
R. Saunders	Cavendish, Cambridge (UK)
P. Scheuer	Cavendish, Cambridge (UK)
R. Schilizzi	Radiosterrenwacht, Dwingeloo (The Netherlands)
R. Schlickeiser	MPIfR, Bonn (Germany)
A. Sillanpää	Tuorla Observatory, Piikkio (Finland)
K. Sokolov	Institut of Radioastronomy, Kharkov (SU)
H. Sol	Observatoire de Paris, Meudon (France)
R. Spencer	NRAL, Jodrell Bank (UK)
P. Véron	Observatoire de Haute Provence, Saint-Michel (France)
M. de Vries	Sterrekundig Instituut, Utrecht (The Netherlands)
S. Wagner	Landessternwarte, Heidelberg (Germany)
A. Witzel	MPIfR, Bonn (Germany)
A. Zdziarski	Copernicus Astronomical Center, Warszawa (Poland)

Importance of Plasma Physics Processes in Extragalactic Radio Sources

V. FEDORENKO

Ioffe Physical Technical Institute, St Petersbourg, Russia

1 INTRODUCTION

The observed non-thermal emission from the extragalactic radio sources (ERS) is an evidence of high-energy particles, produced in these objects. On the other hand, the depolarization of radio emission indicates the presence of a high-temperature, dilute plasma. From the radio maps we conclude that this plasma is strongly violent due to such moving agents like jets. Therefore, the plasma should be turbulent, and the existence of high-energy particles can in principle be explained by some mechanism of their acceleration. Hence, the situation reminds, to some extent, the problem of cosmic ray origin. However, it is even more difficult and uncertain, because our knowledge of ERS is much more poor than that of our Galaxy. The term *particle acceleration* normally means that we want to produce high-energy particles with *isotropic* velocity distribution in the rest reference frame of a thermal plasma (like in the case of cosmic rays). Perhaps, this concept is adequate in the case of extended radio lobes or some compact ERS. Here we can apply the theoretical methods, developped in the cosmic-ray physics. On the other hand, the production of *jets* and *beams* in ERS is also a problem of *particle acceleration*. However, in the latter case it is necessary to organize a *strongly-collimated* motion of charged particles. The latter requires the working-out of some special models, taking into account gravity, rotation and magnetic fields.

There exist many uncertainties in this problem. Usually it is believed that the observed large-scale radio jets are the continuation of the parsec-jets, or in other terms, the beams [3]. The superluminal expansion of the beams indicates their *relativistic velocities*. However, this can be due only to *a small fraction* of the flowing plasma, whereas the dominant massflow of the jets is non-relativistic [19,21]. Therefore, we must be careful and distinguish between the jets and the beams. I hope that there will be a lot of discussion on this subject during the Colloquium.

The temporal frame of my talk does not allow me to elucidate all problems of ERS, which can be treated with the help of the plasma theory. In this talk I will limit myself to the discussion of the relativistic particle acceleration with emphasis to the diffusive shock acceleration mechanism.

2 POSSIBLE MECHANISMS OF PARTICLE ACCELERATION IN EXTRAGALACTIC RADIO SOURCES

Accepting the synchrotron hypothesis, we conclude that the power law radio spectra of ERS are produced by relativistic electrons with *power energetic spectra* :

$$N(E) \sim E^{-\gamma} \quad (1)$$

where $2 \leq \gamma \leq 3$ for a great number of ERS [3]. Therefore, one can conclude that the theory must produce some *universal mechanism* of particle acceleration [14]. Here we must be careful, because to some extent, it is a matter of a credo.

In the case of plasma turbulent acceleration like Fermi-II type, it seems impossible to achieve such an universality, since the spectral index in eq (1), depends on the combination of some unrelated physical parameters. The same is true for any type of plasma turbulence [14].

Only in the case when one considers the back-reaction of accelerated particles on the turbulence, which accelerates them, one can anticipate to obtain some universal spectra. Actually, this idea has been realised by Kaplan and Tsytovich in their theory of Plasma Turbulent Reactor [14]. The applicability of this theory to ERS has been carefully analysed by Norman and ter Haar [18]. They found that the former cannot work in the extended ERS and very questionably in the compact ERS.

In the years 1977-78 the cosmic-ray astrophysicists community received a valuable gift: a discovery of the so-called *diffusive shock acceleration mechanism* [2, 4, 7, 15]. A great advantage of the latter was that it naturally produced the *power-law spectra* like eq. (1) for *any sort of relativistic particles*, with

$$\gamma = \frac{\mathrm{r} + 2}{\mathrm{r} - 1} \quad (2)$$

where r is a compression ratio of thermal plasma on the shock transition. In the case of high-Mach number adiabatic shocks, we have $r = 4$, so $\gamma = 2$ according to eq. (2). Strong shocks are very typical in astrophysical objects, so it seemed very plausible that the desired universal mechanism of particle acceleration has been found. However, the subsequent development of the theory of diffusive shock acceleration posed many questions, so that now we cannot claim that this mechanism really works in the remote astrophysical objects.

3 CURRENT STATUS OF THE DIFFUSIVE SHOCK ACCELERATION MECHANISM

There exist several exellent reviews on the present state of the theory of the considered mechanism e.g. [6, 22]. Here I would like to make a brief outline of some problems of the theory.

One of them, still unsolved, is the problem of *the efficiency of the acceleration*, i.e. the fraction of energy of the onflowing plasma on the shock wave, which is going to the energy of accelerated particles.

In other terms, this is called the *injection problem* [6, 22]. There has been several attempts to solve it by introduction of the so-called *scattering operator*, which uniformily described the structure of shock transition and the scattering of accelerated particles. The latter is a necessary condition of the particle acceleration [6, 22]. However, such a concept seems to be questionable, since in the case of collisionless shock waves, which are intensively studied by the plasma physicists, the situation is much more complicated [9]. Particularly, the boosting of the suprathermal particle distribution function, i.e. the *tails*, is regulated by turbulent acceleration inside the shock transition, which is not described by the scattering operator. It is evident that the process of boosting of the tails is just the injection mechanism for diffusive shock acceleration [22]. Generally, there exist some gap between the studies of collisionless shocks and the diffusive shock acceleration.

The scattering operator should be calculated self–consistently with the process of particle acceleration and the formation of the shock transition, rather than assumed to be given [22]. Some attempts in their direction have been undertaken in papers [5, 12, 16, 17, 23]. The idea was that in the shock upstream region the accelerated particles could themselves generate the Alfven waves, which scatter them. Of course, in this case some mechanism of particle injection is necessary. Therefore, we can imagine such a self-sustaining regime of particle acceleration. One problem is that we do not know how our system comes to such a regime, and whether it is stable or not. The level of the Alfven wave turbulence explicitely depends on the number of accelerated particles, which is regulated by the injection mechanism [12]. This level can become so high that the calculations within the context of the weak plasma turbulence theory break down [12, 16, 23]. Therefore, we are not sure that our current calculations of the above-mentioned regime are correct. There also arises a problem of MHD-turbulence in the shock downstream region. Here the turbulence probably can be produced due to the so-called *firehose instability* [1]. However, we still do not have reliable calculations of its nonlinear stage, so that we cannot say anything about its level and spectrum.

The latter is rather pitiful, since the efficiency of the particle acceleration strongly depends on the properties of the MHD-turbulence in the up and downstream regions. Particularly, the time τ_a of boosting of the particle spectrum (1) from some injection energy E_i to some maximal energy E_{max} can be estimated so [6, 22]:

$$\tau_a(E_{max}) \sim \frac{D_1(E_{max}) + D_2(E_{max})}{u_{sh}^2} \quad , \tag{3}$$

where D_1 and D_2 are the up and downstream diffusion coefficients, u_{sh} the shock wave speed. The values of D_1 and D_2 are normally unknown and should be calculated from the properties of MHD-turbulence in the shock-acceleration region. Above we have seen that the reliable calculations of them are still absent. Therefore, we cannot give reliable estimations for the maximal energy of the accelerated particles with power spectra like eq. (1).

Another problem is account of the back-reaction from the accelerated particles on the structure of the shock transition and on the efficiency of the acceleration. There exists some progress in this direction [6, 22], but since the injection problem is unsolved, we cannot describe this effect self-consistently. Notice, that if the shock transition is substantially broadened by accelerated particles (the so-called mixed shocks), their spectrum can be non-powerlaw (cf. eqs. (1), (2)), and the above mentioned universality is lost.

Hence, our initial euphoria concerning the universality and simplicity of the diffusive shock acceleration mechanism, now has gone. We see that we encountered with a complicated, strongly nonlinear system, in which all processes are mutually connected and cannot be treated separately. Otherwise we cannot say whether this mechanism is working or not. Much yet should be done in the theory.

4 APPLICATION OF THE DIFFUSIVE SHOCK ACCELERATION MECHANISM TO THE EXTRAGALACTIC RADIO SOURCES

In such a situation, what we can say about the action of this mechanism in ERS? One plausible candidate can be the hot spots of the radio galaxies [3]. If the velocities of the jets are nonrelativistic (but supersonic), the mechanism can produce the power spectra of accelerated particles according to eqs. (1) and (2). Additionally, we must hope that we are lucky, and the mechanism actually works. There exist some indications that the efficiency of conversion of the energy of the jets into relativistic particles is not smaller than 10^{-2} [13]. Then, according to [5, 12], we can calculate the level of MHD-turbulence in the upstream region of a (probable) shock wave and to estimate D_1(E) in eq.(3). Let us also assume that $D_2(E) \leq D_1(E)$. Then, equating $\tau_a(E)$ to the electron synchrotron life-time $\tau_{syn}(E)$, we can estimate E_{max} for the accelerated electrons. Such a procedure yields $E_{max}^{(e)} \sim 10^6\ m_e c^2$ with an uncertainty of order of 10 [10]. Therefore, we can have a (power) spectrum of electrons in the range $10 \leq E^{(e)}/m_e c^2 \leq 10^6$, which is sufficient for the synchrotron emission of the hot spots from radio to optical bands [3]. It is noticeable that the level of the small-scale MHD-turbulence in the acceleration region is relatively small, i.e. $W_T \geq 10^{-5} W_B$, where W_B is the energy density of the mean magnetic field. Notice that the latter estimations of $E^{(e)}_{max}$ and W_T are compatible with those deduced from the analysis of the radio-spectra cutoffs of the hot spots [20]. Similar procedure with accelerated protons yields $E^{(p)}_{max} \geq 10^3\ m_p c^2$. The latter means that hot spots cannot be sources of ultra-high energy cosmic rays [10].

If the jets are ultrarelativistic, the diffusive shock acceleration becomes inefficient, since the particles, convected by the flow, cannot intersect the shock transition many times. However, in this case the so-called drift acceleration mechanism can work. This question will be elucidated in the talk of Dr. J. Kirk.

Another sites in ERS where the particle shock acceleration can work, are the accretion disc termination shock [22] and shock waves, produced by interactions of jets with optical clouds in active galactic nuclei [8]. All these models are very badly worked-out and suffer from all the difficulties, mentioned above.

5 CONCLUSION

In the talk I mentioned briefly some problems of ERS, which can be studied with the help of the plasma theory. In the following I concentrated on the particle acceleration within the context of the diffusive shock mechanism [2, 4, 7, 15].

I conclude that the present state of theory of this mechanism does not allow to say for sure that it works in ERS. However, such sites like hot spots, are very plausible candidates for this. The latter encourages me to continue the research. On the one hand, it is desirable to determine better such physical parameters of ERS, like the jet velocities, their composition, the structure of the magnetic field, the plasma density and its temperature, the spectrum of accelerated particles. On the other hand, it is necessary to develop a self-consistent, nonlinear theory of diffusive shock acceleration, and to verify its predictions on such available objects like interplanetary shocks.

Other mechanisms of particle acceleration in ERS also should not be neglected. However, one should bear in mind the problem of the universality of the particle spectra mentioned above.

REFERENCES

[1] Achterberg A., Blandford R.D., 1986, Mon. Not. Roy. Astro. Soc., **218**, 551.
[2] Axford W.I., Leer E., Scadron G.,1977, Proced XV Int., Cosm. Rag. Conf., **11**, 132.
[3] Begelman M.C., Blandford R.D., Rees M.D., 1984, Theory of Extragalactic Radio Sources. Rev. Mod. Phys., **56**, N° 2, 255.
[4] Bell A.R., 1978, Mon. Not. Roy. Astr. Soc., **182**, 147.
[5] Bell A.R., 1978, Mon. Not. Roy, Astr. Soc., **182**, 443.
[6] Blandford R.D., Eichler D., 1987, Phys. Reports, **154**, 1.
[7] Blandford R.D., Ostriker J.P., 1978, Astrophys. J., **221**, 429.
[8] Blandford R.D., Königl A., 1979, Astrophys. Letters, **20**, 15.
[9] Collisionless Shocks-Proceed, Int. Conf. held in Balatonfüred, Hungary, in 1987.
[10] Fedorenko V.N., 1988, Astrophysics (translation from Astrofizika), **28**, N° 1, 71.
[11] Fedorenko V.N., 1990, Preprint N° 1442 A. E. Ioffe Physical Technical Institute, Leningrad. Also: Astrophysics & Space Physics Reviews 1991 (in press).
[12] Fedorenko V.N., Fleischmann G.D., 1988, Sov. Astron., **32**, 398.
[13] Fedorenko V.N., Zentsowa A.S., 1986, Sov. Astron., **30**, 24.
[14] Kaplan S.A., Tsytovich V.N., 1973, Plasma Astrophysics, Pergamon Press, Oxford.
[15] Krymskii G.F., 1977, Sov. Phys. Doklady, **324**, 1306.
[16] Lagage P.O., Cesarsky C.J., 1983, Astron. & Astrophys., **125**, 249.
[17] Lee M.A., 1984, J. Geophys. Res., **88**, 6109.
[18] Norman C.A., ter Haar D., 1975, Physics Reports, C 17, 307.
[19] Pelletier G., Roland J., 1989, Astron. & Astrophys., **224**, 24.
[20] Roland J., Pelletier G., Muxlow T., 1988, Astron. & Astrophys., **207**, 16.
[21] Sol H., Pelletier G., Asseo E., 1989, Mon. Not. Roy. Astr. Soc., **237**, 411.
[22] Völk H.J., 1988, Proceed. of XX Int. Cosm. Ray. Conf., **7**, 157.
[23] Völk H.J., Zank L.A., Zank G.P., 1988, Astron. & Astrophys., **198**, 274.

SIGMA observations of the galactic center and some bright AGNs

F. LEBRUN

Service d'Astrophysique, CEN-Saclay, 91191 Gif sur Yvette Cedex

1 THE SIGMA TELESCOPE

The SIGMA experiment is a french hard X-ray/soft gamma-ray telescope onboard the soviet GRANAT satellite launched in december 1989. The four days, very excentric (2000-200000 km), GRANAT orbit allows three consecutive days of observation outside the radiation belts. The observation program has been shared almost equally between galactic and extragalactic targets.

The SIGMA experiment features basically a coded mask and a position-sensitive detector, it is able to record 15–20 arcmin resolution sky images at different energies in the range 35-1300 keV. The energy resolution varies from 15% at 60 keV to 9% at 511 keV. A typical one day observation is divided in 5–6 successive images. The telescope can detect at a 3σ level a 25 milliCrab source between 40 and 170 keV in one day observation. A more detailed description of the experiment and its performances can be found in Paul et al.(1990).

The main and obvious advantage of a soft gamma-ray imaging experiment over collimated spectrometers is its ability to measure truly a source spectrum, without contamination from the instrumental background, the diffuse emission or other nearby sources. This point is very well illustrated by a recent discovery based on coded-mask X-ray observations of the Virgo cluster. Hanson et al.(1990), analysing data from the Spacelab-2 mission found that the hard X-ray emission was not coming from M87 but from the nearby Seyfert 2 galaxy NGC 4388. Another less obvious, but important, advantage is that the energy response matrix is nearly diagonal (at least up to a few hundred keV) compared to the more triangular energy response-matrix of collimated experiments. This is because most of the source photons scattered in different parts of the experiment before their detection are removed in the image deconvolution process (see Barret and Laurent 1991). One drawback is that the sensitivity is reduced when the spatial resolution of the detector is significant compared to the mask element size, which is the case for SIGMA at low energy ($E < 100$ keV).

The data processing includes image corrections for detector distortions and non-uniform background, deconvolution for the mask pattern and a final flat-field correction to remove the residual background non-uniformity.

2 AGNS OBSERVATIONS

2.1 NGC 4151

This source has been observed six times in summer and at the fall of 1990. The total observing time amounts to 3 10^5 s. The source is visible (3–4 σ) in each observation and within the (limited) statistical accuracy does not show evidence of strong variability (no more than a factor of three). The overall photon spectrum is rather steep with a power law index of -3 ± 1 in contrast with the canonical hard spectrum observed in X-rays. The almost contemporaneous hard X-rays Art-P observations indicate that the source was in a high state and imply the existence of a break near 40 keV.

Our observations match the june 1978 HEAO-1 A4 observation (Baity et al. 1984) and possibly the may 1980 MISO observation (Bassani et al. 1986). However, they differ markedly from most of the others which indicate a continuous power law shape from a few keV to hundreds of keV with a typical index of about -1.5 (Baity et al. 1984, Maisack and Yaqoob 1991, Perotti et al. 1991, Maisack et al. 1989). This result establishes the existence of a soft state of NGC 4151 in which the source may stay a significant fraction of its time. If the source was in such a state during the COS-B and SAS-2 observations, no further (MeV) break would be required to explain their non-detection (Pollock et al. 1981). These results have been published in more details by Ballet et al. 1991 and Jourdain et al. 1991.

2.2 CEN A

This source has been observed successfully only once in july 1990 for about 6 10^4 s, the results of this observation are reported in more details by Ballet et al. (1991). The spectrum is very hard with a power law index of about -1.2. The very low 10 keV intensity observed by Art-P extends this hard spectrum to X-ray energies. Our 100 keV flux of 3 10^{-5} ph $cm^{-2}s^{-1}keV^{-1}$ is rather low, midway between those measured by HEXE in 1987 and 1988 (Maisack et al. 1989), but much lower than that observed by HEAO-1 A4 (Baity et al. 1981).

2.3 3C 273

This source has been observed four times in summer and at the fall of 1990. The source is barely visible in each observation (2–4 σ). The total observing time is 2.2 10^5 s. A spectrum based on all observations can be fitted with a -1.5 index power-law in agreement with previous hard X-ray measurements (Maisack et al. 1989). The 100 keV intensity of 1.7 10^{-5} ph $cm^{-2}s^{-1}keV^{-1}$ is typical of this source. The Art-P observations indicate that the source was also in an ordinary state in X-rays and are compatible with the hard spectrum found in the SIGMA domain. The spectrum steepening necessary to connect these data to the high energy gamma-ray observations of COS-B (Bignami et al. 1981) has to occur above 500 keV. For more details, the reader is referred to Ballet et al. 1991 and Bassani et al. 1991.

3 OBSERVATIONS OF THE GALACTIC CENTER

The case for a black hole in the center of the Galaxy, as in AGNs, is supported by the presence of the compact radio source $SgrA^*$ and the unusual time-variable 511 keV electron-positron annihilation radiation. However, only a hundred solar-mass black hole is necessary to account for these observations (Allen and Sanders 1986) as well as for the UV emission which may originate in an accretion disc around a black hole. The case for a really massive black hole in the center ($SgrA^*$) is mainly supported by the gas dynamics, which shows an increase of velocities inside $\sim$ 2 pc and the presence of a ring of gas outside these 2 pc which may be interpreted as an accretion disc around a very compact object of about $10^{6.5} M_{\odot}$ (Duschl 1990).

Since its discovery (Haymes et al. 1975), the 511 keV annihilation line has been tracked by a number of satellite and balloon experiments. High resolution germanium spectrometers revealed a narrow ($<$ 3.2 keV) line centered at 511 keV (Leventhal et al. 1978) and variable (Riegler et al. 1981, 1985, Leventhal et al. 1980, 1982, 1986). A significant continuum contribution at lower energy has been detected and interpreted as originating from orthopositronium annihilation. A hard tail has also been detected with an intensity correlated with that of the line. The 130° field of view NaI detector onboard SMM has also detected an annihilation line in the general direction of the galactic center but found the line rather stable around 2 10^{-3} ph $cm^{-2}s^{-1}$ during the seven years of observation. In a review of all these results, Lingenfelter and Ramaty (1989) have set up a picture in which the stable emission observed by SMM is dominated by a diffuse contribution (line and orthopositronium) from the galactic disc while the emission observed by the narrowly collimated experiments is dominated by the line and hard tail emission of a variable compact object close to the galactic center ($< 4^{\circ}$), probably a black hole. However,it is not necessarily a massive black hole and therefore may not reside in the nucleus of the Galaxy.

Recent high angular-resolution hard X-ray observations (Skinner et al. 1987, 1990, Kawai et al. 1988) have revealed several point sources around $SgrA^*$ but only very weak emission at the galactic center. A balloon born gamma-ray telescope with a coded-mask aperture giving an angular resolution of 1.1° found that only one of the hard X-ray sources is visible up to 200 keV and suggested to identify it with 1E 1740.7−2942 (Cook et al. 1991).

This source was detected by SIGMA up to 300 keV during its first observation of the galactic center and coincides indeed (within 2') with the Einstein source (Cordier et al. 1990).

The galactic center has been observed by SIGMA at three different epochs: in march–april, september–october 1990 and finally in february–march 1991. The total observing time approaches 6 10^5 s. During seven different observations performed on march 24, april 4,8,12,17 and september 19,29 1990 the spectrum of the source was found stable and very similar to that of the black hole candidate Cygnus-X1 in its low (or hard) state. Moreover, if the source is assumed to be at the galactic center distance, its luminosity between 30 and 300 keV of $\sim 2\ 10^{37}$ erg s^{-1} is similar to that of Cygnus-X1.

On october 10–11, a 4 10^4 s observation revealed a harder spectrum, with an electron temperature for a comptonized disc model of about 55 keV to be compared with the 32 keV found for previous observations ("nominal" state).

On october 13–14, the source spectrum exhibited a large bump between 300 and 600 keV. An image in the 330–570 keV interval shows a $\sim 6\sigma$ excess at the position of 1E 1740.7−2942. The next observation, a few hours later on october 14–15, as well as a subsequent one on october 18 indicated that the source had returned to its "nominal" state.

A series of four observations (2.2 10^5 s in total) performed between the 23 and 28 of february 1991 did not show any excess at the position of 1E 1740.7−2942 which implies a flux decrease by nearly an order of magnitude. On march 22, the source was detected between 40 and 80 keV at a very low level (~ 1/3 "nominal" state). The next day began a strong solar activity, but the first unperturbed twelve hours of observation do not show any excess at the source position. After the decrease of the solar activity, the observations of april 1 and 4 did not show the source. Clearly, we are faced with a new long-term low-state of the source. These results have been reported in Cordier et al. (1990), Bouchet et al. (1991) and Sunyaev et al. (1991).

4 DISCUSSION

These states are reminiscent of those observed towards Cygnus-X1, and the almost identical luminosity and spectrum of their nominal states suggest a similar nature for these two objects, i.e. a stellar mass black hole. However, at hard X-rays, the variability of 1E 1740.7−2942 is more important, greater than a factor of 5, than that of Cygnus-X1 which does not exceed a factor of 2. Moreover, the γ^1 (hard) state of Cygnus X-1 does not ressemble any of the 1E 1740.7-2942 states.

The relation of 1E 1740.7−2942 with the variable narrow 511 keV line observed toward the galactic center by the collimated experiments is still unclear. The feature observed by SIGMA is much broader, at lower energy and lasted at most a few days while the narrow 511 keV line was observed during several years. It seems however difficult to attribute it to another source unseen by SIGMA, e.g. $SgrA^*$, since the 1E 1740.7−2942 high energy bump should have something to do with annihilation. Moreover due to the relatively poor energy resolution of SIGMA, a narrow 511 keV line at a level $< 2\ 10^{-3}$ ph $cm^{-2}s^{-1}$ would not be detectable by SIGMA in one day. The most recent measurements of this narrow line by germanium detectors 1988-1989 (Leventhal et al. 1989, Chapuis et al. 1990) indicate a line flux of about $1\ 10^{-3}$ ph $cm^{-2}s^{-1}$. According to Lingenfelter and Ramaty (1989), for a 20° field of view experiment, about 1/3 of it may be attributed to diffuse emission and 2/3 to a variable source. If such a line from the variable source was present during all the SIGMA observations, it should have been detected at a $3\,\sigma$ level in the sum of all observations.

No clear connexion between jets and gamma-ray emission is apparent on the basis of the present results. Indeed, NGC 4151, NGC 4388 and 1E 1740.7−2942 emit gamma-rays but do not exhibit jets, while 3C 273 and CEN A have jets and emit gamma-rays. M87 and $SgrA^*$ show a jet but do not emit gamma-rays. Why are these last two objects, which are good candidates for harbouring a massive black hole do not emit gamma-rays while less good candidates do (1E 1740.7−2942, NGC 4388) is probably a key question.

The detection of a feature near 511 keV from 1E 1740.7−2942, variable on a day scale shows the need for very sensitive imaging spectrometers in this energy range. The results of the observations of the brightest AGNs indicate that a search for this kind of features in extragalactic sources is unfortunately clearly out of the scope of the SIGMA experiment. This kind of search would require experiments more sensitive by nearly an order of magnitude.

REFERENCES

Allen, D. and Sanders, R. 1986, Nature, 319, 191
Baity, W.A. et al. 1981, Ap.J. 244, 429
Baity, W.A. et al. 1984, Ap.J. 279, 555

Ballet, J. et al. 1991, in Physics of AGNs (Wagner, Editor), Springer Verlag
Barret, D. and Laurent, P. 1991, Nucl.Instr.Meth. A307, 512
Bassani, L. et al. 1986, Ap.J. 311, 623
Bassani, L. et al. 1991, Int.Cosmic Ray Conf., Dublin OG.3.3.10
Bignami, G.F. et al. 1981, A&A 93, 71
Bouchet, L. et al. 1991, Ap.J.(Letters), in press
Chapuis, C. et al. 1990, in Gamma-Ray Line Astrophysics, AIP:NY, in press
Cook, W. et al. 1990, in XVIII COSPAR Symp. ME.4.1.3
Cordier B. et al. 1990, Adv.Sp.Res. 11, 8, 169
Duschl, W.J. 1989, M.N.R.A.S. 240, 219
Hanson, C.G. et al. 1990, M.N.R.A.S. 242, 262
Haymes, R.C. et al. 1975, Ap.J. 201, 593
Jourdain, E. et al. 1991, Int.Cosmic Ray Conf., Dublin OG.3.3.12
Kawai, N. et al. 1988, Ap.J. 330, 130
Leventhal, M. et al. 1978, Ap.J.(Letters) 225, L11
Leventhal, M. et al. 1980, Ap.J. 240, 338
Leventhal, M. et al. 1982, Ap.J.(Letters) 260, L1
Leventhal, M. et al. 1986, Ap.J. 302, 459
Leventhal, M. et al. 1989, Nature 339, 36
Lingenfelter, R.E and Ramaty, R. 1989, Ap.J. 343, 686
Maisack, M. et al. 1989, in Proc. 23rd ESLAB Symp., Bologna, ESA SP-296, 975
Maisack, M. and Yaqoob, T. 1991, A&A, in press
Paul, J. et al. 1990, Adv.Sp.Res. 11, 8, 289
Perotti, F. et al. 1991, Ap.J. 373, 75
Pollock, A.M.T. et al. 1981, A&A 94, 116
Riegler, G.R. et al. 1981, Ap.J.(Letters) 248, L13
Riegler, G.R. et al. 1985, Ap.J.(Letters) 294, L13
Skinner, G.K. et al. 1987, Nature 330, 544
Skinner, G.K. et al. 1990, in Gamma-Ray Line Astrophysics, AIP:NY, in press
Sunyaev et al. 1991, Ap.J.(Letters), in press

Synchrotron modelling of the galactic center jet

M. POHL

Max-Planck-Institut für Radioastronomie, Auf dem Hügel 69, 5300 Bonn 1, FRG

1. THEORY

Several processes can provide monoenergetic spectra of relativistic particles. Reconnection leads to a relativistic beam of electrons, that is unstable with respect to the generation of waves, which scatter the electrons and isotropize the particles distribution function. In the case of accreting galactic centers the very long escape lifetimes cause together with the interplay of Fermi acceleration and radiative energy losses the build of pile-up spectra (Schlickeiser, 1984; Achatz et al., 1990).

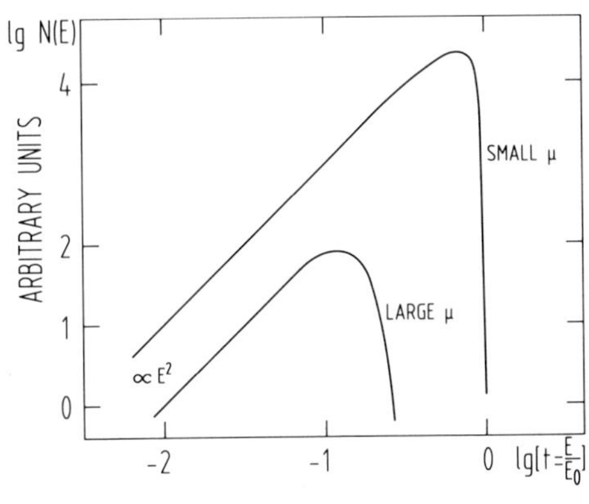

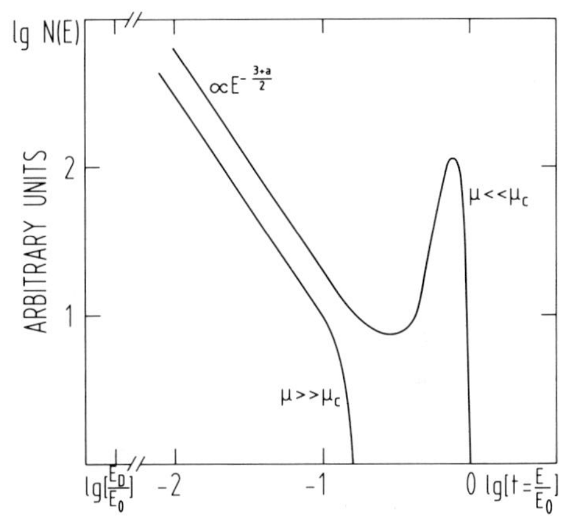

Fig.1-2 . The resulting electron spectra for small energies (dominated by adiabatic losses) left and high energies (dominated by radiative losses) right. μ is the spatial coordinate (=0 at the source position).

We have calculated the steady-state energy spectra in accelerated outflows from monoenergetic sources. The one-dimensional continuity equation of relativistic electrons in these environments includes terms of injection, propagation by convection and diffusion, and adiabatic and radiative losses. We have used the solutions by Lerche and Schlickeiser (1981), which were obtained under some assumptions for the spatial variation of parameters. The asymptotic energy spectra were converted to synchrotron spectra by folding them with the synchrotron emissivity. The resulting spectra depending on the spatial coordinate μ are shown in Fig.1-4.

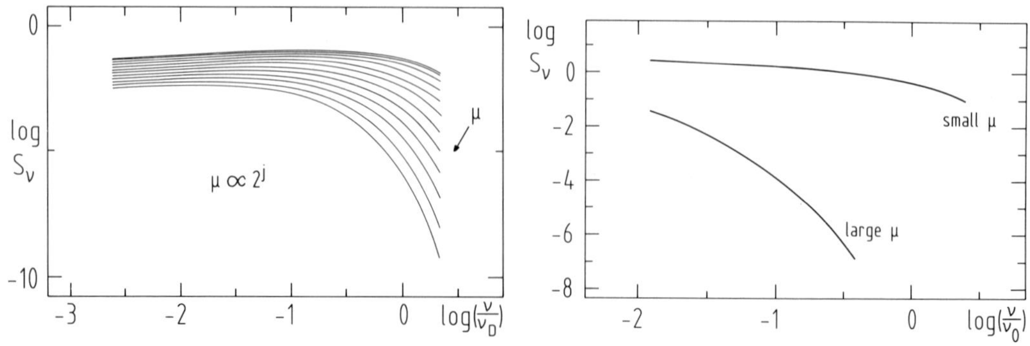

Fig.3-4 . The corresponding synchrotron spectra to the energy spectra shown in Fig.1-2 with the same asymptotic regions. In Fig.3 the dependence of the spectra on the spatial coordinate μ is shown with the distance increasing by factors of two.

2. OBSERVATIONS

These findings may be applied to the so called galactic center jet (Sofue et al., 1989a,b; Reich, 1990), a radio feature extending from the galactic center out to kpc distances. The spectral index is constant ($\alpha \simeq 0.3$) over a large frequency range near the galactic center. This behaviour corresponds to the model spectrum shown in Fig.3. At medium distances the spectral index decreases to roughly $\alpha \simeq -0.6$ and decreases further at large distances. The difficulties in the background elimination of the data hamper the determination of spectral index variations with the distance from GC. Adiabatic losses seem to be the most important process to influence the electron energy spectrum in the jet, i.e. the jet is strongly accelerating from GC to heights of up to 4kpc. Furthermore, the fast decay of the high-frequency emission in the jet can within the framework Lesch, H., Schlickeiser, R.: (1990) A&A 233, 391
strength decreases with distance from GC. An equipartition field strength of 135 μG implies a timescale for adiabatic losses of less than $3 \cdot 10^4 y$, so we can roughly estimate the ultimate jet speed to be $\approx c/10$.

3. REFERENCES

Achatz, U., Lesch, H., Schlickeiser, R.: (1990) A&A 233, 391
Lerche, I., Schlickeiser, R.: (1981) ApJ Supp. 47, 33
Reich, W.: (1990) Proc. IAU 140, ed. R.Beck et al., p.369
Schlickeiser, R.: (1984) A&A 136, 227
Sofue, Y., Reich, W., Reich.P.: (1989a) ApJ 341, L47
Sofue, Y., Reich, W., Reich.P.: (1989b) Proc. IAU 136, ed. M.Morris, p.233

Active Galactic Nuclei observed during the ROSAT All Sky Survey

W. BRINKMANN

Max-Planck-Institut für Extraterrestrische Physik, D - 8046 Garching, BRD

1 INTRODUCTION

The German X-ray astronomy satellite ROSAT was launched from Cape Canaveral on June 1 1990, 17:48 local time by a Delta-II rocket into a circular near-Earth orbit of 580 km altitude and 53° inclination. On June 16 the X-ray telescope saw first light during a set of pointings at regions in the Large Magellanic Cloud. After a detailed calibration and verification phase the all-sky survey started with both the X-ray telescope and the Wide Field Camera co-aligned on July 30, 1990.

With a total mass of the satellite of about 2.4 tons, a length of 4.5 m and a diameter of 2.4 m ROSAT is the biggest scientific satellite built so far in Western Europe. The X-ray telescope (XRT) working between 0.1 to 2.4 keV, consists of a 4-fold nested Wolter type I mirror assembly of 2.4 m focal length and a maximum aperture of 835 mm. The angular resolution is estimated to be 3.6 arcsec, the residual microroughness of the mirror surfaces is 2.8 Å (*Aschenbach* [1]).

The focal plane instrumentation, developed and built at the MPE, consist of two redundant Position Sensitive Proportional Counters (PSPC's), which are operated in gas flow mode (*Pfeffermann et al.* [2]). The PSPC provides spatial and spectral resolution over the full field of view of 2° which vary slightly with photon energy. The angular resolution is limited by the PSPC to about 25 arcsec on-axis and degrades to a few arcminutes for field angles larger than about 20 arcmin due to the geometric aberrations of the telescope. The energy resolution is $\Delta E/E = 0.41/\sqrt{E_{keV}}$. The effective collecting area including the PSPC efficiency is about 220 cm^2 at 1 keV for on-axis rays and decreases slightly with increasing field angle.

The very low background of the PSPC, measured in orbit during a pointing on β Leonis, implies that pointed observations with a typical resolution element of 25 arcsec diameter are not background limited for exposures less than 10^5 sec. For the survey the resolution element is about 2 arcmin in diameter, and PSPC observations become particle background limited only for times exceeding 10^4 sec.

The third detector, the High Resolution Imager (HRI) built at the SAO Cambridge, USA, is a microchannel plate detector with an intrinsic spatial resolution of about 20 μm. With the HRI in the focus of the telescope the angular resolution is dominated by the point spread function of the mirror assembly.

The second telescope is the Wide Field Camera (WFC) furnished by the United K-ingdom. The telescope is a 3-fold mirror assembly of the Wolter type I arrangement with a focal plane turrent from which one of two identical microchannel plate detectors can be selected. A set of 8 filters defines the spectral wavelength passbands which cover the range from about 17 to 210 eV. The telescope has a focal length of 525 mm, a maximum mirror diameter of 576 mm and a field of view of 5° diameter.

A comprehensive and more detailed description of ROSAT and the instruments can be found for instance in the papers by *Trümper* [3] and in *Pounds and Wells* [4] for the WFC.

The survey is performed by scanning the sky along great circles in ecliptic longitude at a rate of about 360° in 96 minutes, the orbital period of the satellite. Thus, the sky is imaged in form of 2° wide stripes, corresponding to the field of view of the XRT/PSPC. Through the motion of the Earth around the Sun the great circle is advanced by about 1° per day completing the all-sky coverage in half a year. Both at the north ecliptic pole and the south ecliptic pole all scan circles from every orbit overlap with each other. Thus the total exposure of a target ranges between $\sim$ 500 secs for objects near the ecliptical equator up to a few 10^4 secs for circumpolar targets. A source is visible for between typically 3 days up to half a year for objects in the innermost 1° of the polar caps.

After completion of the survey the pointing observations of the guest observer programme were started in February 1991. About 30 % of the observation time will be spent on observations of AGN, indicating the enormous interest of the scientific community and ROSAT's potential for the study of these objects.

In the next chapter I will give a 'preview' of results from the survey, related to AGN, going briefly from detailed studies of well known objects to a discussion of the large class of previously unknown objects - their number, properties and contribution to the cosmic X-ray background. I will restrict the presentation to a more qualitative form as even now, near the end of the AO-1 phase, many of the obtained results are still preliminary, awaiting the final 'fine tuning' of the data reduction programmes.

2 DETAILED STUDIES OF KNOWN AGN

Starting already in the pre - launch phase of ROSAT more than 20 international collaborations were set up to study AGN from the survey. Targets are single objects, samples (classes) of objects, certain sky fields (ESO key project, VLA NEP observation, etc.), and simultaneous observations with IUE and Ginga as well as with ground based instruments. Furthermore, about 60 MPE internal projects were defined to maximize the scientific output from the survey data.

The primary objectives of these projects are the study of spectral and flux variability for objects near the ecliptical pole (with exposures for up to half a year in the survey), simultaneous multi-wavelength observations from the radio to X-ray range and the investigation of the X-ray properties of well selected samples.

2.1 The ROSAT - IUE/Ginga programmes

In an attempt to study various X-ray sources simultaneously over a wide energy band with ROSAT and Ginga (and, in most of the cases with IUE as well) 24 AGN were selected with long ROSAT exposures and sufficiently high X-ray fluxes. First results of this programme have been presented elsewhere ([5], [6]).

Most important are simultaneous spectral fits to strong sources for an intercalibration between the two instruments. For example, the combined ROSAT/Ginga spectra for Mkn 501 are in excellent agreement with each other [6]. The close similarity of the parameters in both fits is an important confirmation of the correctness of the calibration of the ROSAT PSPC. This means in particular, that the steep power law slopes of $\alpha \approx 2.3$ generally found for ensembles of extragalactic sources are intrinsic to at least a large class of the objects in the soft X-ray band - perhaps one of the biggest surprises of the ROSAT survey.

2.2 Temporal behaviour of NEP sources

As mentioned above, sources close to the ecliptic poles are seen in the survey up to half a year, 16 times per day for about 30 secs each allowing the study of temporal variations on time scales from $\sim$ hours to months.

Amongst the AGN a group of 16 objects, with distance closer than 2° to the NEP has been observed simultaneously with ROSAT from the radio to the UV wavelengths. First results for Kaz 102, KUV 18004+684 (constant flux level), and KUV 18217+64 (changes on time scale $>$ 10 days) have been presented by *Schaeidt et al.* [7].

3 THE AGN CONTENT OF THE SURVEY

Pre-launch studies of number - count relations predicted that about 40 % of all

ROSAT sources would be AGN, i.e., about 30 000 to 40 000 objects would be newly detected in X-rays. The identification and subsequent classification of these sources clearly represents a task of really 'astronomical' magnitude.

Using the first ROSAT source positions *Hasinger* [8] made a comparative study of these objects with the SIMBAD catalogue. From a total of 1892 ROSAT sources 869 could be identified, the large majority of them (646) with stars, only 32 with AGN.

Plotting the logarithmic ratio of X-ray flux to optical flux against the visual magnitude of the optical counterpart shows a very distinct class separation, very similar to that found from the Einstein EMSS (Stocke et al., 1983 [9]). The optically brighter objects are mainly stars whereas, at magnitudes $m_v > (17 - 18)$, the majority of the sources seem to be AGN. This impression is substantiated by the results of the first deep pointed observations.

Hasinger et al. [10] analysed several fields around sources observed extensively in the ROSAT calibration and verification phase and at the NEP.

The accumulated flux of the isolated point-like sources makes up a fraction of about 40% of the total flux contained in the deepest exposure image. Further, the cumulative number-count relation shows a significant flattening for X-ray fluxes less than about 3×10^{-14} erg cm^{-2} s^{-1}, which could mean that the number density of sources starts to decrease with increasing distance or that the X-ray luminosity shows a decrease because of evolutionary effects.

The photon spectrum of the resolved sources in the 1 - 2 keV range is parallel to that of the residual background. Both show a power law slope of -2.3, suggesting that the steepening of the X-ray background in this energy range [11] is due to extragalactic sources. The excess diffuse galactic emission starts to dominate the XRB only below 1 keV.

Similar conclusions were reached by Shanks et al. [12]. During the PV phase a 30 ksec exposure of a faint QSO survey field yielded 39 sources in the innermost 40' diameter of the PSPC detector with X-ray fluxes down to $> 2\ 10^{-14}$ erg cm^{-2} s^{-1}. This corresponds to a surface density of X-ray sources of $\sim$110 deg^{-2}. From the 30 sources identified in optical follow up observations 24 turned out to be quasars, giving at X-rays a sky density of $\sim$70 QSO / deg^2.

These first studies suggest that the AGN content of the ROSAT All Sky Survey is at least as high as the predicted fraction of 40%. However, the enormous amount of data will require quite a time to be analyzed and checked and follow up observation at other wavelengths will very likely take years before a substantial fraction of these sources will be firmly identified.

4 EXPLORING "THE KNOWN"

So far, only about 200 of the more than 5000 catalogued AGN are known to emit at X-ray and other wavelengths. A correlation of the ROSAT source list with existing optical and radio catalogues is expected to increase this number considerably.

During the pre-launch phase some efforts were undertaken to collect all available informations on AGN in a "master-catalogue". It is based on the Veron-Cetty - Veron catalogue of quasars [13], supplemented with various other compilations like the Hewitt-Burbidge QSO catalogue, the Kühr catalogue of radio sources, the EMSS, etc., and additional data from current publications.

The source positions from this "master-catalogue" are routinely compared with point sources, found by the Standard Analysis System (SASS, [14]). From the first approximately 22% of the processed ROSAT data one can make the prediction that we will find $\approx$ 800 of the catalogued quasars. This is a factor of $\sim$2 improvement in the number of known X-ray quasars.
We further expect to see $\approx$ 92% of all catalogued Seyfert 1 galaxies, again a factor of two improvement in the number of X-ray emitting objects.
Due to their intrinsically low luminosity only very few Seyfert 2 galaxies will be seen in the survey and less than about 45% of the known BL Lacs will be detected.

The majority of the detected sources is too weak to allow proper spectral fits, in general only hardness ratios can be given. From these one can, however, determine the slope of an assumed power law and the value of the interstellar absorption. Likewise, one can perform three band fits with assumed galactic absorption.

Figure 1 shows the photon index α for the first $\approx$ 1000 survey AGN as function of the total count rate. The majority of sources has only few counts in total and the error in the determined power law index is typically of the order of $\pm$1.0. The 'average' power law index has a value of 2.18 $\pm$0.95, similar to that found by *Hasinger et al.* [10] in their analysis of the deep field observations.

No obvious differences in the power law index for the different subclasses of AGN could be found yet. Power laws from spectral fits to the strongest ROSAT sources often show even steeper slopes but it is presently not clear whether selection effects play a role and how the spectral fits are affected by the parameter settings of the Standard Data Analysis System.

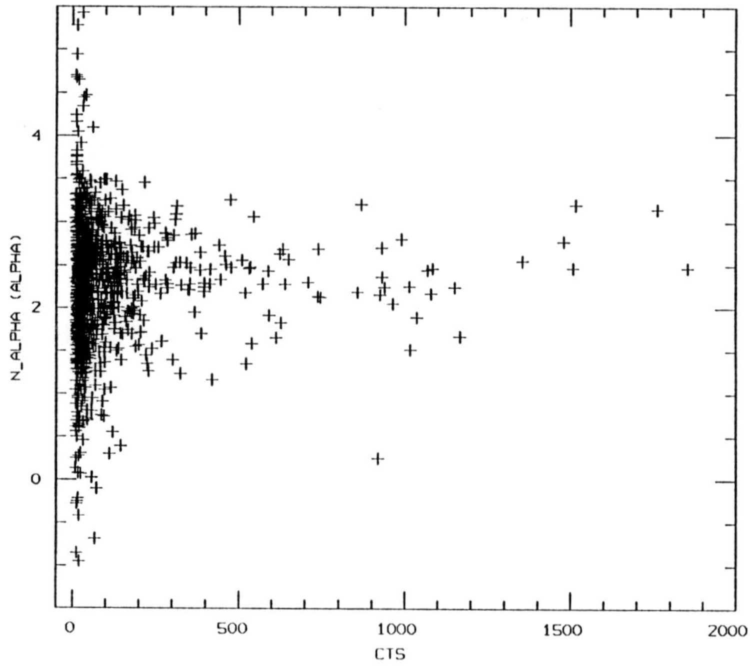

Figure 1: Photon power law index as function of the total count rate for the first $\approx$ 1000 survey AGN, determined from their hardness ratios (Courtesy of N. Schartl)

5 THE CONDON-BRODERICK-SEIELSTAD 6 CM SURVEY

Many AGN show distinctive properties in the radio band, as for example BL Lacs or radio loud quasars. Therefore, a programme was initiated to correlate a point source catalogue generated from the Condon-Broderick-Seielstad 6cm survey of the northern hemisphere [15] with the ROSAT survey. This point source catalogue comprises a total of > 140000 sources with 6cm - fluxes > 20 mJy. For objects found in both energy bands, follow up observations at the 100 m Effelsberg radio telescope were organized, giving radio fluxes at 11 cm, 6 cm, and 2.8 cm. The better radio positions are then used in optical identification programmes to determine the nature of the objects.

More than 1500 sources are expected to be found in the survey data. Figure 2 shows a plot of the 6 cm radio flux versus the ROSAT count rate for the first 501 objects found. 136 of these objects were previously "known", very often only as an entry in another radio survey, i.e., without any optical identification. Those identified are mainly classified as QSO, radio galaxies, BL Lacs, and clusters of galaxies. Only less than a handfull of the sources were known previously as X-ray sources.

Most of the unknown objects populate the lower, left domain of the diagram, "known" sources are generally either X-ray or radio brighter. The hardness ratios of these two subgroups are slightly different and both are different again from the typical hardness ratios found for the AGN of the "master catalogue". There seems to be a tendency for these radio selected X-ray sources to show harder spectra than the "average" AGN.

However, much more detailed studies invoking the whole data set will be necessary to confirm these first trends and to assess their physical significance.

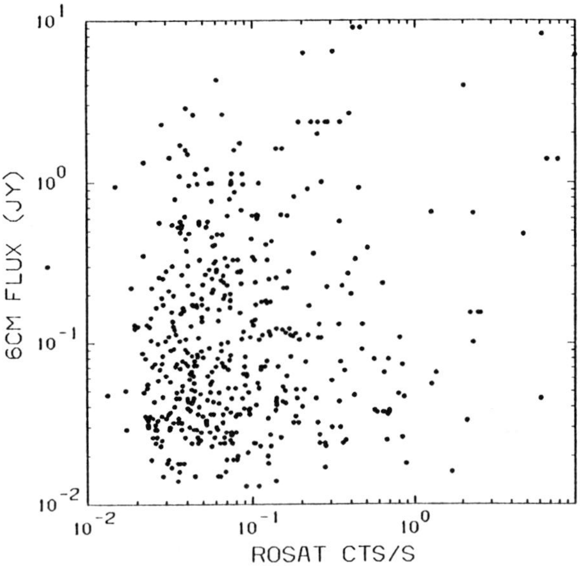

Figure 2: 6 cm radio flux versus X-ray count rate for the first 501 ROSAT / Condon objects

So far, more than about 250 candidates have been re-observed at the Effelsberg telescope and more than 90% of them were positively detected. 44% of them showed a flat or even inverted radio spectrum, 25% had a flat radio spectrum and were polarized, and about 10% had a flat radio spectrum and were extended. So far only about 10 of the candidates could be identified optically.

A detailed discussion of the properties of these sources and a source catalogue will be given elsewhere.

6. CONCLUDING REMARKS

This first review can only give a very preliminary account of the enormous data base available for the study of AGN after the ROSAT All Sky Survey. Most of the data analysis has only just started and will require some time before definite results are available. More observations, in the ROSAT pointed phase, during forthcoming missions, and with ground-based instruments will be necessary to explore the full scientific contents of these data.

ACKNOWLEDGEMENT:

The ROSAT project is supported by the Bundesministerium für Forschung und Technologie (BMFT). I like to thank my colleagues at the Max-Planck-Institut für Extraterrestrische Physik, B. Aschenbach, T. Boller, H. Fink, G. Hasinger, S. Molendi, N. Schartl, and S. Schaeidt for fruitful discussions.

REFERENCES

[1] B. Aschenbach: Appl. Optics **27**, 1404 (1988)

[2] E. Pfeffermann, U.G. Briel, H. Hippmann, G. Kettenring, G. Metzner, P. Predehl, G. Reger, K.H. Stephan, M.V. Zombeck, J. Chappell, S.S. Murray: Proc SPIE **733**, 519 (1986)

[3] J. Trümper: Adv. Space Res. **2**, No. 4, 241 (1983);
J. Trümper: Phys. Scripta **T7**, 209 (1984);
J. Trümper: Phys. BL. **46**, Nr. 5, 137 (1990)

[4] K.A. Pounds, A.A. Wells: COSPAR Proc. *Adv. Space Research*, The Hague (1991)

[5] F. Makino: paper presented at the 28th Yamada conference "Frontiers of X-ray Astronomy", Nagoya (1991)

[6] H.H. Fink: Proceedings of the Heidelberg conference "Physics of AGN", eds. W.J. Duschl, S. Wagner, Springer Verlag (1991).

[7] S. Schaeidt et al.: Proceedings of the Heidelberg conference "Physics of AGN", eds. W.J. Duschl, S. Wagner, Springer Verlag (1991).

[8] G. Hasinger: private communication (1991)

[9] J. Stocke et al.: ApJ **273**, 458 (1983)

[10] G. Hasinger, M. Schmidt, and J. Trümper: Astron. Astr. **246**, L2 (1991)

[11] D. McCammon and W.T. Sanders: ARA&A **28**, 657 (1990)

[12] T. Shanks et al.: Nature (in press) (1991)

[13] M.-P. Veron-Cetty and P. Veron: A Catalogue of Quasars and Active Nuclei (4th Edition), ESO Scientific Report No. 7 (1989)

[14] W. Voges: The Standard Analysis Software System for ROSAT (in preparation)

[15] J.J. Condon, J.J. Broderick, and G.A. Seielstad: A.J. **97**, 1064 (1989)

The Soft X-ray Excess in *Einstein* Quasar Spectra.

JEAN-LOUIS MASNOU
DARC, UPR176, Observatoire de Paris, Section de Meudon, FRANCE

BELINDA J. WILKES, MARTIN ELVIS, JONATHAN C. McDOWELL
Harvard-Smithsonian Center for Astrophysics
60, Garden St., Cambridge, MA 02138 USA

KEITH A. ARNAUD
Goddard Space Flight Center, Greenbelt, MD 20771 USA

1 Introduction

Recent soft X-ray observations have shown that Active Galactic Nuclei (AGN) commonly show strong excess emission above the extrapolation of their higher energy slopes. Information is generally limited to a single discrepant point at low energies when compared with the slope at higher energies (1-10 keV) (e.g. Turner & Pounds 1989). We have selected the highest signal-to-noise Imaging Proportional Counter (IPC) observations from the Wilkes & Elvis (1987) quasar sample, combined with higher energy (2-10 keV) Monitor Proportional Counter (MPC) data where it is useful, to analyse data extending from 0.1-10 keV. These data combined with improved Galactic N_H values allowed two-component spectral fits and thus a well-constrained estimate of the observed flux in the soft excess. We summarize and discuss below the results of our analysis.

2 Analysis and results

Our study has been performed on a signal-to-noise limited sub-sample of 14 quasars. Detailed analysis (Masnou et al. 1991) allowed us to estimate the flux of any detected excess and improve constraints on the spectral parameters at higher X-ray (> 0.6 keV) energies. We have found a significant soft excess in 8 of the 14 quasars, confined in all but one case (PG 1211+143) to below 0.6 keV. The excess is usually strong, being 1-6 times as strong as the high energy component at 0.2 keV. In 3C 273, multiple observations show that the excess is variable. Studies of lower luminosity Seyfert 1 galaxies (Turner & Pounds 1989, Kruper et al. 1990) show a similar 50% detection rate. This number is likely to increase if a significant fraction of the hydrogen in our galaxy is ionized (Reynolds 1991). The high energy slopes determined from the broken power law fits are consistent with those reported earlier. We confirm that a range of slopes is present in the low energy X-ray region, 0.6-10 keV, with the slopes for radio-loud and radio-quiet quasars being ~ 0.5 and 1.0 respectively. We have also estimated the soft X-ray excess above an IR-X-ray power law as a check of our results. The IR-X-ray spectral slope was estimated from data at 1-2 mm (Elvis et al. 1991a) and 0.6 keV. An excess was found above 1σ for 7 (1 radio-loud and 6 radio-quiet quasars) of the 8 quasars which show an excess above the high energy X-ray power law. 3C273, no longer shows an excess. For the other objects, the value of the soft excess relative to the IR-X-ray power law agrees with the value relative to the higher energy X-ray power law. We have found no evidence for a correlation between the strength of the blue bump in the UV range (McDowell et al. 1989) and that of the X-ray soft excess, measured by its ratio to the high energy component at 0.2 keV. This might be due to the large uncertainties in the soft excess measurement.

3 Discussion

A synchrotron self-Compton origin for the α_E~0.5 X-ray component in radio-loud quasars is generally accepted. Shastri (1991) suggested that the flat-spectrum X-ray component is associated with the radio emission and would originate from a jet relativistically beamed. For Seyfert 1 galaxies and radio-quiet quasars, a model has been proposed following *Ginga* observations (e.g. Zdziarski et al. 1990, Pounds et al. 1990, Zdziarski 1991) in which a non thermal initially hard primary spectrum originates in a compact region, constantly filled by relativistic electrons or pairs. This spectrum is reprocessed by soft thermal photons from nearby cold material (e.g. accretion disk or clouds). It predicts α_E~1.0 at low energies, ≤ 10 keV, as seen for our sample of radio-quiet quasars and a tail to the black-body spectrum, which could contribute to the X-ray soft excess observed by us. Moreover, Compton reflection of the cold material adds a broad hump of emission extending from ~10 keV to high energy (200 keV), undetectable in our case. Czerny & Elvis (1987) have reproduced the strong soft excess in PG1211+143 using an electron scattering atmosphere surrounding the disk. Direct application of this model to our sample predicts only 3 quasars with a soft excess implying a broader range for the parameters. Elvis et al. (1991b) infer high optical depths of the emitting plasma from variability of soft excess emission for 7 AGN, including PG1211+143 and 3C273.

4 Conclusion

Two component spectral fits to IPC/MPC quasar data show that the soft X-ray excess dominates the X-ray flux below 0.3 keV in half our sample. The fluxes in the excess at 0.2 keV cover the range 1-6 times that in the high energy component at the same energy. The high energy tail of the blue bump remains the most attractive candidate. However, a power-law tail to the black-body spectrum as predicted in the framework of the reflection model (Zdziarski et al. 1990) is also consistent with our data. Large uncertainties in line-of-sight absorption (Galactic or intrinsic to the quasar) and the lack of simultaneous UV data preclude more detailed modeling.

References

Czerny, B. and Elvis, M. 1987 ApJ **321**, 305
Elvis M. et al. 1991a ApJS (*in preparation*)
Elvis, M., Giommi, P., Wilkes, B.J. and McDowell, J.C. 1991b ApJ (*in press*), Center for Astrophysics Preprint No 3224
Kruper, J.S., Urry, C.M. and Canizares, C.R 1990 ApJS **74**, 347
Masnou, J.L., Wilkes, B.J., Elvis, M., McDowell J.C. and Arnaud, K.A. 1991 Astronomy & Astrophysics (*in press*), Center for Astrophysics Preprint No 3261
McDowell, J.C. et al. 1989 ApJL **345**, L13
Pounds, K.A., Nandra, K., Stewart, G.C., George, L.M. and Fabian, A.C. 1990, Nature **344**, 132
Reynolds, R.J. 1991 ApJ **372**, L17
Shastri, P. 1991 MNRAS **249**, 640
Turner, T.J. and Pounds, K.A. 1989 MNRAS **240**, 769
Wilkes, B.J. and Elvis, M. 1987 ApJ **323**, 243
Zdziarski, A.A. et al. 1990 ApJ **363**, L1
Zdziarski, A.A, 1991, in the Proceedings of this Conference

Physical Parameters in the Nucleus of AGN from X and Gamma Rays

G. GHISELLINI

Osservatorio Astronomico di Torino, Pino Torinese, Italy

ABSTRACT
The most compact regions of AGN are the sites of photon densities so large that they become optically thick to γ-rays, which collide with softer photons to produce electron positron pairs. The created pairs then reprocess the entire spectrum and can explain why the X-ray spectra of Seyfert galaxies have a common shape. Even if the density in the nucleus is dominated by pairs, it is unlikely that the pairs can form an outflowing wind with sufficient density to fuel the inner (parsec) jet of radio sources, due to the efficient pair annihilation close to the center. The magnetic field in the X-ray emitting region is bound to be below the equipartition value with the radiation energy density, to let the relativistic particles (responsible for the X-ray emission) not to overproduce, by synchrotron radiation, the observed IR-optical flux. These severe limits on the magnetic field can be avoided if the synchrotron radiation is absorbed by free-free absorption in small and dense clouds. These clouds, reprocessing the absorbed power, could contribute to the UV bump.

1 THE IMPORTANCE OF ELECTRON–POSITRON PAIRS

A large fraction of the luminosity of a typical AGN is emitted as X and γ-rays. Since the variability at X-ray energies is extremely rapid, this high energy emission must originate at few Schwarzschild radii, making the nucleus of AGN extremely dense in high energy photons. Collisions involving photons are therefore very likely, and can be the most important ingredient in understanding the formation of the high energy spectrum.

The probability of γ-rays collisions with softer photons to produce $e^{\pm}$ pairs is proportional to the compactness l, defined as $l \equiv L\sigma_T/(Rm_ec^3) \sim 4000(L/L_E)(3R_s/R)$, where L/R is the luminosity to size ratio, L_E is the Eddington luminosity, R_s is the Schwarzschild radius, σ_T is the Thomson cross section, m_e is the electron mass and c is the speed of light. In a source exceeding $\sim 1\%$ of its Eddington luminosity, all γ-rays collide with softer photons to produce $e^{\pm}$ pairs.

Assume that high energy electrons (or pairs) are continuously injected throughout the source at a rate corresponding to a compactness l. These electrons produce, by Compton scattering off soft photons, a hard primary spectrum. The γ-rays so produced can be absorbed colliding with softer photons, producing relativistic electron–positron pairs which in turn contribute to the emission, mainly at X-ray energies (see the review by Svensson 1990 and references therein).

The self consistent X-ray spectrum steepens from $\alpha \sim 0.6$ when pairs are only marginally important (for $l = 1$) to $\alpha \sim 1$ when the pair production process saturates (all γ-rays get absorbed). For $l > 30$, up to 10% of the injected electron luminosity is converted into rest mass of pairs, and ultimately into the annihilation line, broadened by scattering and by fast motions within the pair clouds.

Until very recently, it was thought that the pair models could *not* explain the observational data : the X-ray spectrum was fitted with a single power law of energy index 0.7, which in the pair models corresponds to moderate values of the compactness, for which the γ-ray spectrum continues unbroken to large energies. This was inconsistent with the constraints posed by the γ-ray background.

However, the improved spectral resolution of the Japanese X-ray satellite *Ginga* allowed the decomposition of the 1–30 keV spectrum of bright AGN into a power law and a Compton reflection hump. The spectral index of the power law component is $\alpha_x \sim 0.9$, and in general is steeper by a factor $\Delta\alpha \sim 0.2$ than that derived with a single power law fit (e.g. Pounds 1989).

Zdziarski *et al.* (1990) suggested that the source of the power law component is a compact region lying close to an accretion disk. The compactness $l > 60$ of this region means that pair production is saturated, and that the emerging X-ray flux is a power law of index $\alpha_x \sim 0.8 - 1$.

Impinging onto the accretion disk, the hard X-rays are Compton down-scattered, and the soft X-rays are absorbed. The reflected spectrum then peaks at 20–30 keV. An example is shown in Fig. 1. As can be seen, the spectral index of the sum is $\alpha_x \sim 0.7$, and the overall spectrum turns over at a few MeV, in full agreement with present observations. Note that the model is easily testable. In fact we predict that the spectrum above ~ 30 keV should steepen, at least in the most compact sources, because of the steepening of the incident spectrum (due to down-scattering by cold pairs) and because of the vanishing contribution of the reflected component. Spectral variability is allowed, but with small $\Delta\alpha$, as the luminosity of the compact source varies. The γ-ray spectrum of the compact source has $\alpha_\gamma \sim 1.8$, similar to the shape of the γ-ray background. It is therefore likely that the origin of the γ-ray background is emission from nearby AGN.

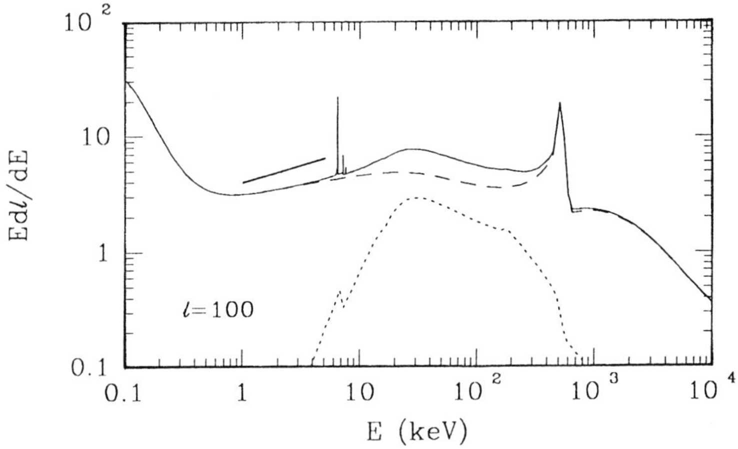

Fig. 1: *Spectrum from combined nonthermal pair cascades and Compton reflection, for a compactness* $l = 100$. *The dashed curve is the incident pair cascade spectrum. The dotted curve is the reflected spectrum, and the solid curve is the total observed spectrum. For comparison an* $\alpha = 0.7$ *power law is also shown as a thick solid line. For clarity, the iron* K_α, K_β *and nickel* K_α *fluorescence lines are shown in the total spectrum only. From Zdziarski et al. (1990).*

2 ARE PARSEC–SCALE JETS MADE BY ELECTRON–POSITRON PAIRS ?

We have seen in the previous section that $e^{\pm}$ pair production play a key role in our understanding of the high energy phenomena in AGN. Can $e^{\pm}$ pairs produced in the nucleus also supply the radio jets ? We (Ghisellini *et al.* 1991) try to answer this question by comparing the *maximum number flux* of $e^{\pm}$ outflowing from the nucleus with the *minimum number flux* of particles needed to emit the synchrotron radiation we observe from the VLBI core of radio sources.

Quite independently on the particular pair production model, in fact, there is a limit on the number of pairs escaping a compact source, due to the efficient annihilation they suffer at the base of the outflow. This limit could be evaded if the pairs were hot, since the annihilation cross section is a decreasing function of the temperature. But a large density of hot pairs results in a catastrophic Compton cooling, and if a strong acceleration mechanism prevented pairs to cool, the resulting saturated Comptonization would destroy the power law shape of the X-ray spectrum. Outflowing pairs are therefore cold, and annihilate efficiently. At a couple of source radii, the $e^{\pm}$ density is very weakly dependent on $e^{\pm}$ density at the base of the outflow : the larger the initial density, the larger the annihilation rate.

The density of the surviving pairs has been obtained by Guilbert & Stepney (1985), assuming a pure pair wind driven by radiation pressure until radiation is trapped in the wind. We then ask if these pairs are a large fraction of the total mass of the matter flowing in the VLBI jets. To answer this, we must estimate the particle density in the jet. We then apply the standard synchrotron self-Compton theory (Jones, O'Dell & Stein 1974) to derive the density of the emitting particles, assumed to have a power law energy distribution $N(\gamma) = N_0\gamma^{-2\alpha-1}$ between γ_{min} and γ_{max}, where α is the thin synchrotron spectral index ($\alpha = 0.75$ is here assumed). To derive the particle density, we take into account beaming effects, due to relativistic bulk motion of the plasma. Note that : i) γ_{min} is a critical parameter, since, for $\alpha > 0$, most of the particles are at energies close to γ_{min}; ii) the derived density N_0 is a lower limit on the actual density, due to the possible presence of other, thermal, non emitting, particles.

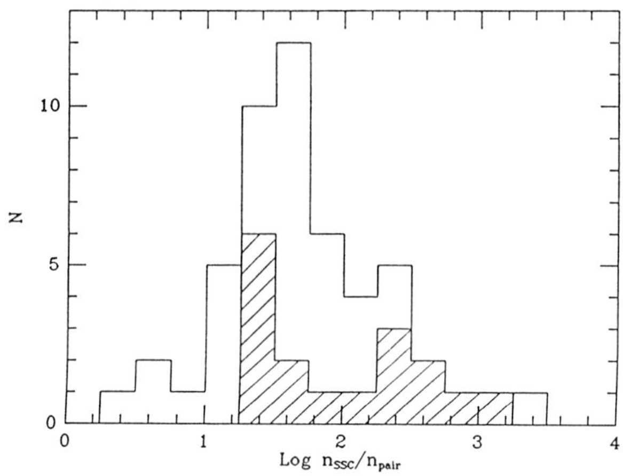

Fig. 2: *Histogram of the logarithm of the ratio n_{SSC}/n_{pair}. The dashed part of the histogram refers to the radio sources listed in Pearson & Readhead (1988). From Ghisellini et al. (1991).*

We derive the density of particles needed to produce the synchrotron emission we observe, n_{SSC}, and compare it with the predicted (maximum) density of $e^{\pm}$ (n_{pair}) able to reach the observed radio region. Fig. 2 shows the histogram of the ratios n_{SSC}/n_{pairs} for a sample of 51 sources, comprising BL Lac objects, HPQs (data in Madau, Ghisellini, & Persic 1987 and references therein), and sources from the 3C catalogue for which VLBI data are available (Pearson & Readhead 1988). The ratio plotted n_{SSC}/n_{pairs} has been calculated assuming $\gamma_{min} = 1$. If this is the case, then $e^{\pm}$ pairs cannot fuel the inner jet of radio sources. The jets would have to be made by "normal" (electron proton) matter.

However, pairs could be "rescued" if the power law energy distribution $N(\gamma)$ has a low energy cutoff at $\gamma_{min} \sim 100$. If this is the case, then an efficient acceleration mechanism, inhibiting pairs to cool, is required. On the other hand, if the plasma is made by normal matter, and polarization is observed, one requires just the same values of the low energy cutoff, to avoid Faraday depolarization (Wardle 1977). The conclusion is therefore that, no matter the nature of the plasma in the jet, there must be an efficient acceleration mechanism at work.

3 LIMITS ON THE MAGNETIC FIELD IN THE NUCLEUS

Up to now, estimates of the value of magnetic field in AGN have been made for the radio emitting components, which refer to linear scale of ~ 1 parsec, or larger. However, Celotti, Ghisellini & Fabian (1991) have recently estimated the value of the B-field in the X-ray emitting region of AGN, with a very simple method. Here I summarize the basic points.

Assume that relativistic electrons (and pairs) scatter soft photons (of the UV bump) to X-ray energies. Knowing the energy density of the soft photons allows us to derive scattering optical depth of the particles. Estimating the size of the region (from variability) we can then derive the number density of the X-ray emitting particles.

If a magnetic field is present, then the particles, besides producing high energy photons by Compton scattering, would radiate in the IR-optical by synchrotron. The larger the B-field, the larger their contribution to the total observed IR-optical flux. Therefore we derive an upper limit to the magnetic field by requiring that they do not overproduce the observed IR-optical spectrum. This limit also depends on the assumed maximum energy $\gamma_{max} m_e c^2$ of the particles, which controls the maximum extent in energy of the synchrotron spectrum. In this way we define a forbidden area in the $B - \gamma_{max}$ plane.

We have applied the method to the two Seyfert galaxies NGC 4051 and NGC 6814, the radio loud quasar 3C 273, the OVV quasar 3C 279, and the radio quiet quasar H1821+643. Basically, we selected sources with well-defined X-ray variability timescales and with evidence for a "blue bump" component.

Surprisingly, we derive that equipartition between the magnetic and the radiation energy density is not possible in the relevant range of γ_{max}. As as example, Fig. 3 shows the plane $B - \gamma_{max}$ for 3C273, for which the derived ratio of the magnetic and the radiation energy densities turned out to be $U_B/U_{rad} \sim 10^{-3}$.

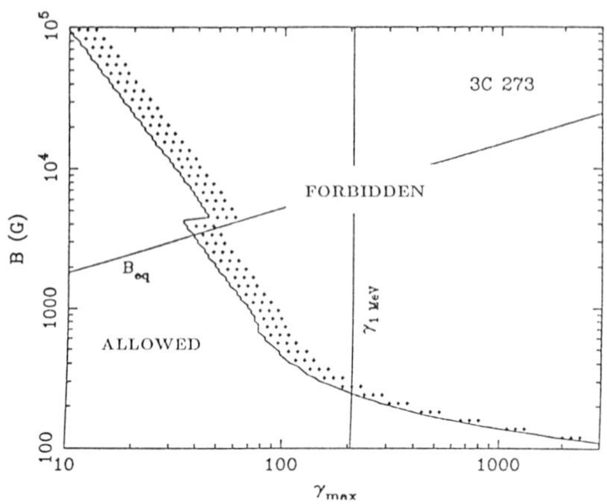

Fig. 3: *The maximum allowed magnetic field versus the maximum Lorentz factor of the electron distribution for 3C 273. The vertical continuous line represents the value of the Lorentz factor needed to produce 1 MeV photons by scattering UV photons. The line labelled B_{eq} shows the magnetic field in equipartition with the radiation energy density. From Celotti et al. (1991).*

Our results can have very important consequence on those models using a dominant magnetic field to accelerate the particles at high energies. Therefore we have explored alternative pictures, and the most promising of them is outlined in the following.

To find our limits we have assumed that the X-ray emission region is directly visible in the optical and UV bands. This would not be the case if the region is blanketed in sufficient cold gas ($T \sim 10^5$ K) that free-free absorption occurs throughout the optical and UV wavebands.

In order the soft X-rays not to be photoelectric absorbed, the gas can be distributed in small dense clouds with a large covering fraction, total column density $N_H \sim 10^{21}$ cm^{-2}, density $n \sim 10^{18}$ cm^{-3}, and size $d \sim 10^3$ cm. The implied thermal pressure of the clouds is then so high that they would rapidly disperse at the internal sound speed if not confined. The intracloud pressure is proposed to be due to a strong magnetic field (Rees 1987), its value being consistent with the equipartition magnetic field required by the X-ray variability.

The power reprocessed by these dense and small blobs can be high, and can contribute to the EUV bump, together with the contribution of other relatively cold gas, of column density of at least 10^{24} cm^{-2}, required to explain the strength of the observed fluorescent iron line (see e.g. George & Fabian 1991).

REFERENCES

Celotti, A., Ghisellini, G. & Fabian, A.C., 1991. *Mon. Not. R. Astr. Soc.*, in press

George, I.M. & Fabian, A.C., 1991, *Mon. Not. R. Astr. Soc.*, **249**, 352

Ghisellini, G., Celotti, A., George, I.M. & Fabian, A.C., 1991, in preparation

Guilbert, P.W. & Stepney, S., 1985, *Mon. Not. R. Astr. Soc.*, **212**, 523

Jones, T.W., O'Dell, S.L. & Stein, W.A., 1974, *Astroph. J.*, **188**, 353

Madau, P., Ghisellini, G. & Persic, M., 1987, *Mon. Not. R. Astr. Soc.*, **224**, 257

Pearson, T.J. & Readhead, A.C.S., 1988, *Astroph. J.*, **328**, 114

Pounds, K.A., 1989, in *AGN and the X-ray background*, 23rd ESLAB Symposium, ESA SP-296, p. 753

Rees, M.J., 1987, *Mon. Not. R. Astr. Soc.*, **228**, 47P

Svensson, R., 1990, in *Physical Processes in Hot Cosmic Plasmas*, eds. W. Brinkmann *et al.* (Kluwer Academic Publishers, The Netherlands), p. 357

Wardle, J.F.C., 1977, *Nature*, **269**, 563

Zdziarski, A.A., Ghisellini, G., George, I.M., Fabian, A.C., Svensson, R. & Done, C., 1990, *Astroph. J.*, **363**, L1

Models of Central Engines of AGNs with Electron-Positron Pair Production

ANDRZEJ A. ZDZIARSKI

N. Copernicus Astronomical Center, Bartycka 18, PL-00-716 Warsaw, Poland
and
NASA Goddard Space Flight Center, Code 666, Greenbelt, MD 20771, USA

1 INTRODUCTION

In this contribution, I will review selected theoretical models of the central regions of active galactic nuclei (hereafter AGNs) with the emphasis on the models and scenarios that predict copious production of $e^{\pm}$ pairs. The pairs can possibly be transported away from the central regions and give rise to winds, beams, and jets. Mechanisms for spatial transport of pairs away from the central engine will not be discussed here. The spectral and time variability predictions of the models and comparison with AGN observations will be discussed briefly.

The models will be classified based on their assumed geometry. One-zone models (with and without reflection), spherical accretion, and disk accretion will be considered. Two further classification criteria are whether particles in the sources are thermal or nonthermal, and whether the available power is supplied mainly to protons or to $e^{\pm}$.

It is assumed that the reader is familiar with both thermal and nonthermal pair production in compact sources. For a current review, see, e.g., Svensson (1990, hereafter S90). Compact source models with pairs in which the primary power is supplied to protons have recently been reviewed in Zdziarski (1991).

We will use here the dimensionless accretion rate, $\dot{m} \equiv \dot{M}c^2/L_{\rm E}$, the accretion efficiency, $e \equiv L/\dot{M}c^2$, the dimensionless electron temperature, $\Theta_e \equiv kT_e/m_e c^2$, the 'proton optical depth', $\tau_p \equiv n_p \sigma_{\rm T} R$, and the total optical depth, $\tau_{\rm T} \equiv (n_+ + n_-)R\sigma_{\rm T}$, where L is the luminosity, $\dot{M}$ is the mass accretion rate, $L_{\rm E}$ is the Eddington luminosity, R is the source size, $n_{\pm}$ is the $e^{\pm}$ density, n_p is the proton density, and $\sigma_{\rm T}$ is the Thomson cross section. The major parameter determining the amount of pair production is the compactness,

$$\ell \equiv \frac{L}{R}\frac{\sigma_{\rm T}}{m_e c^3} = \frac{2\pi}{3}\frac{m_p}{m_e}\left(\frac{L}{L_{\rm E}}\right)\left(\frac{3R_{\rm S}}{R}\right) \simeq 3600\left(\frac{L}{L_{\rm E}}\right)\left(\frac{3R_{\rm S}}{R}\right), \qquad (1)$$

where $R_{\rm S} \equiv 2GM/c^2$ is the Schwarzschild radius. If the luminosity L is emitted mostly in the soft γ-ray range, the source becomes optically thick to pair production in photon-photon collisions when $\ell \gtrsim 10$ (Guilbert, Fabian, & Rees 1983).

2 ONE-ZONE MODELS AND REFLECTION MODELS

In this section, selected models that treat the source as a single homogeneous zone are discussed. This source may be isolated, be irradiated by some external radiation, or be in the vicinity of optically thick cold matter.

2.1 Thermal Pair Equilibrium Plasma Clouds

In this class of models, an astrophysical compact source is approximated as a cloud of plasma of a mildly relativistic T_e in which the rate of $e^{\pm}$ pair production equals the pair annihilation rate (pair equilibrium). The case without magnetic fields or external photon sources is treated by Svensson (1984; see also Lightman 1982 and Svensson 1982). The case with copious external photons or weak magnetic fields is treated by Zdziarski (1985). We will discuss here isolated thermal pair sources only briefly, as they have been a subject of previous reviews (e.g., Svensson 1986; S90).

The condition of pair equilibrium imposes constraints on the parameter space ℓ–Θ_e–τ_p (Svensson 1984; Zdziarski 1985). For $\Theta_e \sim 1$, $\tau_p \lesssim 10$, and $\ell \gtrsim 1$, $n_+ \gtrsim n_p$ can be obtained. The most important consequences of pair equilibrium are that there exist the maximum Θ_e at a given τ_p, and the maximum ℓ at given Θ_e (and $\tau_p \to 0$). For sources with external soft photons, there is an additional dependence (Zdziarski 1985) on the energy spectral index of the power law spectrum from thermal Comptonization, $\alpha = \alpha(\Theta_e, \tau_{\rm T})$ (where α is defined by $F_E \propto E^{-\alpha}$). Those constraints can be compared to observational data on compact sources in AGNs. It is found that the observed AGN X-ray spectral indices of $\alpha \simeq 0.7 \pm 0.15$ (Rotschild *et al.* 1983; Turner and Pounds 1989, hereafter TP89), the X-ray luminosities and variability timescales (constraining ℓ), and the maximum observed photon energies (constraining Θ_e) are consistent with Compton upscattering of the AGN UV photons by thermal pair equilibrium plasma (Zdziarski 1985; Zdziarski & Lightman 1987). On the other hand, there is no apparent reason why spectra from thermal $e^{\pm}$ pair plasmas should have the energy spectral index $\alpha \sim 0.7$. Also, the AGN X-ray spectra either maintain a constant slope or steepen with the increasing flux, which is not explained by this model. A test excluding the thermal model would be to observe a fast-varying power law spectrum extending to such high energies that the implied ℓ and Θ_e violate the pair equilibrium constraints.

An important extension of the thermal model has been proposed by Haardt & Maraschi (1991). In their model, the thermal source forms a corona above the surface of cold matter (e.g., an accretion disk). Dissipation occurs mostly in the corona. The disk is irradiated by the downward flux of the corona. Most of the downward flux is absorbed and reradiated as blackbody radiation, and a fraction of 0.1–0.2 of the downward flux is Compton-reflected (Lightman & White 1988; White, Lightman & Zdziarski 1988). The reflected spectrum peaks at ~ 30 keV because photons with lower energies are bound–free-absorbed and photons at higher energies scatter preferentially forward with the reduced Klein-Nishina cross section and are ultimately absorbed deep in the medium (Lightman and White 1988). The observed X-ray spectrum is the sum of the direct component from the corona and the Compton-reflected component from the cold matter. The observed spectrum is substantially harder below ~ 30 keV because of the presence of the reflected component (Lightman and White 1988).

Haardt & Maraschi (1991) consider only the optically thin case, with $\tau_{\rm T} < 1$, and neglect higher-order reflections. They find the corona temperatures of $T_e \sim 10^9$–10^{10} K and the spectral indices of $\alpha \sim 0.7$–1, which can explain AGNs with softer X-ray spectra. For $\tau_p \ll 1$ and large enough ℓ, $n_+ \gtrsim n_p$ in the corona.

2.2 Thermal Sources with Hot Protons

Sikora & Zbyszewska (1985) consider an inner, quasi-spherical zone of a accretion flow assuming steady state, Coulomb heating of electrons by protons of constant temperature, T_p ($> T_e$), $e^{\pm}$ pair equilibrium, and Comptonized bremsstrahlung as the dominant radiative process. The last assumption requires that the magnetic field strength is much below equipartition. They find that such a source can in principle account for the soft γ-ray spectra observed from strong γ-ray emitting AGNs (e.g., NGC 4151, MCG 8-11-11; Bassani and Dean 1983). However, a rather low compactness for the innermost region of the accretion flow, $\ell \lesssim 10$, is required. An effective shielding from the UV and X-ray radiation emitted outside the γ-ray source (e.g., by an accretion disk) is required as well.

Stability of thermal plasma clouds with electrons and pairs heated by protons is studied by Björnsson & Svensson (1991a; see there for earlier references). They consider a source with a constant power supplied to the protons of a constant density, and neglect pair advection and outflow. They include energy conservation for both electrons and protons (rather than to assume a constant T_p). They find a small linearly-unstable region, with $10 < \ell < 25$ and $\tau_p < 2$.

Dermer (1988) considers a magnetized thermal two-temperature source, with the magnetic field energy density in equipartition with the radiation energy density. The T_p is taken as a free parameter and T_e follows from the balance between Coulomb heating and radiative losses. He considers π^+ production in collisions of thermal protons. Then e^+ from π^+ decay radiate soft photons in the synchrotron process. Those soft photons are then Comptonized by thermal electrons, forming a power law spectrum extending to the X-ray regime. He finds that the presence of the soft photons from π^+-decay positrons has a stabilizing effect on the spectral index of the Comptonization spectrum. The X-ray spectral index turns out to be close to $\alpha = 0.7$, as observed in AGNs (e.g., TP89). Dermer (1988) does not give the obtained values of T_e. Thus it is difficult to assess the importance of thermal $e^{\pm}$ pair production in the model.

Some effects are not taken into account in the model. Irradiation of the source by the UV bump radiation, prominent in AGNs, would provide additional electron cooling and soften the X-ray spectra. Also, γ-rays from π^0 decay would be pair-absorbed in typical AGN conditions. The resulting pairs would outnumber the positrons from π^+ decay and produce more synchrotron soft photons than included in the model. This would have a further cooling and softening effect.

2.3 Nonthermal Pair Plasma Clouds

The crucial feature distinguishing nonthermal pair models is production (or acceleration) of particles with nonthermal distribution at relativistic energies (constituting the primary injection). If the relativistic particles are $e^{\pm}$, they radiate via the synchrotron and Compton processes. The photon energies extend to the γ-ray regime. Those photons produce $e^{\pm}$ pairs if the source is compact. The pairs constitute a secondary injection, radiate γ-rays which again produce new pairs, and a pair cascade develops. Isolated nonthermal pair plasma clouds will be treated here only briefly as they have been reviewed in the past (e.g., in S90).

Svensson (1987), Lightman & Zdziarski (1987), Done & Fabian (1989) and others have studied extensively the nonthermal pair model. The main finding of those studies is that for monoenergetic injection of $e^{\pm}$ at a large Lorentz factor and for the range of $1 \lesssim \ell \lesssim 10^2$, the X-ray spectral index is $\alpha \simeq 0.7 \pm 0.15$, which is in a general agreement with the AGN observations (see S90 for more discussion).

An important extension of this model was the inclusion of the presence of cold matter (Zdziarski *et al.* 1990), prompted by the discovery of reflection features in the AGN spectra (e.g., Pounds *et al.* 1990). The nonthermal source is assumed to be above the surface of cold matter. The nonthermal source can, e.g., be located near the rotation axis of an accretion disk, or it can form a corona above the disk surface. The inclusion of the reflected component makes the total spectra harder (see §2.1). Thus, the intrinsic X-ray spectrum emitted by the nonthermal cloud is now required to be softer, with $\alpha_{\rm intr} \simeq 0.9$ for typical AGNs. This intrinsic index corresponds to saturated pair cascades (Svensson 1987; S90) and it is stable over a large range of compactnesses, $30 \lesssim \ell \lesssim 10^3$. The larger compactnesses implied by this model are in good agreement with the observations of fast varying X-ray spectra as well as with the constraints imposed on the AGN γ-ray spectra by the cosmic γ-ray background (Done & Fabian 1989).

Zdziarski and Coppi (1991) have shown that the nonthermal model (with or without the reflection component) explains the soft X-ray excesses below ~ 1 keV seen in a large fraction of AGNs (TP89). The soft excess is due to Comptonization of the UV photons by the thermalized pairs present in the nonthermal source (see, e.g., S90).

2.4 Nonthermal Sources with Relativistic Protons

Zdziarski (1986) considers a nonthermal source in which protons are accelerated to relativistic energies at the steep rate $\propto \gamma^{-\Gamma}$, $\Gamma = 2.4$ (chosen to match the observed IR–X-ray AGN spectra, see below). The process of pion production in collisions of the relativistic protons with the thermal protons of the accretion flow is considered. This process gives rise to a nonthermal power law $e^{\pm}$ production rate with the same index as that of the primary protons. There is a kinematic low-energy cutoff from the π rest mass in the production rate at $\gamma_{\rm min} \sim 10^2$. The $e^{\pm}$ lose energy in the synchrotron and Compton processes. The magnetic field energy density is taken to be in equipartition with the radiation energy density. The process of $e^{\pm}$ energy loss gives rise to a broken power-law steady-state electron distribution, with the break at $\gamma_{\rm min}$. The near-IR and optical spectra are dominated by synchrotron radiation from the steeper part of the electron distribution and have $\alpha = 1.2$. The UV bump is assumed to come from another source and it provides additional seed photons for Comptonization. The hard X-ray spectrum is due to Comptonization by the entire electron distribution and its spectral index turns out to be $\alpha \sim 0.7$-0.9, in a qualitative agreement with the AGN observations.

The model cannot, however, reproduce spectra with $\alpha \lesssim 0.7$, common in AGNs (e.g., TP89). Also, the model predicts that the IR and optical variability time scales should be qualitatively similar to that of X-rays, which has not been observed in AGNs (however, a fast optical variability has been recently detected in NGC 7469, Dultzin-Hacyan 1991).

Sikora *et al.* (1987) consider acceleration of protons to relativistic energies at a flat power law rate, with $\Gamma < 2$. They point out the importance of collisions of relativistic protons with soft photons for the proton energy loss and production of $e^{\pm}$ pairs and γ-rays. The possible processes are photo-pion production, $p\gamma \rightarrow p\pi$ or $n\pi$, and production of $e^{\pm}$ pairs, $p\gamma \rightarrow pe^+e^-$. Note that the former process can produce neutrons. Proton-photon energy loss dominates energy loss via inelastic proton-proton collisions at high enough energies. For example, for $\dot{m} = 0.1$ and a power law spectrum of soft photons with $\alpha = 1$, the proton-photon processes dominate at

Lorentz factors $\gamma \gtrsim 10^5$. If the soft photon spectrum is a power law, the photo-pion production dominates over proton-photon pair production for $\alpha \lesssim 1$. Both processes give rise to production of nonthermal relativistic $e^{\pm}$ pairs with power law distributions with two different normalizations and cutoff energies, and with the spectral index equal to that of the accelerated protons.

Kazanas & Ellison (1986b), Sikora, Begelman & Rudak (1989), Kirk & Mastichiadis (1989), and Atoyan (1990) point out the importance of generation of relativistic neutrons in *pp* and *pγ* collisions. In contrast to protons, neutrons are not scattered by chaotic magnetic fields and can travel ballistically. The maximum Lorentz factor of the accelerated protons is determined by the balance of the acceleration rate in a shock with the energy loss rate and can be as high at $\gamma_{\max} \sim 10^8$ (Sikora *et al.* 1987). The maximum Lorentz factors of escaping neutrons is somewhat less ($\gamma_{\max} \sim 10^7$), as they have to be able to escape the central source without undergoing a pion-production reaction (Sikora *et al.* 1989). The decay time of neutrons in the source frame is about $10^3\gamma$ s. Thus, the escaping neutrons can travel as far as 100 pc before decaying. Upon decay, they transfer their energy to the ambient matter, which can, e.g., generate large scale winds and inhibit accretion (Sikora *et al.* 1989; Begelman, Rudak & Sikora 1990). The neutrons escaping to large distances in AGNs can also give rise to a detectable flux of VHE and UHE γ-rays from decay of pions produced in hadronic interactions (Mastichiadis & Protheroe 1990).

Begelman *et al.* (1990) consider in some detail hadronic-electromagnetic cascades generated by proton acceleration, taking into account the neutron production. They consider an example in which the dominant electron energy loss process is synchrotron radiation. In the cascades, protons produce pions and $e^{\pm}$ pairs, the pions decay into γ-rays and $e^{\pm}$, the $e^{\pm}$ radiate synchrotron photons and Compton upscatter soft photons, and the γ-rays produce further $e^{\pm}$ pairs.

Stern, Svensson & Sikora (1990) and Stern & Svensson (1991) consider an isolated compact source with constant power used to inject monoenergetic relativistic protons. The injected power gets converted into soft radiation by an $e^{\pm}$ cascade. If the Lorentz factor of the injected protons is low enough, the source becomes unstable and developes limit cycles with the proton and the photon components interchangeably dominating. This may cause periodic large amplitude variations of the luminosity from a compact object. A possibly stabilizing effect of external radiation on the nonthermal source has not been investigated.

3 SPHERICAL BLACK HOLE ACCRETION

In spherical, as well as disk, accretion, a major fraction of the available power is supplied to ions, as the available gravitational energy per particle is proportional to its mass. The maximum number of pairs per accreting proton is $\lesssim 10$ (Svensson 1987; S90).

Spherical accretion can be hydrodynamic (see §§3.1, 3.2) or collisionless (see §§3.3, 3.4). In the former case, one assumes that accreting protons interact effectively with each other, with the mean free path much shorter than the distance to the black hole, and behave like fluid, accreting on a time scale of the order of the free fall time scale. The short mean free path can be due to, e.g., chaotic magnetic fields. In the latter case, proton interactions are ineffective, and the protons orbit the black hole. In that case, the accretion time scale is determined by the rate of the proton energy loss and may be much longer than the free fall time scale.

The accretion efficiency, e, is in some models self-consistently solved for, based on the values of the velocity of the accretion flow and the effectiveness of assumed radiative processes (§§3.1, 3.3, 3.4). This efficiency can be very small, $e \ll 0.1$ (§3.1). On the other hand, dissipative processes related to the presence of chaotic magnetic fields (§3.2) can increase the efficiency to $e \sim 0.1$.

3.1 Thermal Adiabatic Hydrodynamic Accretion

Steady-state adiabatic hydrodynamic spherical accretion including effects of $e^{\pm}$ pairs is studied by Park & Ostriker (1989) and Park (1990a, b). Their models incorporate relativistic hydrodynamics, radiation moment equations, bremsstrahlung and Comptonization as the main radiative process, and take into account the possible presence of a shock and effects of preheating. The accretion efficiency is solved for. Only adiabatic heating of the electrons and protons occurs. It is assumed that magnetic fields are sufficient to fluidize accretion flows, but small enough not to give rise to any substantial synchrotron radiation.

Self-consistent models by Park (1990a, b) predict $n_+ \ll n_p$. They are discussed in some detail in Zdziarski (1991). High-temperature solutions for two-temperature accretion showing $n_+ \sim$ a few n_p are found by Park & Ostriker (1989) at $3 \lesssim \dot{m} \lesssim 40$. The efficiencies and luminosities are relatively high, $e \sim 10^{-2}$ and $L/L_{\rm E} \sim 10^{-2}$. The two different states of the flow at a given $\dot{m}$ can be regarded as two different pair equilibrium states of relativistic thermal plasma (§2.1).

One should note, however, that the high n_+ solutions are not obtained by integration from the accretion radius, in contrast to the models of Park (1990a, b). Rather, they are obtained assuming the density and temperature at $10^3 R_{\rm S}$. Unlike the low n_+ models of Park (1990a, b) they cannot be connected to the steady-state outer part of the flow and require suppressing in some way the preheating disruption (Park 1990b). Park & Ostriker (1989) argue that any related instability will have a time scale much longer than the dynamical time scale at $10^3 R_{\rm S}$ and the inner part of the flow can be considered time-independent.

Park (1990b) considers high-energy γ-ray production from pions produced by hot thermal ions in two-temperature flows. He finds the γ-ray luminosity to be a significant fraction of the total luminosity only for $\dot{m} \lesssim 10^{-3}$, which corresponds only to very weak sources. Thus, pion production in hydrodynamic thermal accretion appears not to be a process relevant to cosmic γ-ray sources. This is due to the flow temperature being too low for efficient pion production to proceed.

The models of Park (1990a, b) predict luminosities many orders of magnitude below the Eddington limit, which is very unlikely for AGNs. On the other hand, the pair-abundant model by Park and Ostriker (1989) gives $L/L_{\rm E} \sim 10^{-2}$, which is a likely value for AGNs. Also, the model predicts $T_e \sim 10^{10}$ K, which can in principle account for X-rays and γ-rays observed from AGNs. However, the spectra from accretion flows in those models are essentially those of Comptonized bremsstrahlung, which typically have the spectral index $\alpha \lesssim 0$. This is much too hard for typical AGNs, with the average X-ray energy spectral index of $\alpha \simeq 0.7$.

3.2 Dissipative Hydrodynamic Accretion

Mészáros (1975) and Maraschi *et al.* (1979, 1982) suggest that the accretion gravitational energy can be efficiently converted into heat by turbulence and magnetic dissipation. The efficiency can be then as high as $e \sim 0.1$. The magnetic field in equipartition with the gravitational energy and $T_e = T_p$ are assumed. The thermal cyclo-synchrotron, bremsstrahlung and multiple Compton scattering processes are then fast enough to radiate away almost all the thermal energy. The adiabatic heating is only about $\sim 5\%$ of the dissipative turbulent heating.

The maximum temperatures of the flow of $T_e \sim 10^9$–10^{10} K at the accretion rates $2 < \dot{m} < 20$ are obtained in the model. A near-isothermal high temperature zone at $R \lesssim 10^2 R_S$ appears in the models. The spectra at infinity are nearly power laws due to multiple Compton scattering of the self-absorbed cyclo-synchrotron radiation. The power laws extend from the infrared to soft γ-rays. The typical spectral indices obtained by Maraschi *et al.* (1982) are $\alpha \sim 0.5$–1, in agreement with the typical X-ray spectra of AGNs. The obtained spectra are, however, in disagreement with the AGN IR and UV spectra, which are more complex than simple power laws.

Pair production is not taken into account in those models. Pairs will be present, however, in the conditions of $\Theta_e \sim 1$ and $\tau_T \sim 1$ (see §2.1), characteristic of this class of models. As a near-isothermal inner zone appears in the model, one can use the pair-equilibrium study of Zdziarski (1985) to infer the average T_e and n_+. Such calculations are done in Zdziarski (1988), where the solutions have abundant pairs ($n_+ \lesssim 5n_p$) for $0.2 \lesssim \dot{m} \lesssim 5$ at the inflow velocity of 1/2 of that of the free fall. An effect of pairs is to lower the temperature and to increase the optical depth. The emitted spectra are now harder (of $\alpha \sim 0$) but cut off at lower energies. The total luminosity is not affected by the pairs, as the dissipation of gravitational energy is related here to turbulent and magnetic processes and its efficiency is assumed *a priori*. No instability due to pairs is expected. The pairs reduce the Eddington luminosity, but they also cause a reduction in the actual luminosity by increasing the trapping radius in the spherical flow (Zdziarski 1988).

Begelman, Sikora, & Rees (1987) and Tritz & Tsuruta (1989) consider dissipative quasi-spherical accretion flows with $T_p > T_e$ and pair production (without integrating the flow dynamic equations). Tritz & Tsuruta (1989) assume the presence of external photons and no magnetic fields. Begelman *et al.* (1987) and Tritz & Tsuruta (1989) find a critical accretion rate, $\dot{m}_{\rm crit} \sim 1$, above which the hot protons (e.g., in ion-supported torii, Rees *et al.* 1982) will cool within some radius on the inflow time scale due to Coulomb interactions with ionization electrons and $e^\pm$ pairs. Then, the flow is expected to collapse to a geometrically thin accretion disk. This effect would occur at lower $\dot{m}_{\rm crit}$ and larger R if pairs were present. However, Tritz & Tsuruta (1989) find only few pairs in these two-temperature accretion flows if the pairs are produced thermally.

Lightman, Zdziarski & Rees (1987) and Tritz & Tsuruta (1989) consider spherical dissipative accretion flows with nonthermal pair production. Lightman *et al.* (1987) find abundant pairs ($n_+ \lesssim 10n_p$) for $0.03 \lesssim \dot{m} \lesssim 30$. The pairs lower the Eddington luminosity and increase the trapping radius, similarly to the thermal case (Zdziarski 1988). Tritz & Tsuruta (1989) find that the abundant (nonthermally-produced) pairs in two-temperature flows decrease the $m_{\rm crit}$ at which an ion-supported flow collapses.

3.3 Thermal Collisionless Accretion

In this class of models, one typically assumes that the accretion flow becomes collisionless downstream of a standing shock, which appears at some $R \gg R_S$ (Mészáros & Ostriker 1983). This effect is due to heating the downstream protons to a near-virial temperature, at which they become collisionless. For this to be the case, the magnetic field must be many orders of magnitude below its equipartition value. The downstream protons orbit the black hole in a star-like manner. The accretion rate is then equal to the rate of proton cooling due to Coulomb collisions with electrons. Typically, $T_e \ll T_p$. As the time scale for proton cooling may be much longer than that of the free fall, the obtained efficiencies can also be much larger, $e \sim 0.1$.

The heated electrons can achieve relativistic temperatures. They radiate via bremsstrahlung, synchrotron, Compton and π^0 decay processes, and the spectra can extend to the X-ray and γ-ray band. Mészáros & Ostriker (1983) do not include $e^{\pm}$ pair production in their model.

The effects of pairs are considered by Moskalik & Sikora (1986). They consider a time-dependent source at a constant T_p and a constant matter inflow rate from external region. The proton accretion rate equals their cooling rate. The rate of change of the $e^{\pm}$ pair density equals the difference between the pair production rate due to photon-photon, photon-particle, and particle-particle interactions and the pair annihilation rate (pair advection and escape are neglected). They find that there is a narrow range of the values of the proton inflow rate at which there is no stable equilibrium of the source. For the case they consider with $kT_p = 60 m_e c^2$ (of the order of the proton virial temperature at $R = 10 R_S$), the unstable range corresponds to the average luminosity close to $10^{-2} L_E (10 R_S/R)$, or, equivalently, to $\ell \simeq 10$. Note that the proton temperature is not determined self-consistently here, but assumed, which may cause the absence of the instability in a more realistic accretion flow model (see below; Babul *et al.* 1989).

In the unstable range, the source exhibits a limit cycle behavior with recurrent flares. The flare timescale equals approximately the diffusion time, $\tau_T R/c$, and the recurrence period is found to be about $10^2 R/c$. During the recurrence period, the energy density in the protons is gradually accumulated, and during the flash, there is a pair production runaway and the accumulated energy gets transferred to $e^{\pm}$ pairs and, almost instantaneously, to their radiation.

Babul *et al.* (1989) also extend the model of Mészáros & Ostriker (1983) by including the effects of pairs (although in a very simplified manner). They consider supercritical accretion rates, for which solutions with a shock are found. The shock exists typically at $R \sim 10^2 R_S$. Downstream of the shock, there exists a collisionless transition region of the kind discussed in Mészáros & Ostriker (1983). The thickness of the transition region is assumed to equal the mean free path for proton cooling. In the transition region, $n_+ \lesssim 0.2 n_p$, $L \sim 0.1 L_E$, and no flaring is found. This L, and $R \sim 10^2 R_S$ correspond to the same compactness, $\ell \sim 10$, at which the flaring instability was found by Moskalik & Sikora (1986). However, the typical proton temperatures found by Babul *et al.* (1989) are $\sim 5 m_e c^2$ (of the order of the virial temperature at $10^2 R_S$), much less than the value of $60 m_e c^2$ assumed by Moskalik & Sikora (1986).

3.4 Nonthermal Collisionless Accretion

This class of models is similar to that discussed above in §3.3 except that a mechanism producing nonthermal, relativistic particles is invoked. This can be first order Fermi acceleration in an accretion shock (Protheroe & Kazanas 1983; Kazanas & Ellison 1986a). The shock is supported by the pressure of collisionless nonthermal protons downstream the shock. The magnetic field is assumed to be strong enough to confine the protons in the shock region, but small enough not to fluidize them. These requirements roughly correspond to the gyroradius of the most energetic protons being smaller than the shock radius but larger than the Schwarzshild radius. The latter condition corresponds to the rather high maximum Lorentz factor of the accelerated protons, $\gamma \gtrsim 10^8 (B/1\mathrm{G})(M/10^9 M_\odot)$.

The relativistic protons lose energy through pion production in collisions with ambient protons, provided the accretion rate is large enough. The balance between the relativistic proton energy loss rate and the injection rate determines the energy density downstream and the position of the shock (Kazanas & Ellison 1986a). The final products of pion decay are γ-rays, $e^\pm$, and neutrinos. Their production rates are calculated by Protheroe & Kazanas (1983) and Stecker *et al.* (1991). The primary γ-rays and $e^\pm$ give rise to an $e^\pm$ cascade, which is the process determining the form of radiated photons. The X-ray and γ-ray spectra from this process are qualitatively similar to those of AGNs and the cosmic γ-ray background (Kazanas & Protheroe 1983).

4 DISK ACCRETION

4.1 Cold Disks

The standard cold disk model of Shakura & Sunyaev (1973) assumes that the disk is geometrically thin and that the rate of angular momentum transfer is proportional to the total pressure times the constant viscosity parameter, α_v, with $\alpha_v \lesssim 1$. The pressure in an inner region of the disk is typically dominated by radiation, and $T_e = T_p$. When the disk is effectively optically thick in the vertical direction, its local radiation field can be approximated as modified blackbody. Then, the temperature of a disk is always nonrelativistic, for both stellar mass as well as supermassive black holes, and there is no $e^\pm$ pair production.

However, the innermost region of the disk is effectively optically thin, although the scattering optical depth is still very large. Then the modified blackbody approximation is no longer valid. Rather, one needs to explicitly consider the radiation production mechanisms, with bremsstrahlung and Comptonization being the major processes. Maraschi & Molendi (1990) study the temperatures and radiation spectra assuming that the inner part of the disk is homogeneous and isothermal in the vertical direction. They find that if $\dot{m} \sim e^{-1}$ and $\alpha_v \sim 1$, the maximum disk temperatures can be as high as $T \sim 10^9$ K. For disk accretion, $e \sim 0.1$. Maraschi & Molendi (1990) do not include effects of $e^\pm$ pairs although they notice that they would be present at $T \gtrsim 10^9$ K. One of the effects of pairs would be to limit T, as pair production increases the medium density (§2.1).

Maraschi & Molendi (1990) find that the temperatures of $T \gtrsim 10^8$ K, which are necessary for this model to account for the AGN X-ray spectra, are obtained only for large accretion rates, $\dot{m} \gtrsim 0.5e^{-1}$ and for $\alpha_v \sim 1$. However, when $\dot{m} \sim e^{-1}$, the assumption of geometrical thinness breaks down. The effects of the increase in disk thickness on the temperature and spectra are not investigated in that paper. Thus, it is not clear whether the results of the paper are valid for $0.5 \lesssim \dot{m}e \lesssim 1$. Even if this is so, the required narrow range of $\dot{m}$ limits the applicability of this model to AGNs.

4.2 Hot Disks

A solution to disk accretion in which the conditions for the presence of copious $e^\pm$ pairs are satisfied is that of the hot, effectively optically thin disk of Shapiro, Lightman

& Eardley (1976). In this model, protons and electrons are thermal but may have different temperatures, $T_p \sim 10^{11}$ K, and $T_e \sim 10^9$ K. As in dissipative thermal spherical accretion (§3.2), the protons are heated by some dissipative processes and cooled through Coulomb collisions with the electrons. The electrons radiate away the received energy.

The effects of $e^\pm$ pairs in such disks are studied by Kusunose & Takahara (1988, 1989), Tritz & Tsuruta (1989), White & Lightman (1989, 1990), and Björnsson & Svensson (1991b, c). White & Lightman (1989) consider several kinds of steady-state thermal accretion disks. They find $e^\pm$ pairs may be important in two-temperature Comptonized bremsstrahlung disks. For the accretion rates $\dot{m} > \dot{m}_{\rm crit} \lesssim 1$, there are no pair-equilibrium solutions in an annulus centered on $R \simeq 5R_{\rm S}$. In the stable regions of R and $\dot{m}$, there are two solutions, one with high n_+, and one with $n_+ \ll n_p$. The high n_+ solution appears unphysical, as $T_p > T_{\rm virial}$ in a large range of R, as well as it is unstable to pair runaway (White & Lightman 1990). The low n_+ solution is stable (White & Lightman 1990). If the electrons and protons are effectively coupled and $T_e = T_p$, the critical accretion rate is lower and the electron temperatures are higher (White & Lightman 1989). The disk at radii at which no pair-equilibrium solutions exist is likely to collapse to a cold disk (White & Lightman 1990).

Two-temperature disks with a supply of external soft photons are considered by White & Lightman (1989) and Tritz & Tsuruta (1989). They find that solutions with unsaturated Comptonization of soft photons (with $y = 1$, where $y \sim \Theta_e[\tau_{\rm T} + \tau_{\rm T}^2]$ is the Compton parameter) are stable to pair runaway and have $n_+ \ll n_p$. No gas-pressure dominated solutions are found for the case with $T_e = T_p$ (White & Lightman 1989).

Björnsson & Svensson (1991b) have shown that the equations describing effectively thin accretion disks can be rewritten as two relations between the accretion disk parameters: $\dot{m}$, α_v, and R; and the pair-equilibrium parameters: the *local* compactness, ℓ, and the local proton optical depth, τ_p. The results of pair equilibrium studies can be then directly mapped onto the disk parameter space. Each solution from the pair plasma studies then corresponds to a specific $\dot{m}$, α_v, and R. This constitutes a set of generic disk solutions for given α_v. Björnsson & Svensson (1991b) apply their method to the case of disks where the only source of photons is optically thin bremsstrahlung.

Björnsson & Svensson (1991c) give some results for the case of Comptonized bremsstrahlung disks. That specific disk solution can be obtained from the generic solution of Björnsson & Svensson (1991b). They find that pairs become important for $\dot{m} \gtrsim 1$. For $\alpha_v < \alpha_{\rm crit} \sim 1$, infinite radial gradients appear when pairs become important and the standard disk assumptions break down. For $\alpha_v > \alpha_{\rm crit}$, fully self-consistent disk solutions with $n_+ \sim n_p$ exist.

Eilek & Kafatos (1983) consider the effects of pion production in two-temperature hot accretion disks around Kerr black holes. Relativistic $e^\pm$ are produced in this model via $\pi^\pm$ decay and via photon-photon pair production of γ-rays from π^0 decay. Note that those $e^\pm$ are nonthermal although the plasma in which they are produced is thermal. The authors also note that the nonthermal pairs can be seeds for an effective nonthermal acceleration.

Eilek & Kafatos (1983) sketch the spectra from such a disk around a canonical Kerr black hole for a few values of y, α_v and $\dot{m}$. For $\alpha_v \ll 1$ and $\dot{m} \sim 1$ they find T_p high enough for efficient pion production. The γ-ray luminosity from π^0 decay can reach then $\sim$ 10% of the total luminosity. The effects of energy loss of nonthermal pairs and of nonthermal pair cascades (see §2.3) are neglected in their treatment. Those effects are rather important for reliable determination of the X-ray and γ-ray spectra from hot disks in the Kerr metric, and more work on the model appears desirable.

Eilek & Kafatos (1983) suggest that such a disk might be an efficient source of relativistic pairs. However, in the parameter range in which the pair production via pions is effective, the compactness $\ell \gg 1$, and the pairs lose energy, thermalize, and finally annihilate before having the chance to leave the source (see, e.g., S90).

Tritz & Tsuruta (1989) consider accretion disks with nonthermal pair production and Comptonization of external soft photons (with the Compton parameter $y = 1$). They find high equilibrium pair densities for $\alpha_v \sim 1$ and $\dot{m} \sim 1$. Some effects of the pairs are to inflate the disk, lower n_p, increase T_p, and decrease T_e.

5 SUMMARY AND DISCUSSION

Electron-positron pair production in theoretical models of the central engines of AGNs has been reviewed. The theoretical models have been divided into categories based on the source geometry (one-zone and reflection models, spherical and disk accretion) and on the particle distribution (thermal or nonthermal).

Pairs are found to be important in several models. Abundant pairs can be present in isolated one-zone sources, either thermal or nonthermal (see S90 for a review). Such pair sources can be located in the vicinity of cold material, e.g., forming a corona above the surface of an accretion disk (Zdziarski *et al.* 1990; Haardt and Maraschi 1991). The observed radiation is then the sum of the direct component from the pair source and the reflected one from the disk, which can account for the reflection features in AGN spectra (see, e.g., Pounds *et al.* 1990).

The presence of pairs in isolated sources in which the available power is supplied to protons may lead to instabilities in both thermal (Björnsson & Svensson 1991a) and nonthermal (Stern & Svensson 1991) cases. The instabilities are likely to disappear in the presence of copious external soft photon source.

Abundant pairs can also be present in spherical and disk accretion flows. A high pair density solution is found for hydrodynamic adiabatic spherical accretion (Park & Ostriker 1989). The solution may be unstable on long time scales due to preheating. Abundant pairs are present in quasi-spherical accretion flows with turbulent dissipation and either thermal or nonthermal pair production for some flow parameters (Begelman *et al.* 1987; Lightman *et al.* 1987; Zdziarski 1988; Tritz & Tsuruta 1989). An effect of pairs is then to reduce both the Eddington luminosity and the actual luminosity (due to an increase in the trapping radius; Lightman *et al.* 1987; Zdziarski 1988). Another effect is to enhance a rapid cooling of protons in the inner parts of the flow, which is likely to lead to a collapse of the quasi-spherical flow to an optically thick geometrically thin disk (Tritz & Tsuruta 1989).

Pairs may be abundant and lead to a limit cycle instability in thermal collisionless spherical accretion with $T_p \gg T_e$ and with a shock (Moskalik & Sikora 1986). However, the study of Babul *et al.* (1989) indicates that the shock may occur at a rather large radius ($R \sim 10^2 R_S$), at which T_p is too low to initiate efficient pair production and the instability.

Pairs are unimportant in Shakura and Sunyaev (1973)'s accretion disks, except for $\dot{m} \sim 1$ and $\alpha_v \sim 1$ (Maraschi & Molendi 1990). They are important and may lead to instabilities in hot, Comptonized bremsstrahlung disks (White & Lightman 1989, 1990; Björnsson & Svensson 1991c). Pair densities are negligible in hot thermal disks with a copious external source of soft photons (e.g., from the outer regions of the disk; White & Lightman 1989; Tritz & Tsuruta 1989). Pairs are found important in soft-photon cooled disks with nonthermal pair production (Tritz & Tsuruta 1989).

In general, high pair densities are more likely in nonthermal models than in thermal models. The former *assume* efficient acceleration of particles, which can then easily initiate pair production. This contrasts thermal models, in which the electron distribution is exponentially cut off at high energies and the temperature adjusts itself to the heating, cooling, and pair production processes. This adjustment has then a limiting effect on the rate of pair production.

I thank Chris Done and Roland Svensson for their valuable comments. This research has been supported in part by NASA grants NAS5-30442 and NAG5-1570, a grant from the Polish Research Council, and a grant from Institut d'Astrophysique de Paris.

REFERENCES

Atoyan, A. M. 1990, preprint.
Babul, A., Ostriker, J. P., & Mészáros, P. 1989, *Ap. J.*, **347**, 59.
Bassani, L. & Dean A. J. 1983, *Space Sci. Rev.*, bf 35, 367.
Begelman, M. C., Rudak, B., & Sikora, M. 1990, *Ap. J.*, **362**, 38.
Begelman, M. C., Sikora, M., & Rees, M. J. 1987, *Ap. J.*, **313**, 689.
Bell, A. R. 1978, *M. N. R. A. S.*, **182**, 443.
Björnsson, G., & Svensson, R. 1991a, *M. N. R. A. S.*, **249**, 177.
Björnsson, G., & Svensson, R. 1991b, *Ap. J. (Letters)*, **371**, L69.
Björnsson, G., & Svensson, R. 1991c, in *Relativistic Hadrons in Cosmic Compact Objects*, eds. A. A. Zdziarski & M. Sikora (Heidelberg: Springer), p. 53.
Dermer, C. D. 1988, *Ap. J. (Letters)*, **335**, L5.
Done, C., & Fabian, A. C. 1989, *M. N. R. A. S.*, **240**, 81.
Dultzin-Hacyan, D. 1991, poster presented at the Conference on *Testing the AGN Paradigm*, College Park, MD.
Eilek, J. A., & Kafatos, M. 1983, *Ap. J.*, **271**, 804.
Guilbert, P. W., Fabian, A. C., & Rees, M. 1983, *M. N. R. A. S.*, **205**, 593.
Haardt, F., & Maraschi, L. 1991, *Ap. J. (Letters)*, **380**, L51.
Kazanas, D., & Ellison, D. C. 1986a, *Ap. J.*, **304**, 178.
Kazanas, D., & Ellison, D. C. 1986b, *Nature*, **319**, 380.
Kazanas, D., & Protheroe, R. J. 1983, *Nature*, **302**, 228.
Kirk, J. G., & Mastichiadis, A. 1989, *Astr. Ap.*, **213**, 75.
Kusunose, M., & Takahara, F. 1988, *Publ. Astr. Soc. Japan*, **40**, 435.
Kusunose, M., & Takahara, F. 1989, *Publ. Astr. Soc. Japan*, **41**, 263.
Lightman, A. P. 1982, *Ap. J.*, **253**, 842.
Lightman, A. P. & White, T. R. 1988, *Ap. J.*, **335**, 57.
Lightman, A. P. & Zdziarski, A. A. 1987, *Ap. J.*, **319**, 643.
Lightman, A. P., Zdziarski, A. A., & Rees, M. J. 1987, *Ap. J. (Letters)*, **315**, L113.
Maraschi, L., & Molendi, S. 1990, *Ap. J.*, **353**, 452.
Maraschi, L., Perola, G. C., Reina, C., & Treves, A. 1979, *Ap. J.*, **230**, 243.
Maraschi, L., Roasio, R., & Treves, A. 1982, *Ap. J.*, **253**, 312.
Mastichiadis, A., & Protheroe, R. J. 1990, *Astr. Ap.*, **246**, 279.

Mészáros, P. 1975, *Astr. Ap.*, **44**, 59.
Mészáros, P., & Ostriker, J. P. 1983, *Ap. J. (Letters)*, **273**, L59.
Moskalik, P., & Sikora, M. 1986, *Nature*, **319**, 649.
Park, M.-G. 1990a, *Ap. J.*, **354**, 64.
Park, M.-G. 1990b, *Ap. J.*, **354**, 83.
Park, M.-G., & Ostriker, J. P. 1989, *Ap. J.*, **347**, 679.
Pounds, K. A., Nandra, K., Stewart, G. C., George, I. M., and Fabian, A. C. 1990, *Nature*, **344**, 132.
Protheroe, R. J., & Kazanas, D. 1983, *Ap. J.*, **265**, 620.
Rees, M. J., Begelman, M. C., Blandford, R. D. & Phinney, E. S. 1982, *Nature*, **295**, 17.
Rotschild, R. E., Mushotzky, R. F., Baity, W. A., Gruber, D. E., Matteson, J. L. & Peterson, L. E. 1983, *Ap. J.*, **269**, 423.
Shakura, N. I., & Sunyaev, R. A. 1973, *Astr. Ap.*, **24**, 337.
Shapiro, S. L., Lightman, A. P., & Eardley D. M. 1976, *Ap. J.*, **204**, 187.
Sikora, M., Begelman, M. C., & Rudak, B. 1989, *Ap. J. (Letters)*, **341**, L33.
Sikora, M., Kirk, J. G., Begelman, M. C., & Schneider, P. 1987, *Ap. J. (Letters)*, **320**, L81.
Sikora, M., & Zbyszewska, M. 1985, *M. N. R. A. S.*, **212**, 553.
Stecker, F. W., Done, C., Salamon, M. H., & Sommers, P. 1991, *Phys. Rev. Lett.*, **66**, 2697.
Stern, B. & Svensson, R. 1991, in *Relativistic Hadrons in Cosmic Compact Objects*, eds. A. A. Zdziarski & M. Sikora (Heidelberg: Springer), p. 41.
Stern, B., Svensson, R. & Sikora, M. 1991, in *Variability of Active Galactic Nuclei*, eds. H. R. Miller & P. J. Wiita (Cambridge University Press) p. 229.
Svensson, R. 1982, *Ap. J.*, **258**, 335.
Svensson, R. 1984, *M. N. R. A. S.*, **209**, 175.
Svensson, R. 1986, in *Radiation Hydrodynamics in Stars and Compact Objects*, eds. D. Mihalas & K.-H. A. Winkler (Heidelberg: Springer), p. 325.
Svensson, R. 1987, *M. N. R. A. S.*, **227**, 403.
Svensson, R. 1990, in *Physical Processes in Hot Cosmic Plasmas*, eds. W. Brinkmann *et al.* (Kluwer Academic), p. 357 (S90).
Tritz, B., & Tsuruta, S. 1989, *Ap. J.*, **340**, 203.
Turner, T. J., & Pounds, K. A. 1989, *M. N. R. A. S.*, **240**, 833 (TP89).
White, T. R., & Lightman, A. P. 1989, *Ap. J.*, **340**, 1024.
White, T. R., & Lightman, A. P. 1990, *Ap. J.*, **352**, 495.
White, T. R., Lightman, A. P., & Zdziarski, A. A. 1988, *Ap. J.*, **331**, 939.
Zdziarski, A. A. 1985, *Ap. J.*, **289**, 514.
Zdziarski, A. A. 1986, *Ap. J.*, **305**, 45.
Zdziarski, A. A. 1988, *M. N. R. A. S.*, **233**, 739.
Zdziarski, A. A. 1991, in *Relativistic Hadrons in Cosmic Compact Objects*, eds. A. A. Zdziarski & M. Sikora (Heidelberg: Springer), p. 27.
Zdziarski, A. A., & Coppi, P. S. 1991, *Ap. J.*, **376**, 480.
Zdziarski, A. A., Ghisellini, G., George, I. M., Svensson, R., Fabian, A. C., & Done, C. 1990, *Ap. J. (Letters)*, **363**, L1.
Zdziarski, A. A., & Lightman, A. P. 1987, in *Variability in Galactic and Extragalactic X-ray Sources*, ed. A. Treves (Milano: Associazione per l'Avanzamento dell'Astronomia) p. 121.

On Pair Creation in a Black Hole Magnetosphere

V.S.BESKIN, YA.N.ISTOMIN and V.I.PARIEV

P.N.Lebedev Physical Institute, USSR Academy of Sciences,
117924, Leninsky prosp., 53, Moscow

1 CRITICAL CHARGE DENSITY

If the nature of AGN and quasars is connected with magnetically dominated wind, the "central engine" may be similar to pulsar magnetosphere (Phinney,1983; Thorne,Price and Macdonald,1986), where in the QS regime the charge density ρ_e must be equal to ρ_c. Here

$$\rho_c = -\frac{1}{4\pi c}\cdot \mathrm{div}[[\vec{\Omega}^F,\vec{r}],\vec{B}] \approx -\frac{\vec{\Omega}^F\vec{B}}{2\pi c} \tag{1}$$

($\vec{\Omega}^F$ = const) is the Goldreich–Julian (1969) charge density. General relativity changes this relation in a following manner (Macdonald and Thorne,1982)

$$\rho_c = -\frac{1}{8\pi^2 c}\nabla\left[\frac{(\Omega^F-\omega)}{\alpha}\cdot\nabla\Psi\right];\quad \vec{B} = \frac{[\nabla\Psi,\vec{\phi}]}{2\pi\sqrt{g_{\phi\phi}}} - \frac{2I(\Psi)}{\alpha\sqrt{g_{\phi\phi}}c}\vec{\phi}. \tag{2}$$

For example, near the rotation axis in the axisymmetrical case (Muslimov and Tsygan,1990; Beskin,1990)

$$\rho_c \approx -\frac{(1-\omega/\Omega^F)\vec{\Omega}^F\vec{B}}{2\pi c\alpha}. \tag{3}$$

Here α is the lapse function, the flux Ψ is constant along magnetic field lines, $\Omega^F(\Psi)$ is an angular velocity of a given magnetic field line, $I(\Psi)$ is the total electric current in the tube with magnetic flux Ψ and all values are measured by zero angular momentum observers moving with angular velocity ω ($\Omega^H=\omega(r_+)$).

We see that for $\Omega^F<\Omega^H$ (when the energy flux is directed from the hole to infinity (Macdonald and Thorne,1982; Thorne et al,1986)) in the vicinity of a black hole the "Goldreich–Julian" charge density changes the sign. For example, for the split monopole magnetic field

$$\Psi = \Psi_0\cdot(1-\cos\theta);\qquad \theta < \pi/2 \tag{4}$$

wth $\Omega^F=\Omega^H/2$ (Blandford and Znajek,1977) and for slow rotation the surface $\rho_c=0$ is a sphere with the radius

$$r_0 = 2^{1/3}r_+ \approx 1.26r_+. \tag{5}$$

Consequently, this surface is not so close to the black hole's horizon.

2 PAIR CREATION

As is well known in the theory of radio pulsar, the domain $\rho_c \approx 0$ can be the region of extensive electron–positron pair creation (Cheng, Ho and Ruderman,1986). Really, charge separated flow is impossible through the surface $\rho_c=0$. Hence, in this region the relation $\rho_e=\rho_c$ may be violated, large longitudinal electric field E_p appearing here. Any primary particle accelerated in such a gap gives photons hard enough to produce secondary pairs as a result of a reaction $\gamma+\gamma \to e^++e^-$.

There are two main differences between pulsar and black hole "outer" gap:

i. In the black hole magnetosphere primary particles have energy much smaller than ones in the pulsar magnetosphere as a result of soft photon damping:

$$\gamma_{max} \approx \left[\frac{eE_p}{\sigma_T U}\right]^{1/2} \sim 10^2 - 10^3. \tag{6}$$

Here $U \approx L_{tot}/r^2c$ is the photon density in the vicinity of a black hole. For this reason the curvature radiation (whose energy is proportional to γ^3) is not important near the black hole.

ii. The origin of hard photons in the black hole magnetosphere may be connected only with Inverse Compton (IC) scattering of primary particles on the external soft photons. So, the size and energetics of the gap near the black hole depend upon the energy density and spectrum of an external source. On the other hand, the relation (4) shows that at first step we can neglect the general relativity correction to the energy density U_ν. For this reason in what follows we shall use the isotropic energy density U_ν with frequency dependence of the power

$$U_\nu \propto \nu^{-\alpha}. \tag{7}$$

The height of the gap H is defined by the sum of three lengths

$$H = l_0 + l_1 + l_2. \tag{8}$$

Here l_0 is the acceleration length allowing the particles to gain the energy (6) in the electric field, l_1 is the free path of energetic particle before colliding the background photon with energy ϵ and creating the energetic (IC) photon with $\epsilon' = \gamma^2_{max}\epsilon$, l_2 is the path of the pair generation by energetic and background photons.

The acceleration length is defined by equilibrium between the electric force and the friction force acts on the particle from the background photons

$$l_0 = mc^2/\sqrt{(eE_p\sigma_T U)}. \tag{9}$$

On the other hand, $l_1 \approx l_2$ and

$$(l_1)^{-1} \approx \begin{cases} \int_{2a/3}^{3a/2} \sigma_T U_\nu d\nu; & \gamma_{max} > \epsilon_2/mc^2 \\ \int_{a^2h/\epsilon_2}^{\epsilon_2/h} \sigma_T U_\nu d\nu; & \gamma_{max} < \epsilon_2/mc^2 \end{cases} \tag{10}$$

where $a = mc^2/h\gamma_{max}$ and ϵ_2 is the maximum energy of the background photons. Finally, the longitudinal electric field E_p is defined by the value H

$$E_p \approx \frac{\pi}{4} \cdot H^2 \cdot \nabla\rho_c \Big|_{\rho_c=0}. \tag{11}$$

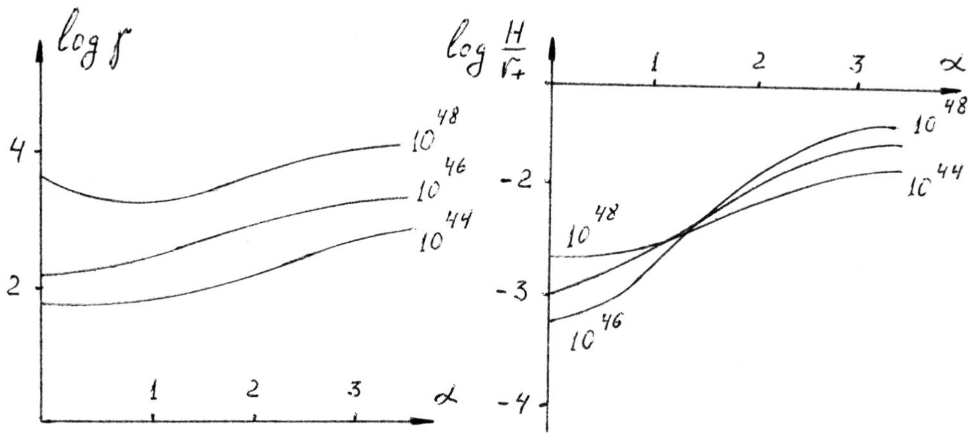

Fig. 1. Dependence of the height H of the gap and the energy of primary particles $mc^2\gamma_{max}$ on the spectral index α for fast rotation. Here $M = 10^8 M_\odot, h\nu_{min} = 10$ eV, $h\nu_{max} = 10$ keV . Digits indicate the total luminosity L in erg/s

In Figure 1 we show the dependences of the height of the gap H and the energy of primary particles $mc^2\gamma_{max}$ on the spectral index α determined from the relations (7) – (11). We see that the dimension H is much smaller than the black hole radius r_+.

Hence, we see that stationary generation of a secondary plasma in the vicinity of a black hole horizon can really take place (Beskin, Istomin and Pariev,1990).

3 SIMPLE MODEL OF THE BLACK HOLE MAGNETOSPHERE

To find the real shape of $\rho_c=0$ surface it is necessary to know the magnetic field structure (that is, the potential Ψ), and also the angular velocity $\Omega^F(\Psi)$. The potential Ψ must fulfill the nonlinear stream equation (Macdonald and Thorne,1982)

$$\nabla\left\{\frac{\alpha}{g_{\phi\phi}}\left[1-\frac{(\Omega^F-\omega)^2 g_{\phi\phi}}{c^2\alpha^2}\right]\nabla\Psi\right\}+\frac{(\Omega^F-\omega)}{\alpha c^2}\cdot\frac{d\Omega^F}{d\Psi}\cdot(\nabla\Psi)^2+\frac{16\pi^2}{\alpha g_{\phi\phi}c^2}\cdot I\cdot\frac{dI}{d\Psi}=0. \tag{12}$$

Actually, the split monopole magnetic field (4) has already given the real structure of the black hole magnetosphere. Indeed, according to Blandford and Znajek(1977), for $\Omega^F=\Omega^H/2\ll c/r_+$ and

$$4\pi I(\Psi)=\Omega^F(2\Psi-\Psi^2/\Psi_0) \tag{13}$$

(Ψ_0 is the magnetic flux through the hole), the monopole field (4) is the solution of the nonlinear stream equation (12) not only at large distances from the hole (Michel,1973), but also in the vicinity of the horizon. In this model the returning current flows along the disk.

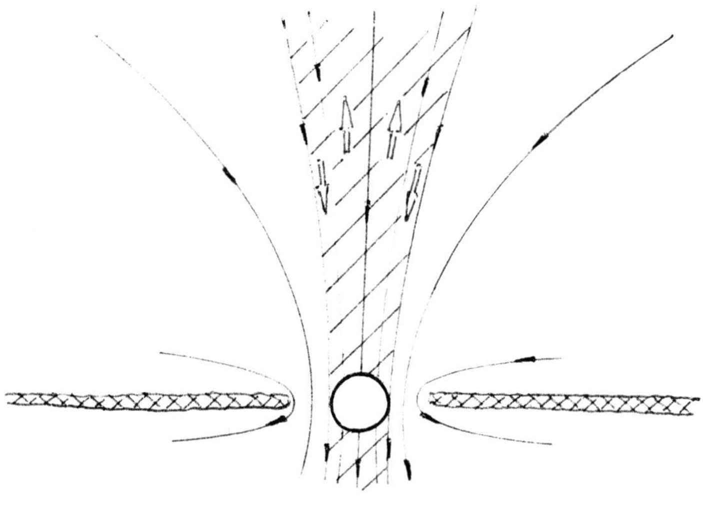

Fig. 2. The simple model of the black hole magnetosphere containing a well conducting disk. The electric current structure corresponding to the relations (21) is also shown

Now let's consider a more realistic model concerning the black hole in the center of a disk with inner radius b (see Fig. 2.). Let the magnetic flux through the disk be Ψ_0, and we assume that magnetic field lines do not penetrate the disk surface. This case corresponds to a very high turbulent level in the disk.

Without the black hole in the center of a disk the potential Ψ is given by the following relation

$$\Psi = \Psi_0\left\{1 - \left\{\frac{1}{2}(1 - \frac{r^2}{b^2}) + \left[\frac{1}{4}(1 - \frac{r^2}{b^2})^2 + \frac{r^2\cos^2\theta}{b^2}\right]^{1/2}\right\}^{1/2}\right\}. \quad (14)$$

Expanding (14) near $r = 0$ and matching every harmonic to black hole horizon, it is possible to find the vacuum magnetic field structure in the presence of the black hole. For example, for $r_+ \ll b$ (and for slow rotation) we obtain

$$\Psi_V = \Psi_0\left[\frac{1}{2}\cdot\frac{r^2\sin^2\theta}{b^2} - \frac{1}{120}\cdot\frac{r^4}{b^4}\cdot F(r)\cdot(4\sin^2\theta\cos^2\theta - \sin^4\theta) + \ldots\right]. \quad (15)$$

Here $F(r) = F(-4,-2,1,1-r_+/r)$ is the hypergeometric function. The flux Ψ_V is the solution of vacuum equation

$$\nabla\left\{\frac{\alpha}{g_{\phi\phi}}\cdot\nabla\Psi\right\} = 0. \quad (16)$$

Hence, near the black hole the main term is the homogeneous magnetic field

$$\Psi_V = \frac{1}{2}\cdot\Psi_0\cdot\frac{r^2\sin^2\theta}{b^2}. \quad (17)$$

Now we shall show that it is possible to construct the solution of the nonlinear stream equation (12) for $\Omega^F(\Psi) \neq 0$ and $I(\Psi) \neq 0$ with the same asymptotics (4) and (17) as in the vacuum case. Indeed, it is easy to check that at large distances $r \gg b$ the monopole field (4) is the solution of the stream equation with

$$4\pi I(\Psi) = \Omega^F(\Psi)(2\Psi - \Psi^2/\Psi_0) \quad (18)$$

for *arbitrary* dependence $\Omega^F = \Omega^F(\Psi)$ (Blandford and Znajek,1977). Assume now that near the horizon, as at large distances, the solution of the stream equation does not differ strongly from the vacuum field (15), (18). It is possible because, as is directly seen from (12), for slow rotation the stream equation in the domain $r_+(\Omega^H r_+/c)^2 \ll r - r_+ \ll c/\Omega^H$ coincides with the vacuum equation (16). Hence, the real solution of nonlinear stream equation (13) differs from vacuum one by the small value $\sim (\Omega^H r_+/c)^2$ (for details see Blandford and Znajek,1977). Then for $\Psi = \Psi_V$ (19) the boundary condition on the horizon

$$\sin\theta\cdot\frac{d\Psi}{d\theta} = \frac{4\pi I(\Psi)}{\Omega^H - \Omega^F(\Psi)} \quad (19)$$

gives the following connection between $\Omega^F(\Psi)$ and $I(\Psi)$

$$I(\Psi) = 2\Psi\sqrt{(1 - \Psi/\Psi*)}[\Omega^{H} - \Omega^{F}(\Psi)]. \tag{20}$$

Here $\Psi* = 0.5\Psi_0 r_+^2/b^2$ is the magnetic flux through the black hole. Combining (18) and (20) we finally obtain

$$\begin{aligned} \Omega^{F}(\Psi) &= \frac{\sqrt{(1 - \Psi/\Psi*)}}{1 - \Psi/2\Psi_0 + \sqrt{(1 - \Psi/\Psi*)}} \cdot \Omega^{H}; \\ 4\pi I(\Psi) &= 2\Psi \cdot (1 - \Psi/2\Psi_0) \cdot \frac{\sqrt{(1 - \Psi/\Psi*)}}{1 - \Psi/2\Psi_0 + \sqrt{(1 - \Psi/\Psi*)}} \cdot \Omega^{H}. \end{aligned} \tag{21}$$

We see that

$$I(\Psi*) = 0. \tag{22}$$

Hence, in this model the returning current flows in the outer part of a tube $\Psi = \Psi*$, the total current being equal to zero. So, as shown in Fig. 2., this model gives the jet with the zero whole current in it. The semiangle θ of this jet is

$$\theta = \arccos(1 - \Psi*/\Psi_0) \approx r_+/b \tag{23}$$

The condition $\Omega^{F}(\Psi*) = 0$ demonstrates that the total electric charge in the jet is equal to zero too. Remember that in such current–carrying jets the energy losses of a central object is due to the Pointing flux inside the jet because $S_r = (c/4\pi)EB_{\phi}$ does not change sign here (Benford,1978; Begelman, Blandford and Rees,1984).

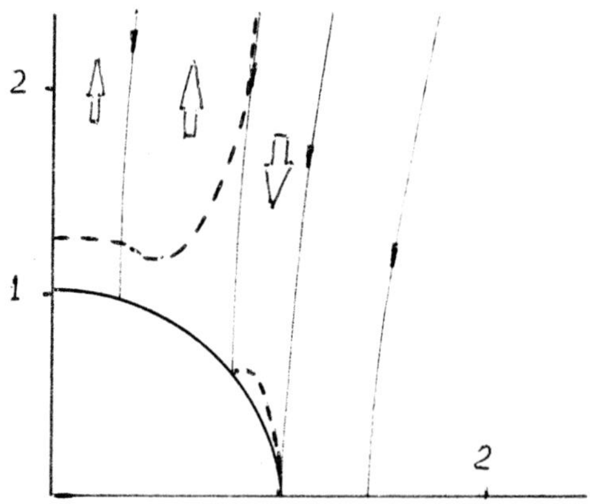

Fig. 3. The structure of the pair creation surface for $r_+/b = 0.25$

4 REAL SHAPE OF THE PAIR CREATION SURFACE

Figure 3 shows the pair creation surface $\rho_c = 0$. We see that every magnetic field line crossing the black hole horizon penetrates the pair creation surface, different branches of this surface corresponding to the domains of inward and outward current j $\propto$ dI/dΨ. It is possible to prove that this is the general result and, for example, on the horizon the point ρ_c = 0 coincides with the point dI/dΨ = 0. Indeed, using the definition (2) and condition $E^{\hat{\theta}} \longrightarrow -B^{\hat{\phi}}$ when $r \longrightarrow r_+$ (Thorne et al,1986), we obtain on the horizon

$$4\pi\rho_e = \frac{1}{\sqrt{\gamma}}\frac{d}{d\theta}\left[\sqrt{\gamma}\cdot\frac{E^{\hat{\theta}}}{\sqrt{g_{\theta\theta}}}\right] = -\frac{1}{\sqrt{\gamma}}\frac{d}{d\theta}\left[\sqrt{g_{rr}}\sqrt{g_{\phi\phi}}B^{\hat{\phi}}\right] = -\frac{2}{\sqrt{\gamma}}\frac{d}{d\theta}\left[\frac{\sqrt{g_{rr}}}{\alpha c}I(\theta)\right]. \qquad (24)$$

So, ρ_c = 0 when dI/dθ = 0. This result (which does not depend upon the hole rotation velocity) is absolutely clear from the physical point of view.

In the vicinity of the horizon, where E $\longrightarrow$ B (and all particles move toward the hole at a velocity close to that of light), the main role is played by the drift component $\vec{j} = c\rho_e[\vec{E},\vec{B}]/B^2 \approx -c\rho_e\vec{r}/r$ which is just proportional to the charge density ρ_e. Similarly, at large distances, where also $\Psi = \Psi(\theta)$

$$4\pi\rho_e = \frac{dE}{\rho d\theta} = -\frac{1}{2\pi c\rho^2}\cdot\frac{d}{d\theta}[\Omega^F(\Psi)\Psi_0\sin\theta] = -\frac{1}{2\pi c\rho^2}\cdot\frac{dI}{d\theta} \qquad (25)$$

Here we have used the definition $\vec{E} = -(\Omega^F/2\pi c)\nabla\Psi$ and the relations (4) and (18). Hence, at large distances the pair creation surface also coincides with the surface j = 0.

The authors are grateful to R.Blandford, A.V.Gurevich, I.D.Novikov and A.G.Polnarev for fruitful discussions.

REFERENCES

Begelman M.C., Blandford R.D. and Rees M. (1984) Rev. Mod. Phys., **56**, 265
Benford G. (1978) M.N.R.A.S., **183**, 29
Beskin V.S. (1990) Soviet Astronomy Letters, **16**, 286
Beskin V.S., Istomin Ya.N. and Pariev V. I. (1991) (in preparation)
Blandford R.D. and Znajek R.L. (1977) M.N.R.A.S., **179**, 433
Cheng K.S., Ho C. and Ruderman M.A. (1986) Ap.J., **300**, 522
Goldreich P. and Julian W.H. (1969) Ap.J., **157**, 869
Lipunov V.M. (1991) Astrophysics of neutron stars. Berlin: Springer – Verlag
Macdonald D. and Thorne K. (1982) M.N.R.A.S., **188**, 345
Michel F.C. (1973) Ap.J., **180**, L133
Muslimov A.G. and Tsygan A.I. (1990) Soviet Astronomy, (in press)
Phinney E.S. (1983) Ph.D. dissertation, University of Cambridge
Thorne K., Price R. and Macdonald D. (1986) Black Holes: The Membrane Paradigm. New Haven: Yale University Press

ABOVE THE ACCRETION DISC & BLACK HOLE: A STARBURST STELLAR CLUSTER?

JUDITH J. PERRY

Institute of Astronomy, Madingley Road, Cambridge CB3 0HA, U.K.

ABSTRACT
We review the consequences of assuming that the AGN phenomenon manifests itself when a nuclear starburst stellar cluster coexists with a massive central object responsible for the continuum.

1 INTRODUCTION

Radio galaxies and quasars manufacture jets and "eye-catching" large scale structures which have fascinated astronomers for decades: the physical processes involved directly in the production and propagation of jets by "the central engine" is the subject of this meeting. During these discussions, we should keep in mind that this central engine is a child of the galaxy within which it resides and that it manages to create the structures under discussion with only a small fraction of its total energetic output. Not only is more than 90% of its energy output in optical, UV and X-ray radiation but only 10% of all galaxies and quasi-stellar objects with "active" nuclei are radio sources. Thus from the radio properties alone we are more apt to learn about plasma physics than about the structure and evolution of active galaxies . An immense amount of information about the structure and dynamics of the interstellar medium (ISM) of the host galaxy is contained within the emission, and sometimes also the absorption, line spectrum. Even at the highest redshifts (~ 4) quasar spectra contain lines from heavy elements with abundances not dramatically different from solar, clear evidence that a generation of supernovae have enriched the nuclear ISM.

There can be little difference between the nuclear ISM (and probably also the nuclear stellar clusters) of radio loud (RL) and radio quiet (RQ) AGN, as there is no *significant* difference between the emission line spectra of the two. Yet a strong correlation appears to exist with galaxy type, with RLs in ellipticals and RQs in spirals. It is unlikely that the central engine can be strongly affected directly by the outer galaxy. Therefore, it seems likely that the features of the central engine controlling its radio properties (probably the angular momentum of the core) are determined simultaneously with the *formation* of the outer structure. Since collisions between galaxies appear to be linked to both starbursts and AGN it may be that the impact parameter and strength of the collision determine both simultaneously. I shall outline some of the attempts my colleagues and I have been making to understand the structure of the nuclear regions of AGN and their link to the starburst **phenomenon.**

What are the primary (non-radio) observations to be explained? The energy source for the luminosity (up to 10^{48-49} erg s^{-1}); the existence of the broad emission line gas, its abundances, excitation and kinematics, including the systematic differences between the low and high ionization emission (the LIL and HIL respectively); the broad absorption lines seen in roughly 5 – 10 % of QSOs (as large a percentage as are radio loud!), and the (short) time variability of the line spectra. Once a self-consistent "steady-state" model explaining these features is developed, its evolution can be addressed. It is worth remembering the stellar analogy: steady-state models preceeded evolutionary models and star formation is still not fully understood. Yet in AGN research we still have no analogue of an "HR" diagram to guide us.

The emission lines contain the most detailed quantitative information available about AGNs (in addition to the continuum). However, the mass of gas *directly* involved in line emission is but a tiny proportion of the gas present. Estimates of the total mass of "cold" gas in the broad emission line region (BELR), are $\lesssim 100\ M_\odot$, even in luminous QSOs. The broad line gas must exist in miniscule, heavy element enriched, cold ($T \sim 10^4$K), fast moving ($v \sim$ 5,000 - 10,000 km s^{-1}) "clouds". The interstellar medium (ISM) in the BELR contains at least $10^4 M_\odot$ of gas (Perry & Dyson, 1985 (hereinafter PD)). Thus the gas we observe directly is a tracer, not a determinant, of the local dynamics. Furthermore, since the *recombination time* is orders of magnitude shorter than any other time scale of interest (flow times, heating and cooling times, etc) the observation of a photoionisation equilibrium emission line spectrum tells us only that clumps of cold moving gas are exposed to the continuum radiation field—it tells us nothing about their origin, lifetime, or dynamical history. The emission line spectrum is thus merely a *snapshot* of the gas. It is not surprising that photoionisation models give good agreement with observed line intensities and ratios independent of the dynamical model in which they are embedded; they cannot, by themselves, yield a global, self-consistent model for activity in galaxies.

In 1982 John Dyson and I first proposed that a young stellar cluster surrounding the central engine (then, as now, assumed to be a massive black hole) would naturally solve many of the outstanding problems in QSO phenomenology: in particular, how the central engine is supplied with fuel and simultaneously, how the emission line clouds are created and maintained. Cloud formation by radiative shocks (Perry & Dyson, 1982, PD) and the two component AGN model of Collin-Souffrin, Dyson, McDowell and Perry (1988, hereinafter CDMP) is the basis of our current work. Ultimately, the only free parameters are the initial mass function (IMF) of the cluster and the ratio of the initial black hole (BH) and cluster masses. The consequences of the interaction between the BH and cluster for the structure and evolution of the system is the subject of my talk today. I concentrate on high luminosity QSOs in what follows, believing them to be the greatest challenge to model builders, mentioning low luminosity AGNs where I believe they may differ from QSOs.

2 STARS AND GAS IN THE BROAD LINE REGION

Assume that a massive black hole exists in the centre of a galaxy: the Keplerian velocity is then, $v_K \approx 660(M_8/r_{pc})^{1/2}$ km s^{-1}. [We use standard notation, *e.g.* $M_8 = M/(10^8 M_\odot)$, $T_7 = T/(10^7\mathrm{K})$, and r_{pc} is the distance from the centre in pc, etc.] The ISM must be optically thin since rapid time variations in the continuum emission are not smeared out: therefore the ISM is at the Compton temperature, $T_c \sim 10^7 T_7$K, and has a sound speed, $c_s \approx 380 T_7^{1/2}$ km s$^{-1} < v_K$. Thus *all* stars within the region move *supersonically* at Mach numbers of $\mathbf{m}_* \sim 1.7(M_8/r_{pc}T_7)^{1/2}$. They create bow shocks in the ISM, and have shocks driven into their outer layers. Therefore *all stellar mass injection into the ISM - whether through stellar winds, supernovae, tidal stripping or stellar collisions - is mediated by shocks,* even in a static ISM. A static ISM is dynamically improbable and observations imply that flow velocities range from 1000 to perhaps 30,000 km s^{-1}. Shock velocities are thus far in excess of v_K. The effects on stellar evolution due to such interactions is still an open question.

Shocks convert kinetic energy into thermal energy; the post-shock gas is heated and compressed: the shocked ISM is heated to temperatures ($T_s \approx 10^8\omega^2$ where $v_{shock} \equiv 3000\omega$ km s^{-1}) in excess of T_c, and is no longer in equilibrium with the radiation field; the gas therefore cools. It flows through the shock before re-expanding, downstream, into the ISM. Its initial post-shock pressure equals the stagnation pressure, $p_{stag} \approx \rho_{ISM} v_{shock}^2$. As the gas accelerates downstream the gas pressure drops, and the ionization parameter, Ξ, consequently increases. ($\Xi = L_{ion}/4\pi r^2 n_e kTc = 2.3 p_{rad}/p_{gas}$. At fixed r, p_{rad} is constant, and changes in p_{gas} are directly reflected in changes in Ξ). If the shock is extended the region within which the gas is subsonic (with respect to the shock) and effectively isobaric is correspondingly large, and the cooling gas can reach catastrophic radiative cooling before becoming supersonic (see PD for details of the flow). The shocked gas cools to $\sim 10^4$K if the cooling time is shorter than the time spent in the high pressure (low Ξ) region. When cold, the gas is in equilibrium with the continuum so that, *although created by radiative shocks, the emitted spectrum is primarily that of photoionized gas* (see discussion below). The final state of the post-shock gas is thus dependent primarily on the size of the shocked region.

The size of the shock in the ISM depends on the size of the obstacle forming the shock: on the effective size of the star, or supernova bubble. The size of the obstacle is self-consistently determined by approximate hydrostatic equilibrium at the contact surface between the shocked ISM and shocked stellar material. A star will be compressed by the shock down to that point in its atmosphere where the pressure is p_{stag}. The bowshock will be very small, and the cooling time for gas from $\gtrsim 10^7$K is so long that the gas cannot cool before it re-enters the ambient ISM.

A supernova, (or a disrupted star) expands until it reaches pressure equilibrium with the shocked ISM. The pressure of the fully expanded bubble $\approx E_{SN}/V$ (where E_{SN} is the energy injected, $\equiv 10^{51}E_{51}$ erg, and V is the volume of the bubble). Hence $V \approx E_{SN}/p_{stag}$ giving $r_{SN} \approx 0.09(E_{51}/p_{0.01})^{1/3}$ pc. (The stagnation pressure of the flowing ISM is $p_{stag} \equiv 10^{-2}p_{0.01}$ dynes, and $p_{0.01} \approx 1.1(n_{ISM}/10^5)\omega^2$). Behind the shock $nT \approx p_{stag}/k$, giving $nT \sim 8 \times 10^{13}p_{0.01}$. Thus *when the stagnation pressure of stars moving through the ISM is roughly equal to the thermal pressure deduced for BELR "clouds", supernovae occurring in the flow will expand to typically 0.1 pc in size.* When "obstacles" this large exist near the continuum source, catastrophic cooling of the gas flowing through the bow shock occurs (PD, Innes & Perry, 1991 *in prep.*), resulting in continuous production of "cold" line emitting gas. The gas fragments into "clouds" with a spectrum of column densities whose maximum is $N_H \sim 10^{22}$ cm^{-2}. There is a natural maximum radial distance from the continuum source within which cooling is effective: the radiation field, and probably the ISM density, decrease $\propto r^{-2}$, and with them the cooling. PD found that $r_{max} \sim L_{47}^{1/2}$ pc, in agreement with observations. I would like to emphasize here that clouds are formed by this process throughout the region $r < r_{max}$.

A conservative estimate for the yield of cold gas by the shocks is found by assuming that only gas entering the shock within $\sim r_{SN}/2$ of the stagnation line cools isobarically. Then $\dot{M}_{\rm cold} \approx (\pi/4)r_{SN}^2\rho_{ISM}v_{ISM}$ per supernova. Its lifetime (between cooling and downstream re-expansion and reheating), is considerably greater than the post-shock cooling time, $t_{\rm life} > t_{\rm cool}$. The mass of cold gas produced per supernova is $\approx \dot{M}_{\rm cold}t_{\rm life}$, and the observed equivalent widths then require a SN rate $\dot{N}_{SN} \lesssim 0.7$ L_{47} yr^{-1} (PD). This *maximum* for the *required SN rate* leads to the *minimum mass starburst stellar cluster* required to explain the BELR of QSOs, as discussed below.

3 STELLAR CLUSTER SIMULATIONS

Evidence has been accumulating recently linking non-axisymmetric perturbations of galaxies with accumulation of gas in their nuclear regions. Hernquist and his collaborators have shown that galaxy interactions can lead to the accumulation of as much, or more than, $10^9M_\odot$ of gas in the nucleus. If that gas fragments and forms a dense stellar cluster, its early history may include nuclear activity. Robin Williams and I (Williams & Perry, 1991) have been studying the evolution of such clusters and have calculated models with accurate stellar evolutionary tracks for a wide range of initial mass functions and several star formation rates, to determine the mass loss and supernovae rates of the cluster. The cluster simulations are combined with two-dimensional (2-D) hydrodynamical simulations to study the flow of the ISM.

The 2-D simulations show that, particularly in the early high mass-loss stage, infall roughly perpendicular to the axis of symmetry accompanied by hypersonic outflow over a wide opening angle about the axis can occur. The flows are complex and have more structure than spherically symmetric simulations could reveal: well known severe restrictions on the parameters necessary to achieve supersonic winds in spherical symmetry are considerably relaxed and a far wider set of parameters give rise to high speed ISM flows. Interestingly, something like half the total stellar mass loss from the cluster, $\dot{M}_{cluster}$, can be accreted, and turned into continuum luminosity.

The simulations, for *e.g.* a Salpeter IMF ranging over 1 - 50 $M_\odot$ with star formation decaying over 10^7 yrs, yield a SN rate, $\dot{N}_{SN}$ yr^{-1} of $\dot{N}_{SN}/L_{Bol} \approx 0.6\mathrm{yr}^{-1}/(10^{47}$ $\mathrm{erg\,s^{-1}})$ after about 5×10^7 yr. A peak luminosity of $L_{Bol} \approx 10^{47}\mathrm{erg\,s^{-1}}$ (assuming that accreted stellar mass loss (including SN ejecta) is converted with 10% efficiency) requires an initial cluster stellar mass of $M_{cluster} \sim 2 \times 10^9 M_\odot$, implying an initial average stellar density, ρ_*, within a volume defined by $r_{pc} \sim L_{47}^{1/2}$ of $\rho_* \sim 4 \times 10^8 M_\odot \mathrm{pc}^{-3}$. Hydrodynamic considerations suggest that at these masses, accretion may be near the Eddington limit, so $M_{BH} \sim 8 \times 10^8 L_{47} M_\odot$. Depending on the IMF, the total luminosity decreases smoothly with time, the SN rate, however, drops precipitously $\sim 10^8$ yrs after onset of the burst. The slow growth of the BH mass and decrease in $\dot{M}_{cluster}$ means that accretion becomes sub-Eddington, as is deduced from observations. At low luminosities the ISM may be thermally unstable (PD). Low luminosity AGNs may well represent these late stages of cluster evolution.

To summarize, we find that if young stellar clusters surround BHs, their mass loss creates an ISM with a characteristic mass flux of $\dot{M}_{ISM} \sim L_{Bol}/(0.1c^2)$. This ISM has a complex flow pattern with accretion and outflow with Mach numbers as high as ~ 30. The concomitant SNae produce the HIL BELR.

4 TWO COMPONENTS: THE HIL AND LIL

Both theoretical **arguments** based on photoionization equilibrium calculations (see CDMP for references) and observations (see *e.g* Espey, *et al.* 1989) have shown that the broad emission spectrum must be divided into at least two parts. The HIL (Lyα, CIII], CIV, etc) emitted by low density gas ($n < 10^9$ cm^{-3}) are systematically blue shifted with respect to the LIL. The LIL (Balmer lines, MgII, CII, FeII, etc.) are emitted by mildly ionized gas of high density ($n > 10^{11}\mathrm{cm}^{-3}$). Collin-Souffrin and her collaborators have shown that the LIL can be emitted in the outer regions of the massive accretion disc which produces, in its inner regions, the "blue bump" (see *e.g.* Rokaki, Boisson & Collin-Souffrin, 1991). In CDMP we showed that the systematic differences *and* the correlations between the HIL and LIL are the natural consequences of the interactions between a nuclear stellar cluster and central BH. We argued there that the HIL is produced in the shocked ISM and the LIL in the accretion disc. The bulk of cold HIL gas produced by the radiative shocks in the ISM moves on outwardly directed *quasi*-radial trajectories. Because the accretion disc obscures some of the farthest side of the AGN, systematic blue shifts are observed. The relative constancy of the ratio of the HIL and LIL intensities results from the fact that the accretion disc radiating the LIL is fed by the same ISM creating the HIL. Radiative shocks in the ISM reprocess the continuum radiation by Compton scattering and the shocked gas emits bremsstrahlung, so the continuum reflected onto the accretion disc is harder than the direct central continuum. Detailed line ratios depend on this reprocessing of the continuum, both for the LIL and the HIL.

An important feature of the observed HIL spectrum is the near constancy of the ionization parameter, $\Xi \sim 0.5$. The HIL is observed when cooled shocked gas is rejoining the flow, so $v_{shock} \approx v_{ISM}$ (PD). Then $\dot{M}_{ISM} = 4\pi r^2 \rho v_{ISM}$. Behind shocks, $nkT \approx \rho_{ISM} v_{ISM}^2$, and thus $\Xi_{shock} \approx L_{ion}/(\dot{M}_{ISM} v_{ISM} c)$. Since the luminosity is produced by accretion from the ISM we can write $L_{ion} = f_{Bol} f_{acc}(0.1)\dot{M}_{ISM} c^2$ where $f_{Bol} \equiv L_{ion}/L_{bol}$, $f_{acc} \equiv \dot{M}_{acc}/\dot{M}_{ISM}$ and $\dot{M}_{acc} \equiv L_{Bol}/(0.1c^2)$. Therefore $\Xi \approx 0.1 f_{Bol} f_{acc}(c/v_{ISM})$. From observations $f_{Bol} \sim 0.1$; the hydrodynamics implies $f_{acc} \sim 1$ and the width of the broad lines implies $v_{ISM}/c \sim 0.02$. Thus $\Xi \sim 0.5$, in agreement with observations.

5 SUPERNOVAE ON THE LINE OF SIGHT: BROAD ABSORPTION?

In the initial interaction of a SN and the ISM, at the same time as a fast shock moves into the ISM, a reverse shock moves into the high speed ejecta, accelerating back towards the explosion centre. Although the ISM is shocked to $T_{shock} > 10^8$K the temperature of the shocked ejecta, T_{se}, is initially low, since it goes through the reverse shock at low relative velocity; T_{se} increases with time. About 10% of the post shocked ejecta has temperatures < a few $\times 10^4$K while still moving at $v > 8{,}000\ \mathrm{km\,s^{-1}}$. If the SN lies on the line of sight, broad absorption lines (BAL) will be observed. Ejecta just outside the BELR are slowed down in $\sim$ 10 yrs, which for $z \sim 2$ is $\sim$ 30 yrs in the observer's frame (Perry & Dyson, *in prep.*). Ejecta moving toward the central source slow down far more quickly than ejecta moving away from the source because of the outwardly directed ISM flow, and expand a smaller distance: therefore only blue-shifted troughs will be observed. Very fast moving cool ejecta does not exist for any appreciable time in the BELR, explaining why the BAL are broader than the BEL. The observed velocity dispersion results from a complex turbulent shell breaking up on scale sizes comparable to the continuum source. The frequency with which BALs are observed is proportional to the covering factor of nuclear cluster SN shells, roughly 10% (PD). The interface between the shocked ambient material and the SNR is Rayleigh-Taylor unstable on time scales less than t_{stop} which may explain the observed BAL variability.

6 SUMMARY

The Starburst-BH model postulates that a young central dense stellar cluster (probably formed by collisions between gas-rich galaxies) coexists with a massive central black hole. The cluster releases mass for a substantial period, by loss-cone disruption, collisions, stellar winds and supernovae. This mass loss creates an optically thin, flowing ISM kept at the Compton temperature by the non-thermal continuum. The flow patterns are complex; there is both accretion and a hypersonic wind. The accretion feeds the BH and accretion disc and is responsible for the continuum; the LIL are radiated by the outer accretion disc. The outflowing wind shock interacts with the stellar cluster. Large bow shocks around supernovae are radiative: they cool, come into equilibrium with the central continuum and radiate the HIL. Although created by shocks and transient, the HIL is primarily photoionized and so responds to variations of the continuum. The structure of the HIL region changes on the timescale for supernova radiative shock evolution, which is of the order of 30 years. BALs originate in supernova ejecta just outside of the BELR. Their frequency reflects the covering factor and relative number of SN in the cluster.

ACKNOWLEDGEMENTS

I am grateful to John Dyson for a decade of rewarding and stimulating collaboration; to Robin Williams, Davina Innes, Suzy Collin and many other colleagues for stimulating discussions which have enhanced my understanding of AGN.

REFERENCES

Collin-Souffrin S., Dyson, J. E., McDowell, J., Perry, J. J. 1988. *M.N.R.A.S*, **96**, 123.

Dyson, J. E. & Perry, J. J., 1982. In: *Third European IUE Conference,* Eds.Rolfe, E., Heck, A. & Battrick, B., ESA Publication, Noordwijk, p. 595.

Espey, B.R., Carswell, R. F., Bailey, J. A., Smith, M. G. & Ward, M. J., 1989. *Ap. J.*, **342**, 666.

Perry, J. J. & Dyson, J. E., 1985. *M.N.R.A.S*, **213**, 665. (herein PD)

Rokaki, E., Boisson, C. & Collin-Souffrin, S., 1991. *Astr. Ap.*, , *in press.*

Williams R. J. R. & Perry, J. J., 1991. In: *Physics of Active Galactic Nuclei,* Eds.Duschl, W. J. & Wagner, S. J., Springer Verlag, Heidelberg, p. *in press.*

Intraday Variability of Active Galactic Nuclei Radio and Optical Observations

STEFAN J. WAGNER[1] AND ARNO WITZEL[2]

[1]Landessternwarte Heidelberg-Königstuhl
[2]Max-Planck-Institut für Radioastronomie, Bonn

ABSTRACT

Some aspects of the intraday variability of compact, flat-spectrum radio-sources are discussed. We show such variations to be a typical phenomenon in this sub-class of AGN. At least in one object a clear and close correlation between the optical and radio variability is observed. Based on this correlation and on other observations extrinsic origins of the variability can be excluded. The extreme changes in flux and spectral behaviour also rule out most of the current models to explain intrinsic variations. Some consequences on the geometrical structure and emission mechanisms are discussed.

1 INTRODUCTION

Intraday variability of extragalactic radio sources was detected accidentally in 1986 (Witzel et al. 1986, Heeschen et al. 1987, Witzel 1990) during a systematic search for the characteristics of the flux density variations of mainly compact radio sources caused by the interstellar medium called flickering (Heeschen 1984, Heeschen and Rickett, 1987). The initial discovery of IDV in the BL Lac object 1749+70 (see fig. 1) which showed a 7 % decrease followed by a comparable increase within 20 hours at 11 cm wavelength, started a number of observational campaigns - both in the radio and in the optical domains - to investigate the characteristics of the phenomenon and eventually find a theoretical interpretation for it.

In this paper we describe the observational results of both radio and optical observations and briefly discuss the theoretical explanations discussed to date.

2 RADIO OBSERVATIONS

In numerous observational campaigns between May 1985 and May 1991 using the 100 m telescope of the MPIfR and the VLA of the NRAO we investigated extragalactic radio sources for IDV at wavelengths between 2 and 90 cm. Radio-polarization information on IDV is available at wavelengths between 3.6 and 20 cm.

Table 1: The sample of flat sources ($\alpha_6^{11} < -0.5$) and their types of variability.

Source	Id	Type	5/85	8/85	12/85	3/88	6/88	12/88	5/89	4/90	7/90	5/91
			11 cm	11 cm	11 cm	6/11 cm	6 cm	6/11 cm	6/11 cm	6 cm	6/11 cm	6/11 cm
0016+73	QSO	vc	0	I	I	I/0	-	-	-	-	-	-
0153+74	QSO	dble	I	0	0	-	-	-	-	-	-	-
0212+73	QSO	jet	0	0	0	-	-	-	-	-	-	-
0615+82	QSO	vc	0	0	I	-	-	-	-	-	-	-
0716+71	BLL	vc	-	II	II	II/II	II	II/II	II/II	-	-	II/II
0836+71	QSO	jet	0	0	0	-	0	0/0	0/0	-	0/0	0/0
0917+62	QSO	c	-	II	II	II/II	II	II/II	II/II	-	-	-
0954+65	BLL	c	-	-	-	-	I	I/II	II/II	-	-	0/0
1150+81	QSO	c	II	I	II	II/II	II	II/II	-	-	-	-
1642+69	QSO	c	-	II	I	II/II	II	II/I	0/0	-	-	-
1749+70	BLL	c	II	I	?	II/I	-	-	II/II	-	I/I	-
1803+78	BLL	c	0	I	II	0/II	II	I/I	0/I	-	I/I	-
1807+69	Gal	jet	-	0	I	-	-	-	-	-	0/0	-
1928+73	QSO	jet	0	I	0	0/0	I	I/I	0/0	-	0/0	-
2007+77	BLL	c	-	I	I	II/II	II	I/I	II/II	-	I/I	-
2021+61	Gal	dble	-	0	0	-	-	-	-	-	-	-
0108+38	Gal	dble	-	-	-	-	-	-	-	0	-	-
0133+47	QSO	c	-	-	-	-	-	-	-	II	-	-
0316+41	Gal	irr	-	-	-	-	-	-	-	0	-	-
0710+43	Gal	dble	-	-	-	-	-	-	-	0	-	-
0711+35	QSO	dble	-	-	-	-	-	-	-	?	-	-
0804+49	QSO	vc	-	-	-	-	-	-	-	II	II/II	-
0814+42	BLL	c	-	-	-	-	-	-	-	II	-	-
0833+58	QSO	?	-	-	-	-	-	-	-	-	-	0/0
0850+58	QSO	c	-	-	-	-	-	-	-	-	-	0/0
0859+47	QSO	jet	-	-	-	-	-	-	-	0	-	-
0906+43	QSO	jet	-	-	-	-	-	-	-	0	-	-
0923+39	QSO	dble	-	-	-	-	-	-	-	0	-	-
0945+40	QSO	jet	-	-	-	-	-	-	-	I	-	-
0954+55	QSO	?	-	-	-	-	-	-	-	-	-	0/0
1418+54	BLL	?	-	-	-	-	-	-	-	-	-	0/0
1624+41	QSO	jet	-	-	-	-	-	-	-	0	-	-
1633+38	QSO	jet	-	-	-	-	-	-	-	0	-	-
1637+57	QSO	vc	-	-	-	-	-	-	-	-	-	0/0
1641+39	QSO	jet	-	-	-	-	-	-	-	0	-	-
1652+39	Gal	jet	-	-	-	-	-	-	-	?	-	-
1739+52	QSO	vc	-	-	-	-	-	-	-	-	-	II/II
1823+56	BLL	c	-	-	-	-	-	-	-	-	-	I/I
2200+42	BLL	jet	-	-	-	-	-	-	-	?	I/I	-
2351+45	QSO	jet	-	-	-	-	-	-	-	0	-	-
2352+49	Gal	dble	-	-	-	-	-	-	-	0	-	-
0235+16	BLL	vc	-	-	-	-	-	-	-	II	-	-
0300+47	BLL	c ?	-	-	-	-	-	-	-	I	-	-
0735+17	BLL	jet	-	-	-	-	-	-	-	II	-	-
0851+20	BLL	c	-	-	-	-	-	-	-	I	-	-
1226+02	QSO	jet	-	-	-	-	-	-	-	0	-	-
1253−05	QSO	jet	-	-	-	-	-	-	-	0	-	-
1308+32	BLL	vc	-	-	-	-	-	-	-	I	-	-
1404+28	Gal	c	-	-	-	-	-	-	-	0	-	-
1502+10	QSO	c ?	-	-	-	-	-	-	-	0	-	-
1611+34	QSO	c ?	-	-	-	-	-	-	-	?	-	-
1741−03	QSO	c ?	-	-	-	-	-	-	-	?	-	-
1749+09	QSO	c	-	-	-	-	-	-	-	I	-	-
1821+10	QSO		-	-	-	-	-	-	-	0	-	-
2214+35	QSO		-	-	-	-	-	-	-	II	-	-
2251+15	QSO	jet	-	-	-	-	-	-	-	0	-	-

- vc : very compact $F_c \geq 0.9$
- c : compact $F_c \geq 0.6$
- jet : prominent jet structure
- dble: compact double sources
- irr : irregular sources

Table 1 lists all sources investigated so far: Col 1 gives the name of the source. Col 2 gives the optical identification (Gal = galaxy, BLL = BL Lac object, QSO = quasar). Col 3 gives information on the structure on the mas scale. (c = compact, vc = very compact, dble = compact double, jet = prominent mas jet). Col 4 - 12 list the variability type found during the observation at the indicated epoch at the respective wavelengths (see header). The symbols 0, I and II refer to the type of variations observed for the object: 0 indicates no significant variations, I indicates significant variability of the flux densities whose structure functions do not exhibit a clear maximum and II denotes significant flux density variations whose structure functions show a clearcut maximum (see fig. 2).
Table 1 is divided in three parts. The upper third contains the sample of sources north of declination 60 degrees, which is best studied, since these sources are circumpolar and thus easily observable from Effelsberg. Moreover during the combined VLBI, X-ray etc observational campaigns for the S 5 sources (e.g. Witzel, 1987, Witzel et al., 1988, Schalinski et al., 1991) a wealth of relevant information has been accumulated for these objects. The middle part of table 1 lists sources with declinations between 35 and 60 degrees. These two upper parts of the table form essentially complete statistical samples of sources selected on the basis of their 5 GHz flux densities ($S > 1$ Jy at 5 GHz) and their radio spectra (radio spectral indices $\alpha > -0.5$ with $S \propto \nu^{\alpha}$). In contrast the sources found in the bottom third of the table 1 are selected due to their a priori known properties which made them good candidates for the appearance of IDV. From the table it is obvious that at any given epoch about a third of all flat spectrum radio sources show IDV and that the repeatability for IDV is high. Also there is a correlation with source size in the sense that the most compact sources have a high probability for displaying IDV. No clearcut correlation with galactic latitude or location on sky is apparent from the table.
The rms error of the flux density determinations with the 100 m telescope at 11 and 6 cm wavelengths varied between 0.25 and 0.5 per cent according to the integration time, the weather conditions etc. (e.g. Witzel, 1991). Fig. 3 shows the distribution of the variability index for flat and steep spectrum sources. The hatched area in fig. 3 a displays sources which exhibit prominent jets on the mas scale. On the basis of these figures it is evident that at any given epoch essentially all very compact sources show variability on a level of at least 2 per cent.

3 OPTICAL VARIABILITY

Optical variations of rapid variability in BL Lac objects has been known since the definition of this class. In spite of being used as a defining criterion (Stein, 1978) and extensive studies in a few objects (e.g. Moore et al., 1982) theoretical investigations have rarely taken intraday variability into account. Miller et al. (1989) have shown that the time-scales in the optical regime can be shorter than 1 h, close to the short

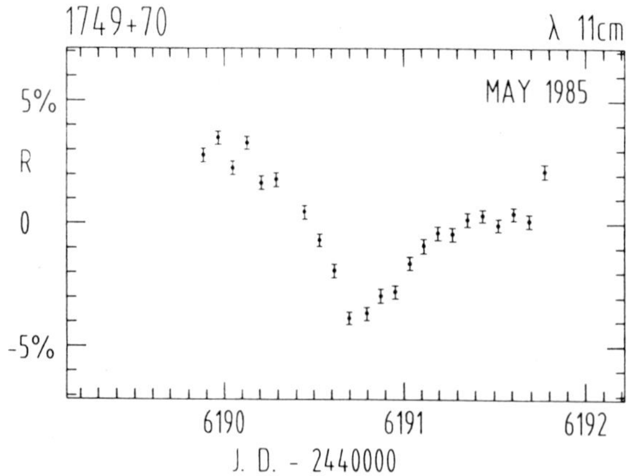

Fig. 1. The intensity variations of the BL Lac 1749+70 at 11 cm wavelength in May 1985, R = residual flux density (see Heeschen et al., 1987).

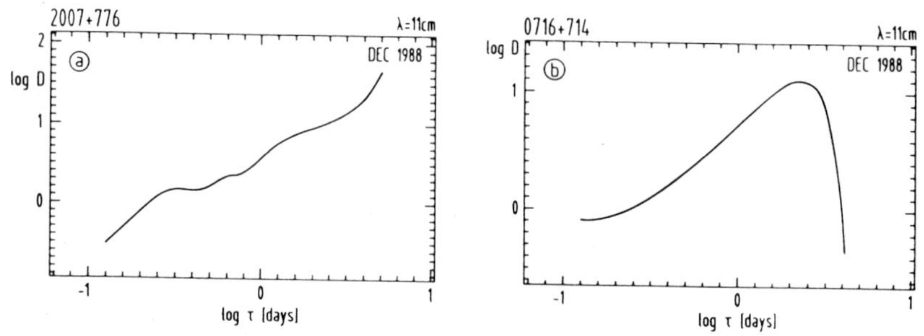

Fig. 2. Typical structure functions of a source showing IDV (right) and an object varying on longer timescales (left) (see e.g. Witzel, 1990).

time-scales observed in the x-ray regime.

We started optical monitoring of the sources which were found to show IDV at radio frequencies in order to resolve the question whether interstellar scintillation may cause the variability. Early statistical investigations (Wagner et al., 1990) showed that out of a pool of sources studied simultaneously at optical and radio frequencies, the same ones varied in both, the radio and the optical regime. Likewise the sets of quiet objects were identical, too.

In ongoing investigations optical monitoring of compact, flat-spectrum radio sources we find a fraction of $\sim$ 60 % of the objects to exhibit IDV at any given epoch. Since long-term monitoring of some of these sources reveals a duty-cycle of $\sim$ 60% as well, IDV is probably a phenomenon occurring in almost every source showing the properties described in Section 2.

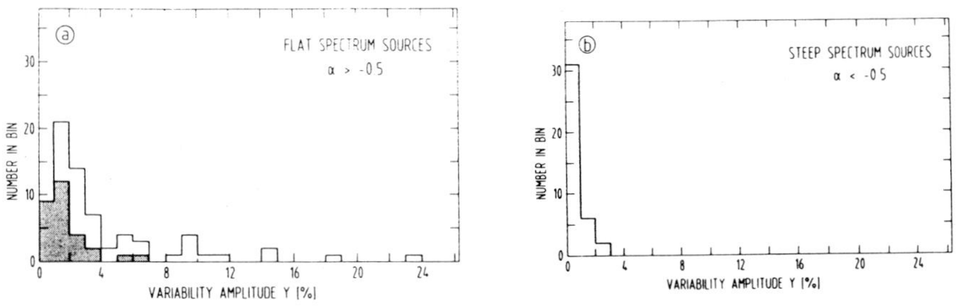

Fig. 3. Distributions of the variability amplitudes Y of flat (a) and steep spectrum sources (b) at 6cm wavelengths. The sources with prominent VLBI jets are shaded in (a).

4 POLARIZATION

The flux density variations described above are in a number of cases accompanied by variations of the polarization characteristics of the AGN.

4.1 Radio Data

In 6 out of 9 cases simultaneous variability has been found. In the quasar 0917+62 an anti-correlation of the total and the polarized flux densities has been observed repeatedly (Quirrenbach et al. 1989, Witzel 1991). Significant changes of the position angle of the polarization (including a 180°-swing at the time of a pronounced minimum of the polarized flux density) have been observed (Quirrenbach et al., 1989). In contrast, in the BL Lac 0716+714 correlated variations between the total and the polarized flux densities have been found during the slow state of its variability (see below for the two states of variations in this object, see fig. 4).
In the BL Lac object 0954+65 no correlation between the two quantities has been established yet.

4.2 Optical Data

Most of the objects showing IDV at optical frequencies are highly polarized (5 % to 35 %). In all of the cases were multiple measurements have been described in the literature, the values deviate from another by much larger amounts than the quoted errors, indicating that all of the sources have variable polarization. This is clear evidence that the variable optical emission is non-thermal. As yet, little is known about the correlation between variations of the total flux and variations of the polarized flux or the polarization angle.

5 CORRELATIONS BETWEEN RADIO AND OPTICAL DATA

A more specific test to the relation of IDV in the two frequency regimes was enabled by a simultaneous long-term campaign. In Feb. 1990 we observed 3 sources for 25 days continuously at the VLA and with two optical telescopes on Calar Alto (2.2 m) and Königstuhl (0.7 m). 0716+714 and 0954+65 were found to vary at radio and optical frequencies. While IDV in 0716+714 exhibits a permanent variation around a slowly changing mean, 0954+65 shows individual flares. Unfortunately, electronic maintenance at the VLA during the two of the major outbursts in 0954+65 prohibits a detailed comparison of the variability characteristics during this particular campaign. In 0716+714, a close match of the variability pattern at optical and radio frequencies is observed (fig. 4). As has been described e.g. in Quirrenbach et al., 1991 and Wagner, 1991, the close correlation is evident from the simultaneous change between a rapid state at high level and a slow oscillation at a lower level on JD 2447930. Furthermore the power-spectra and structure functions in the two regimes are very similar, both exhibiting clear maxima at periodicities of about a day and 6 days during the two states respectively. In both cases there are no differences in the positions and widths of the peaks in the power-spectra (see e.g. fig. 5, which shows the power spectra of 0716+714 in the optical and radio regimes).

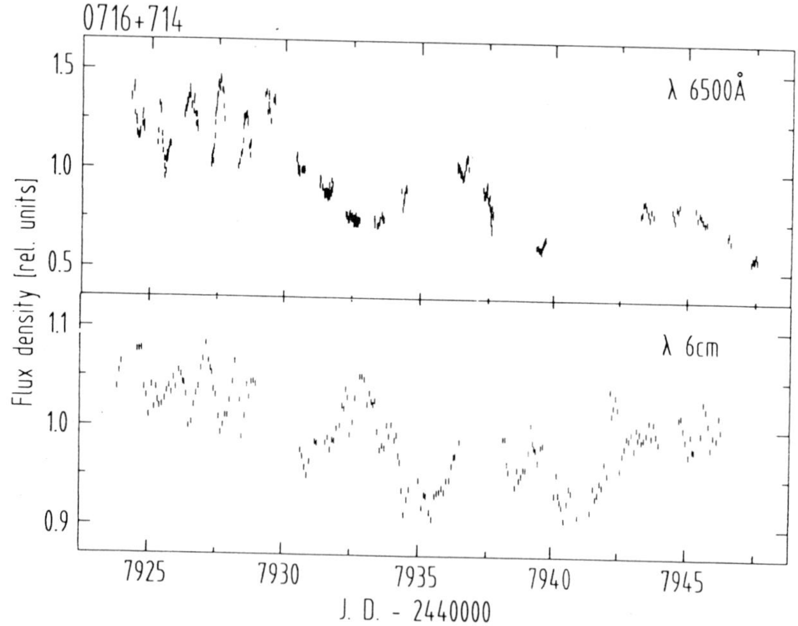

Fig. 4. The variations of the 6cm flux density and the R-band emission of 0716+714 observed in February 1990 (Quirrenbach et al., 1991).

6 STATISTICAL PROPERTIES

In spite of the preferred time-scales, suggested by the clear maxima of the power spectra, it is obvious from the structure functions that significant power resides on all measurable time-scales up to longest ones which can be sampled in the 25 day campaign (see e.g. Quirrenbach et al., 1991). This causes a strong blending of any individual flare rising on the preferred timescale of ~ 1 day. It is possible, however, to determine the skewness of an individual flare from higher-moment structure functions. Fig. 6 displays the differences in intensity dI between any two measurements spaced dt apart. For a sufficiently long campaign the average flux has no long-term trends, forcing the mean $dI(t)$ to zero. However, the distribution of dI in any given interval $(dI(t_1) - dI(t_2))$ will be skewed if the variability occurring on timescales t (with $t_1 < t < t_2$) is asymmetric in time. Asymmetrical light-curves are expected for most intrinsic models, since optical depth effects and the interplay between acceleration and cooling time-scales, which determine the shape of each flare require fine-tuning to produce light-curves with are invariant under time-reversal. Extrinsic mechanisms may usually create lightcurves which are statistical symmetric in time unless the source has an asymmetric luminosity profile.

Although the correlated radio and optical variability in the source 0716+714 seems to rule out any extrinsic explanation at least for this source , in the following we will briefly discuss the main theoretical approaches for the explanation of IDV, which are being discussed to date.

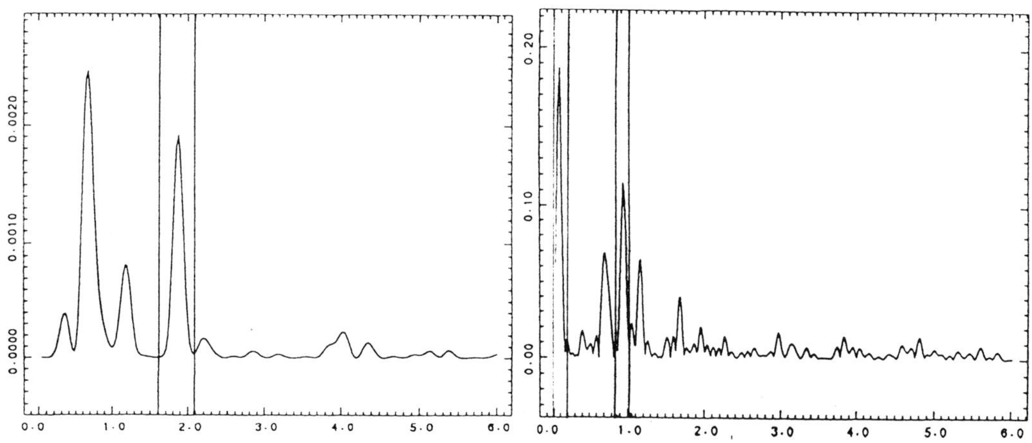

Fig. 5. Power spectra of the variability of 0716+714 during the first week of the monitoring campaign in February 1990. Both at optical (left) and radio frequencies (right) preferred frequencies of about 1 day are obvious. Broken lines indicate time-scales which are strongly influenced by the window function.

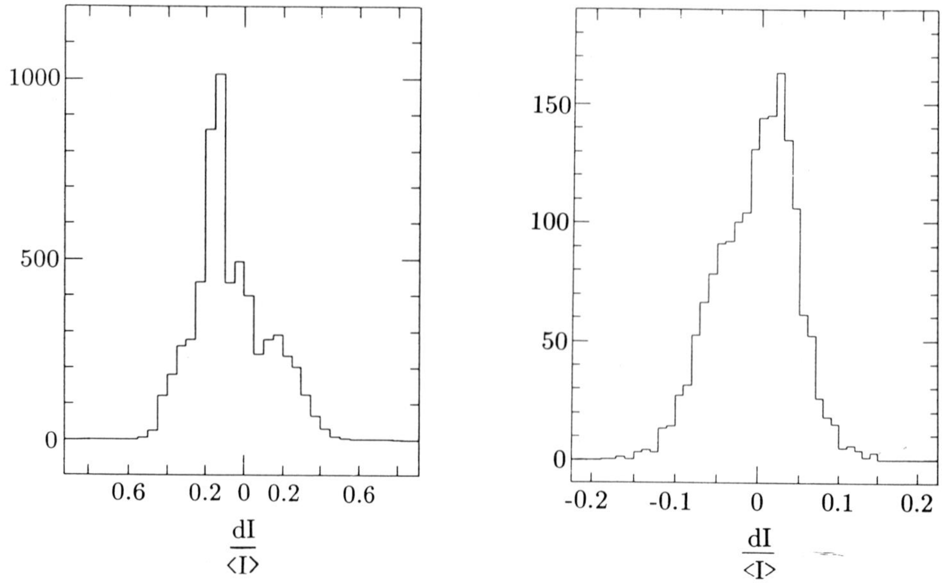

Fig. 6. The statistical distribution of $dI(dt)$ of 0716+714 during February 1990. The histograms of dI in the range of $0.5d < dt < 1.5d$ for the optical (650 nm, left) and radio (6cm, right) are asymmetric, but display different signs of skewness.

7 EXTRINSIC MODELS

The high brightness temperatures invoked from the short variability time-scales in the optical regime (up to $10^{18}K$) have given rise to the suggestion that propagation-induced variability rather than an intrinsic origin dominates the observations. While gravitational microlensing is achromatic, interstellar scintillation only affects observations at lower frequencies. These wavelength dependencies allow us to exclude either of these mechanisms as a dominant effect.

7.1 Microlensing

In the microlensing theory the observed time-scales are related to the relative transverse velocity of either the observer, the lens, or the source by:

$$\Delta t = v^{-1} \ (M \ / \ M_{\odot})^{0.5} \ 5 \ 10^{16} cm$$

In order to explain the observed variability on the very short timescales ($\Delta t < 1d$), the required velocities v_{trans} are relativistic or superluminal if the lenses have $M > 0.01 M_{\odot}$. Since these velocities are only possible in the source-plane, relativistic motions are required. The high duty cycle of the variability implies that *several* knots moving in the source plane are required. Such a scenario requires intrinsic variability, which reduces the potential role of microlensing to an amplification of intrinsic variability.

Another argument against microlensing is given by the facts that the actual autocorrelation functions at radio and optical frequencies have similar widths, which are also comparable to the widths of the cross-correlation peaks but smaller than the observed lags between the radio and optical data. If the variability observed in the two frequency bands is due to microlensing, these observations would require that the optical and radio sources were spatially separated in the source-plane by a distance exceeding their diameters. Since both sources are nonthermal (see Sect. 4), this scenario is very contrived.
Furthermore the statistical asymmetry of the flares (see Sect. 6) contradicts predictions from microlensing. Although individual high-amplification events, like any other microlensing-induced variations may be asymmetric, a symmetric source amplified by microlensing will produce statistically symmetric variability.

7.2 Interstellar Scintillation

Although RISS induced variations were one of the prime movers of the study of rapid variability at radio frequencies and an early explanation until recently, the observed variations pose very strong limits on the parameters of the interstellar medium (see e.g. Wambsganss et al., 1989). It is not clear at the moment whether the observed frequency dependence can be explained by standard models of interstellar scintillation (Schneider, private communication; Rickett, private communication). It is also unlikely that the frequent occurrence of correlated or anti-correlated variability of total and polarized flux (see Sect. 4) can be accommodated in a scenario of RISS-induced variability. In the RISS models these correlations would call for an intrinsically inhomogeneous source whose substructure has varying degrees and directions of polarization, some of them being amplified by the interstellar scattering. Especially changes from correlated to anti-correlated variability could only be explained if the flow pattern varies or if the source changed as well.
Like the microlensing-scenario, RISS induced variations are expected to be symmetric in time for simple geometries, inconsistent with the results described in Sect. 6.

The most convincing argument to exclude RISS as the *dominant* mechanism for the variations described in Sect. 3 is the intimate correlation between radio and optical frequencies. RISS cannot produce any variations at optical frequencies. It would be extremely contrived, if a source which varies in the optical with a preferred time-scale would not vary at cm frequencies intrinsically but at the same time where subject to RISS induced variability giving rise to variations with exactly the same power spectrum as seen in the optical regime.
The close correlation, described in Sect. 5 therefore rules out interstellar variations as most important mechanism for the quasi-periodic IDV. We would like to point out, however, that any source of the high degree of compactness as implied by the rapid intrinsic variability is bound to be subject to RISS-induced variability as well. Indeed a statistical analysis of other flat-spectrum sources exhibits clear indications for a minor contribution of IDV (Witzel and Quirrenbach, 1991).

8 INTRINSIC MODELS

The most dramatic challenge to present-day models of rapid variability is given by the extremely high brightness temperatures suggested from the intraday variability. Unfortunately only limits can be given in the case of 0716+714, which is discussed extensively above, since it was not yet possible to measure a redshift in the featureless spectrum of this BL Lac object. From the absence of any signatures of a host galaxy even on deep images taken under very good seeing conditions, a lower limit to the redshift of $z > 0.3$ was derived. This leads to brightness temperatures of $T_b > 10^{16} K$, comparable to the values found for other objects (e.g. $T_b = 10^{18}$ K in 0917+62 (Quirrenbach et al., 1989)).
Furthermore the regular and quasi-periodic nature of the variability is a challenge for models explaining the variability as caused by some turbulent process (such as the propagation of shocks in turbulent shocks, e.g. Marscher, 1991).
The quasi-periodic nature (i.e. the significant peak in the power spectra) and the regular, albeit asymmetrical shape of the lightcurves both favour a regular geometry for a macroscopic description of the orbit of the radiating material. Following the standard paradigm for compact radio sources and blazars we assume that the sources exhibiting IDV are seen jet-on. Because of the close correlation between cm and optical light curves we assume similar optical depths for the emission regions, which would then both be in the jet. The regular geometry may then be a helical jet with downward-propagating shocks. This scenario follows the original suggestions of Königl and Choudhuri, 1984 and may, by invoking relativistic beaming even produce the apparent high brightness temperatures. If the shock were planar, the time-scale of the variability would not be comparable to the light-travel diameter of the source (i.e. the shock front) but to the cooling length behind the shock. In spite of the success of shock models explaining longer-term variability, it is still unclear, whether

the whole jets are helical, if such shocks could actually be produced and whether they could propagate as a coherent, planar structure through the whole configuration long enough.

A similar scenario are helical paths of plasma blobs moving in straight jets along the field lines carried by the jet from the rotating disk (see Camenzind, these proceedings). In this case the helical structure describes the orbit of the emitter, which due to relativistic beaming, sweeps over the line-of-sight at successive intervals.

We would like to mention that - in the absence of a clear picture about the microphysical emission mechanisms - the origin of the emission in the jet is uncertain. An alternative site for the emission might be the disk directly. As discussed below, coherent radiation mechanisms could explain the observed high brightness-temperatures. In analogy to the radiation produced in solar flares, coherent plasma effects and magnetic reconnection in flares coupled to the accretion disk could also produce quasi-regular signatures, if the radiation would escape from the rotating disk along a preferred line-of-sight.

Irrespective of the overall geometric configuration, the rapid time-scales and high brightness temperatures are unlikely to be caused by relativistic beaming and special geometries alone. In this case a modification of the classical assumptions on the microphysics might be necessary. Following early discussions about the nature of the nonthermal continuum (e.g. Jones, et al., 1974) it was concluded that most of the emission is due to incoherent synchrotron emission. Coherent radiation mechanisms could explain the high brightness temperatures fairly easy (Benford, 1984, Lesch, these proceedings). It is not clear, however, whether any of the suggested mechanisms (magnetic reconnection, collective plasma effects, cyclotron-masers) can produce the observed high energy densities in a sufficiently small volume without violating the tight limits on the optical depths and electron densities. It is also not clear, at present how much the primary, coherent radiation contributes to the observed flux, and where comptonized, re-processed radiation sets in to be the major contributor.

Acknowledgements. It is a pleasure to thank M. Camenzind, J. Heidt, T. Krichbaum, A. Quirrenbach, R. Wegner, and K.-H. Wüllner for their help and for many discussions throughout the project. This work was supported by the Deutsche Forschungsgemeinschaft (SFB 328).

REFERENCES

Benford, G.: 1984, in *"Physics of Energy Transport in AGNs"*, NRAO Proceedings, p. 185

Heeschen, D.S.: 1984, *Astron. J.*, **89**, 1111.

Heeschen, D.S., and Rickett, B.J.: 1987, *Astron. J.*, **93**, 589.

Heeschen, D.S., Krichbaum, T., Schalinski, C., Witzel, A.: 1987, *Astron. J.*, **94**, 1493.

Jones, T.W., O'Dell, S.L., and Stein, W.A.: 1974, *Astrophys. J.*, **188**, 353

Königl, A. and Choudhouri, A.R.: 1985, *Astrophys. J.* **289**, 173

Miller, H.R., Carini, M.T., and Goodrich, B.D.: 1989, *Nature*, **337**, 627

Marscher, A.: 1991, in: *Physics of Active Galactic Nuclei*, ed. S.J. Wagner and W.J. Duschl (Springer Verlag), in press.

Moore, R.L. et al.: 1982, *Astrophys. J.*, **260**, 415

Quirrenbach, A., Witzel, A., Qian, S.J., Krichbaum, T., Hummel, C., Alberdi, A.: 1989, *Astron. Astrophys. (Letters)*, **226**, L1.

Quirrenbach, A., Witzel, A., Wagner, S., Sanchez-Pons, F., Krichbaum, T., Wegner, R., Anton, K., Erkens, U., Haehnelt, M., Zensus, J., Johnston, K. et al.: 1991, *Astrophys. J.* **372**, L71

Schalinski, C.J., Witzel, A., Krichbaum, T.P., Hummel, C.A., Quirrenbach, A., Johnston, K.J.: 1991, in: *Variability of Blazars*, ed. E. Valtaoja and M. Valtonen (Cambridge University Press), in press.

Stein, W.A.: 1978, in: *Pittsburgh Conference on BL Lac Objects*, ed. A.M. Wolfe, p. 1.

Wagner: 1991, in: *Variability of Blazars*, ed. E. Valtaoja and M. Valtonen (Cambridge University Press), in press.

Wagner, S.J., Sanchez-Pons, F., Quirrenbach, A. and Witzel, A.: 1990, *Astron. Astrophys.* **235**, L1

Wambsganss, J., Schneider, P., Quirrenbach, A. and Witzel, A.: 1989, *Astron. Astrophys.*, **224**, L9

Witzel, A., 1987, in: *Superluminal Radio Sources*, ed. J. A. Zensus and T. J. Pearson (Cambridge University Press), p. 83.

Witzel, A.: 1990, in: *Parsec-scale radio jets*, ed. J. A. Zensus and T. J. Pearson (Cambridge University Press), p. 206.

Witzel, A.: 1991, in: *Physics of Active Galactic Nuclei*, ed. S.J. Wagner and W.J. Duschl (Springer Verlag), in press.

Witzel, A., Heeschen, D.S., Schalinski, C., Krichbaum, T.: 1986, *Mitt. Astron. Ges.*, **65**, 239.

Witzel, A., Schalinski, C.J., Johnston, K.J., Biermann P.L., Krichbaum, T.P., Hummel, C.A., and Eckart, A.: 1988, *Astron. Astrophys.*, **206**, 245.

Witzel, A. and Quirrenbach, A.: 1991, in: *Proceedings of the conference "Propagation Effects in Space VLBI"* held in Leningrad, June 1990, ed. N. Kardashev and L. Gurvits, in press.

Formation of Relativistic Jets in Quasars and the Origin of Quasi–Periodicity

M. CAMENZIND

Landessternwarte Königstuhl, D-W–6900 Heidelberg, Germany

Abstract: Relativistic jets are naturally formed in the rapidly rotating magnetospheres of geometrically thin accretion disks around rapidly rotating black holes. Disk fields generated by dynamo effects in the boundary layer between disk and rotating hole immerse the hole into a rotating magnetosphere which is loaded by plasma from the disk. Magnetized winds ejected from the surface of the disk will be collimated by magnetic effects on scales somewhat larger than the light cylinder of the magnetosphere. Non–axisymmetric perturbations dragged along by these jets will rotate and produce quasi–periodic emission by a kind of lighthouse effect.

1 INTRODUCTION

The actual formation of jets in Quasars and BL Lac objects is fundamental for the understanding of the emission from these objects. The formation of relativistic jets is rather difficult to understand, since detailed structural information on the subparsec–scale is still lacking. VLBI observations for 3C 345 show, however, that jets are not pencil–like, but have a radius of the order of a fraction of a light year already on the parsec–scale. This rules out formation of jets by the accretion funnel of geometrically thick disks. The most likely origin of these jets is due to rapidly rotating magnetospheres built up by the accretion disks around supermassive objects. Magnetic fields are advected inwards by the accretion flow and redistributed in the Keplerian disk by strong differential rotation and turbulent diffusion. The central black hole appears immersed into a rotating magnetosphere. A detailed model will be presented, which explains the observed superluminal speeds and accounts for the emission properties of jets.

2 THE MAGNETIC STRUCTURE OF AGN DISKS

According to the standard model, AGNs are powered by accretion disks surrounding a supermassive black hole with masses in the range $M_H \simeq 10^6 - 10^{11}\ M_\odot$. In this respect, each source is specified in terms of 4 parameters: the mass M_H of the central black hole, the Kerr parameter $a_H = J_H/J_{\rm max} < 1$ for the black hole angular momentum ($J_{\rm max} = GM_H/c$), the accretion rate $\dot{M}_{\rm acc}$ for the disk and the angle θ for the disk inclination with respect to the observer's line of sight. Masses and accretion rates can essentially be determined from model fittings of the big blue bump, the angular momentum J_H is not directly observable.

Under moderate accretion rates, $\dot{M} < \dot{M}_{\rm ED}$, the hole will be surrounded by a geometrically thin disk. Radiation from the disk, either as direct thermal emission from its surface due to dissipation of energy within the disk (UV and optical), or as non–thermal reprocessed emission is responsible for most of the observed activity in radio–weak objects. Their X–ray emission is most probably produced in an optically thin corona above and below the disk, or it might be due to inverse–Compton scattering of photons off relativistic electrons in the magnetic corona.

The formation of a magnetic corona around the innermost part of the accretion disk is very probable. Fully turbulent disks are ideal setups for magnetic dynamos (Pudritz 1981; Kuperus 1987; Camenzind 1990). Any small magnetic field in the accreting matter will be amplified in a Keplerian disk by the strong differential rotation. At the same time, the turbulence in the disk will generate a turbulent magnetic field and a turbulent velocity field, which will also generate large–scale magnetic fields over the helicity $\alpha_D(R,z) = L_T^2\, \Omega\, \bar{\alpha}(z)/H(R)$, quite similar to the dynamo action observed in the disks of spiral galaxies. In such turbulent disks, the mean magnetic fields evolve according to the induction equation of mean field electrodynamics

$$\partial_t \vec{B} + (\vec{u}\cdot\nabla)\,\vec{B} = (\vec{B}\cdot\nabla)\vec{u} - (\nabla\cdot\vec{u})\,\vec{B} + \nabla\wedge\left\{\vec{E} - \eta_T\,\nabla\wedge\vec{B}\right\}. \qquad (1)$$

$\vec{B}$ is the mean magnetic field, and $\vec{u}$ the mean velocity field in the disk, $\vec{u} = (u_R, R\Omega, 0)$. η_T is the total magnetic diffusivity including the microscopic one and the turbulent diffusivity. $\vec{E} = \alpha_D\,\vec{B}$ represents the electromotive force induced by correlations between turbulent motions and turbulent magnetic fields.

Any small seed field accreted from the molecular torus at the parsec scale will decay exponentially or be amplified by the dynamo until it reaches equipartition with the turbulent pressure in the disk (Camenzind 1990). In geometrically thin disks, the toroidal component B_ϕ is the dominant component of the magnetic field in the disk, the poloidal component $\vec{B}_p$ is regenerated from the toroidal one via the dynamo effect. A disk dynamo generates a large scale structure for the magnetic fields which is superposed by turbulent small scale structures. This large scale field is essentially time–dependent, and its structure depends largely on the dynamo numbers (Pudritz 1981; Camenzind 1990; 1991c). For typical parameters of geometrically thin accretion disks, $H/R \leq 0.5$, seed fields in the disk are rapidly amplified and form a quadrupolar structure near the inner edge of the accretion disk. When the dynamo effect is included, dipolar fields are in general unstable. Quadrupolar fields are the first modes to be excited in geometrically thin disks (Camenzind 1990; 1991c).

For accretion rates $\dot{m} = \dot{M}/\dot{M}_{\rm ED} > 0.01$ the disks are dominated by radiation pressure, and the growth of the toroidal magnetic fields in the disk is limited by the turbulent pressure, the magnetic field is practically independent of the accretion rate

$$B_D \leq B_{\rm ED}\,(GM_H/c^2R)^{3/4} \quad , \quad B_{\rm ED} = \sqrt{L_{\rm ED}/M_H^2\,c} = 4\times 10^4\,\text{Gauss}\,M_{H,8}^{-1/2}\,. \tag{2}$$

This disk field corresponds to a total poloidal current driven by the differential rotation

$$I_D \leq 3\times 10^{18}\,\text{Ampere}\,M_{H,8}^{1/2}\,(c^2R/GM_H)^{1/4}\,. \tag{3}$$

Quasars with their high accretion rates, $\dot{m} \simeq 0.1$, will therefore have magnetic field strengths given by Eq. (2), and the total current generated in the accretion disk is of the order of 10^{19} Amps. The accretion disks of Seyfert galaxies, having masses $M_H \simeq 10^7\,M_\odot$, will carry higher magnetic fields, but much lower currents. Since the accretion rate in FR 1 radio sources is quite low, $\dot{m} \simeq 0.001$, the magnetic fields are also weaker in these sources; the turbulent pressure scales with the accretion rate. In this way, all AGNs can be classified in terms of their masses M_H and in terms of the strength of the magnetic field in the disk (Camenzind 1991a). The main difference between radio–weak objects (QSOs and Seyfert galaxies) and radio–loud ones (Quasars, radio galaxies and BL Lac objects) must be hidden in a third parameter, which is most probably the angular momentum J_H of the rotating black hole (Camenzind 1991b). Radio–weak objects have a low spin, $a_H < 0.6$, radio–loud objects might still have a high spin, $0.6 < a_H < 1.0$. The role of the angular momentum in the jet production mechanism is, however, still unclear.

3 DISK MAGNETOSPHERES AND THE ORIGIN OF JETS IN AGNS

Axisymmetric magnetospheres around a rotating Kerr black hole can be described in terms of the magnetic flux function $\Psi(r,\theta)$

$$\Psi = \int_\Sigma \vec{B}_p \cdot d\vec{S} \quad , \quad \vec{B}_p = \frac{\nabla\Psi \wedge \vec{m}}{2\pi R}\,. \tag{4}$$

They consist of a family of nested magnetic surfaces, each of which is obtained by taking a single $\vec{B}$–line and rotating it around the rotational axis with angular velocity $\Omega^F(\Psi)$. The toroidal field $\vec{B}_T$ is then responsible for the spiraling.

3.1 The Global Structure of Disk Magnetospheres

The structure of a rotating magnetosphere follows from Ampère's equation with the toroidal current density given by force–balance perpendicular to the flux surfaces

$$\rho_e\vec{E}_\perp + (\vec{j}_T/c)\wedge\vec{B}_P + (\vec{j}_P/c)\wedge\vec{B}_T = \vec{f}_\perp\,. \tag{5}$$

$\vec{f}_\perp$ is the inertial force due to the streaming plasma, which is in general quite thin, so that the force-free approximation is a reliable one, $\vec{f}_\perp \simeq 0$, except for certain critical regions in the magnetosphere. Under force–freeness, the condition $\vec{f}_T = 0$ also implies $\vec{j}_P \parallel \vec{B}_P$, and therefore the total current density

$$\vec{j} = \rho_e \vec{v}^F - \frac{1}{\alpha} \frac{dI}{d\Psi} \vec{B} \quad , \quad \vec{v}^F = \frac{\Omega^F - \omega}{\alpha} R \vec{m} \, . \tag{6}$$

The magnetic structure of a rotating corona follows therefore from the relativistic Grad–Schlüter–Shafranov (GSS) equation (Blandford & Znajek 1977; Macdonald & Thorne 1982; Haehnelt & Camenzind 1991)

$$\nabla \cdot \left\{ \frac{\alpha D}{R^2} \nabla \Psi \right\} + \frac{\Omega^F - \omega}{\alpha c^2} \frac{d\Omega^F}{d\Psi} |\nabla \Psi|^2 + \frac{8\pi^2}{R^2 \alpha c^2} \frac{dI^2}{d\Psi} = 0 \, , \tag{7}$$

$$D = 1 - \frac{R^2 (\Omega^F - \omega)^2}{\alpha^2 c^2} \, . \tag{8}$$

This GSS equation has critical points at the light surfaces of the magnetosphere, $R_{\mathrm{L,out}} = \alpha c / (\Omega^F - \omega)$ and $R_{\mathrm{L,in}} = \alpha c / (\omega - \Omega^F)$, which are identical with the Alfvén surfaces for the force–free condition (Camenzind 1991a). The existence of an inner light surface near the horizon is a consequence of the frame–dragging effect represented by the angular velocity ω of the ZAMOs.

The GSS equation can be solved with the method of Finite Elements for a given flux distribution $\Psi_D(r)$ along the surface of the accretion disk and a given current distribution $I(\Psi)$ (Haehnelt & Camenzind 1991). For a flux distribution localized in the boundary layer of the accretion disk, the rotation of the field lines is approximately constant $\Omega^F(\Psi) \simeq \Omega(r_{\mathrm{ms}})$, where r_{ms} denotes the radius of the marginal stable orbits. This determines then the position of the outer light cylinder ($a_H \simeq 0.8$)

$$R_L = c/\Omega^F \simeq 10 G M_H / c^2 \simeq 1.5 \times 10^{15} \, \mathrm{cm} \, M_H / 10^9 \, M_\odot \, . \tag{9}$$

In general, a certain fraction Ψ_H of the magnetic flux will cover the horizon. Magnetic fields in the disk near the horizon will be dragged onto the horizon by strong advection. The resulting global topology of a force–free rotating magnetosphere, generated by a localized dipolar flux tube in the boundary layer, has a complicated structure (see Haehnelt & Camenzind 1991).

The most striking difference between a Newtonian magnetosphere and the relativistic one is the existence of a critical flux surface Ψ_c, which extends between the inner and outer light surface and ends in a neutral ring on the outer light surface. From there the flux surface closes back to the disk. The magnetosphere of a dipolar flux tube contains an inner closed part bounded by this critical flux surface, and two open parts. The inner one covers the horizon of the hole and is bounded by the outer light surface. It forms the magnetic core of the jet and contains the return currents. The outer part, which is anchored in the accretion disk outside the critical flux surface, must cross the outer light cylinder. Since the magnetic field lines are here strongly bent towards the surface of the disk, plasma of the corona can easily escape along these flux surfaces. Outside the light cylinder, these field lines are collimated by strong pinching forces. The scale for collimation is the light cylinder radius R_L.

3.2 Plasma Motion in the Collimated Region

The open disk magnetosphere is filled up with plasma which is injected near the boundary layer of the disk. Plasma is streaming along the magnetic surfaces, $\vec{u}_p \parallel \vec{B}_p$. The kinematics follow from two additional constants of motion, the total energy E and total angular momentum L (Camenzind 1986)

$$u_t = \frac{E}{\mu}\frac{1 - M_g^2 - \Omega^F L/E}{1 - M_g^2 - \Omega^F\, l(\Omega^F)}, \quad M^2 = \frac{4\pi\alpha^2\mu n u_p^2}{B_p^2}, \quad M_g^2 = \frac{M^2}{g_{tt} + \Omega^F g_{t\phi}}, \tag{10}$$

$$l = \frac{l(\Omega^F)(1 - \Omega^F L/E) - M_g^2\, L/E}{1 - M_g^2 - \Omega^F L/E}, \quad l(\Omega^F) = -\frac{g_{\phi\phi}(\Omega^F - \omega)}{g_{tt} + \Omega^F g_{t\phi}}. \tag{11}$$

The magnetization σ_D of the disk corona given by (Camenzind 1991a)

$$\sigma_D = \frac{\Phi_D^2}{4\pi m_p\, J_D c\, R_L^2} \simeq \frac{\Psi_D^2}{2\dot{M}_w c\, R_L^2} \simeq \pi^2\, \frac{\dot{m}}{\epsilon_H}\, (3GM_H/c^2 R_D)^{3/2} \tag{12}$$

for radiation pressure dominated disks, $0.001 < \dot{m} < 1.0$, determines the poloidal outflow velocity u_p. Physical outflows must pass through the fast magnetosonic point (Camenzind 1989). This is also true for magnetized accretion flows onto black holes. The magnetization parameter is the crucial quantity for the question of the outflow speeds. When $\sigma_D > 1$, the outflow is accelerated to relativistic speeds with Lorentz factors $\gamma \simeq \sigma_D$, for $\sigma_D \leq 1$ only transrelativistic speeds are achieved.

4 KNOT SPECTRA AND INTRADAY–VARIABILITY

The radio–optical spectra of quasars and BL Lac objects have in general two different components (Brown et al. 1989), which are also supported by variability data. Emission in the mm–UV is from one single synchrotron component which becomes self–absorbed at a wavelength of about 1 mm. This component also shows a higher and more rapid degree of variability than the cm–component which is identified with the underlying quiescent relativistic jet and is variable on the timescale of years. A similar decomposition also holds for the archetypical quasar 3C 273 (Courvoisier & Camenzind 1989). The mm-UV component arises from injection, or re–acceleration of high energy electrons related to the general variability seen on timescales of weeks to months.

Detailed monitoring of a few sources in the last years has revealed a quasi–periodic nature of these short–term fluctuations. During the optical activity in 1988, 3C

273 has shown a quasi–periodic fluctuation with a fundamental period $P_{\rm obs} \simeq 13$ days (see Fig. 1 in Courvoisier et al. 1988). This activity precedes the ejection of the superluminal knot C9, whose appearance has been detected in mm–VLBI (Krichbaum et al. 1991; Baath et al. 1991). The actual birth of the knot coincides with a strong X–ray flare (Camenzind 1991a). Superluminal motion can be best explained in this source for an inclination $i \leq 8°$ and $\gamma_{\rm jet} \simeq 10 - 20$. The central mass of this quasar is known from its UV–luminosity, $M_H \geq 2 \times 10^9\, M_\odot$, and the black hole should be rapidly rotating, $a_H \simeq 0.8$. The light cylinder R_L of the innermost part of the disk magnetosphere is therefore extremely compact, $R_L \simeq (5-10)GM_H/c^2 \simeq (2-5)\times 10^{15}$ cm $M_{H,9.5}$. The variability of this source can be explained in terms of knots injected into the jet at a radius $R_0 \simeq 10\, R_L$ with a typical magnetic field strength $B_0 \simeq 1$ Gauss. These knots are rotating with the jet plasma at the fundamental period $P_0 \simeq 660$ days. Due to beaming effects, this period appears shortened, $P_{\rm obs} = (1+z)\, P_0\, (1 - \beta_{\rm jet} \cos(i - \Theta_0)) \simeq 15$ days.

The spectral flux S_ν observed from an unresolved knot C moving in a collimated jet with constant opening angle Θ_0 is strongly amplified by beaming (for the emissivity j_ν we have $j_\nu = D^2\, j_{\nu'}$)

$$D(t) = \frac{1}{\gamma_{\rm jet}(1 - \beta_{\rm jet}\, \mu(t))}, \tag{13}$$

$$S_\nu(t) = \int_C d\Omega\, j_\nu(\nu)\, dl = \frac{1}{d^2}\, D^{2+\alpha}(t)\, (1+z)^{1-\alpha} \int_C dV\, j_{\nu'}(\nu, t)\,. \tag{14}$$

This flux is in general time–dependent, since the Doppler factor D is now a function of t because of the varying angle between the line of sight and the velocity vector of the knot (a kind of light–house effect). In addition, the time–evolution of the relativistic particle distribution also produces a time–dependent effect. In Camenzind & Krockenberger (1991) we discuss a theoretical lightcurve for 3C 273 as derived from our model for $R_0 = 10R_L$ and a jet opening angle $\Theta_0 = 0.07°$. If the opening would be much larger, the flares would be completely damped and no quasi–periodicity would be visible. The stretching of the period by about 15 percent is due to this particular opening of the magnetic surface which guides the optical knot. After a few revolutions, the spectrum of the knot moves into the mm–regime with its potential detectability by mm–VLBI. With high resolution mm–VLBI, this helical motion of the knot should be visible. At a distance of a few mas, this helical motion will be washed out by the diffusion of the relativistic particles, which now fill up the entire jet. This helical motion of newly born knots is clearly visible within a distance of one mas from the core in the case of 3C 345 (Zensus 1990). The high amplitude of the helix ($\simeq 200\,\mu$as) requires a high mass for the central object in 3C 345, $M_H > 10^{10}\, M_\odot$.

A similar quasi–periodicity, but on a shorter time–scale has also been detected in the optical emission of some BL Lac objects (Wagner 1991), and also in the radio regime (Witzel 1991). The jets of BL Lac objects have somewhat smaller Lorentz factors than the jets of quasars, $\gamma_{\rm jet} \simeq 2 - 4$ (see Eq. (12)), but we observe their jets under small inclination angles. Under this condition, the fluctuations in the lightcurve are much weaker, typically of a factor two, and the lightcurve itself is more regular (of a periodic shape). The masses involved in BL Lac objects are smaller (Camenzind 1991a), $M_H \simeq 10^7 - 10^8 \, M_\odot$. This means that the radii of the jets are also at least a factor 10 smaller as compared to the radii of quasar jets. For the same reason, the observed periods are shorter, $P_{\rm obs} \simeq 1.0$ day, corresponding to the intrinsic period $P_0 \simeq 15$ days. For $M_H \simeq 7 \times 10^7 \, M_\odot$, we find $P_F = 2\pi/\Omega^F \simeq 10$ hours, i.e. $R_0 \simeq 10 R_L \simeq 50$ AU. This would be an extremely compact jet comparable to the extension of jets in protostellar objects !

Acknowledgements: This work is partially supported by the Deutsche Forschungsgemeinschaft (SFB 328).

REFERENCES

Baath, L.B. et al. 1991. *Astron. Astrophys.* **241**, L1.

Brown, J. et al. 1989. *Astrophys. J.* **340**, 129.

Camenzind, M. 1986. *Astron. Astrophys.* **162**, 32.

Camenzind, M. 1989. In *Neutron Stars and their Birth Events.* W. Kundt, Ed., 139 - 177. Kluwer, Dordrecht.

Camenzind, M. 1990. In *Reviews of Modern Astronomy.* G. Klare, Ed. Vol. **3**, 234 - 265. Springer–Verlag, Heidelberg.

Camenzind, M. 1991a. In *Variability of Active Galactic Nuclei*, S. Wagner, W. Duschl & M. Camenzind, Eds., 201 - 210, Springer–Verlag, Heidelberg.

Camenzind, M. 1991b. In *Proc. of the TEXAS/ESO–CERN conference on Relativistic Astrophysics*, Brighton 1990, in press.

Camenzind, M. 1991c. In prep.

Camenzind, M. & Krockenberger, M. 1991, submitted to *Astron. Astrophys.*

Courvoisier, T.J.–L. et al. 1988, *Nature* **335**, 330.

Courvoisier, T.J.–L. & Camenzind, M. 1989, *Astron. Astrophys.* **224**, 10.

Haehnelt, M. & Camenzind, M. 1991, submitted to Mon. Not. Roy. Astron. Soc.

Krichbaum, T.P. et al. 1991. *Astron. Astrophys.* **237**, 3.

Kuperus, M. 1987. In *Gravitation in Astrophysics.* B. Carter & J.B. Hartle, Eds. Vol **156**, 195 – 208. Plenum Press. New York, NY.

MacDonald, D. & Thorne, K.S. 1982. *Mon. Not. Roy. Astron. Soc.* **188**, 345.

Pudritz, R.E. 1981. *Mon. Not. Roy. astron. Soc.* **195**, 881 - 896; 897.

Wagner, S. 1991, these Proceedings.

Witzel, A. 1991, these Proceedings.

Zensus, J.A. 1990. In *Parsec–scale radio jets*, J.A. Zensus & T.J. Pearson, Eds., 28-31 Cambridge Univ. Press.

Magnetic Reconnection in Active Galactic Nuclei

H. LESCH

Max-Planck-Institut für Radioastronomie

Auf dem Hügel 69, D-5300 Bonn 1, FRG

ABSTRACT
A model for the emission processes causing rapid variability (less than one day) in active galactic nuclei is developed. Relativistic electron beams escape from reconnection sheets in coronae of accretion disks and excite plasma turbulence with a typical frequency $\omega_{\rm pe} \simeq 5.6 \cdot 10^4 \sqrt{n}$, which depends on the electron number density n. The finite lengths of different beams emerging from different reconnection sheets allow that the waves are **coherently** scattered to frequencies $\simeq \gamma^2 \omega_{\rm pe}$. For Lorentz factors $\gamma \simeq 10^3$ and densities typical for disk coronae $n \simeq 10^6\, cm^{-3}$ (derived from iron line observations) one easily reaches the optical frequency range. The time scale of the variability is then caused by the relaxation of the electron beams. Likewise, this model explains the very rapid variability in the X-ray (less than 10 minutes) by changing the parameters slightly. According to this scenario the higher the variable frequency is, the closer to the central black hole it should originate.

1. INTRODUCTION

The radio (Witzel, 1991) and optical (Wagner, 1991) observations reveal variability in flat spectrum sources on time scales shorter than one day. With the assumption that the radio variability is caused by incoherent synchrotron radiation the apparent brightness temperature $T_{\rm B}^{\rm var} \simeq 10^{18}\, K$ exceeds the Compton limit by six orders of magnitude. The high brightness temperatures pose severe problems for the model of relativistically beamed radiation, since $T_{\rm B}^{\rm var} \propto \gamma_{\rm var}^3 T_{\rm B}^{\rm int}$, that is the radio observations require beaming factors $\gamma_{\rm var}$ of the order of 100, which is a magnitude larger than typical for variable radio sources (Blandford, 1990).

Moreover, the reported light curves of the optical variabilities exhibit an exponential increase of the emission and a very steep decline of the variable signal (for more details see Wagner, 1991). Since the optical emission is polarized and exhibits a power-law spectrum it is interpreted as synchrotron emission. The shape of the light curve is then hard to understand in terms of acceleration models in which the energy of the relativistic electrons is due to stochastic or diffusive shock wave acceleration.

Therefore we like to propose an additional model that might explain the variable part of the continuous emission in active galactic nuclei, by using the results of laboratory experiments of the coherent emission of relativistic electron beams during their propagation through a plasma (see also Benford, 1991).

2. DISSIPATION OF MAGNETIC FIELDS IN CORONAE OF ACCRETION DISKS

We consider as a possible scenario a rotating black hole surrounded by a magnetized accretion disk (Camenzind, 1991). The origin of the magnetic field must be due to the existence of seed fields in the disk which extend to larger scales of a few parsecs to a few hundred of parsecs in galaxies (Begelman et al., 1989). These larger disks and rings represent the reservoir for gas and magnetic flux which is ultimately accreted onto the central black hole.

As the disk plasma accretes in the potential of the central mass, magnetic fields are convected inwards, amplified and finally deposited on the horizon of the black hole. A dynamo in the disk may be responsible for the maintenance and amplification of the magnetic field. The interplay of differential rotation and convective turbulence (ascending and descending turbulent cells) amplifies the magnetic field to a strength where processes like magnetic buoyancy limit further amplification by generating ascending magnetic loops with length scales comparable to the disk height from which the field escapes. An inevitable consequence of convective turbulence and magnetic buoyancy will be to form a hot, magnetically active corona by transferring magnetic energy into the coronal plasma (see Fig. 1).

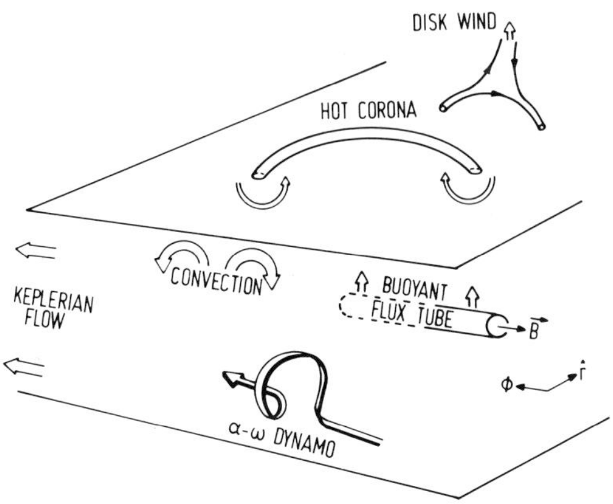

Fig. 1.

The transfer of magnetic energy is related to the process of energy dissipation. The question is how is the magnetic energy dissipated ? The way to do that is by magnetic reconnection, which in its simplest form means that magnetic field lines with different polarity experience each other. It is an inevitable process in turbulent plasmas (Biskamp and Welter, 1990) and it proceeds if magnetic flux within the rising loops encounters the magnetic fields in the corona above the disk. Between these fields with antiparallel directions a plane exists where the magnetic field must be zero (neutral sheet), i.e. strong magnetic field gradients are built up near a neutral sheet. Since magnetic gradients are equivalent to high current densities j, the Joule dissipation rate

$$Q = j^2/\sigma \tag{1}$$

is increasing even if the electrical conductivity σ stays constant. However, if the current density exceeds a critical value plasma instabilities arise which change the plasma dynamics in the neutral layer. The electrical conductivity is proportional to the inverse of the collision frequency. For small current densities only the Coulomb collisions have to be considered. But when plasma waves are generated the wave-particle interactions are much more intense than the Coulomb interactions. The conductivity changes into an anomalous state where the plasma waves reduce σ by increasing the collision frequency. The reduction of σ also increases the dissipation of the magnetic energy in neutral layers.

2.1 Electron Acceleration in Reconnection Sheets

The reduction of the conductivity allows the plasma to move relative to the magnetic field (which is impossible in the case of a "frozen-in" field, where the conductivity is large), i.e. electric fields are built up, which follows from an application of Faraday's law (Lesch, 1990).

The magnetic field lines of opposite polarity merge at a neutral layer where $B = 0$. The merging speed u is typically a fraction of the Alfvèn speed v_{A}. If the pre-merging field is B_0 a particle of charge e and Larmor radius r_{g} finds the magnetic flux $\simeq \pi r_{\mathrm{g}}^2 B_0$ decreasing through its orbit with a rate $-\pi r_{\mathrm{g}}^2 B_0/t_{\mathrm{rec}}$. $t_{\mathrm{rec}} = r_{\mathrm{g}}/u$ is the reconnection time as seen from the orbit. The change of magnetic flux induces an electromotive force emf $\simeq \pi r_{\mathrm{g}} B_0 u/c$ which is equal to the work done per unit charge by an electric field E, emf $\simeq 2\pi r_{\mathrm{g}} E$. One obtains (Lesch, 1991)

$$E \simeq \frac{u}{2c} B_0. \tag{2}$$

This electric field accelerates particles. We note that reconnection prefers the electron acceleration (for the details we refer to Lesch and Pohl (1991)). The maximum energy the electrons can attain depends on the strength of the electric field and on

the length L of the reconnection sheet. For a detailed investigation of the acceleration mechanism it is necessary to determine L and E by taking into account the underlying microscopic plasma processes which lead to the reduction of the electrical conductivity, which is the ultimate driver of the reconnection process.
The resulting maximal Lorentz factor $\gamma_{\rm rec}$ of the electrons is (for the details see Lesch and Pohl (1991))

$$\gamma_{\rm rec} \simeq 10^4 \left[\frac{B}{100\,{\rm Gauss}}\right]^3 \left[\frac{n}{10^6\,{\rm cm}^{-3}}\right]^{-3/2} \left[\frac{T}{10^8\,{\rm K}}\right]^{-1/2} \left[\frac{\epsilon}{0.1}\right] . \tag{3}$$

We have introduced an effiency factor ϵ of the acceleration process. The temperature and the number density were derived from observations of the iron K-line in AGN (Stella and Campana, 1990) and the magnetic field strength is about 20% of the maximal field in an accretion disk at a radius of about 6 Schwarzschild radii, which should correspond to the inner edge of an accretion disk. The length of a reconnection sheet is about one light second. We note that for the calculation of the electron energy and of the length scale it was assumed that lower-hybrid and ion acoustic turbulence determines the anomalous conductivity in the neutral sheet.

We have considered only one neutral layer (which is equivalent to a current sheet) in the disk corona. However, if the whole disk and its corona is in a turbulent state a large number of current sheets might exist. If the mean free path of the electrons is larger than the acceleration length scale, they can reach much higher energies than those given by Eq. (3). The electrons are then accelerated by a mean electric field along a scale length which is given by the **macroscopic** turbulence (in our case the scale of the loops).
Since the electrons are accelerated by an electric field their energy distribution function will be monoenergetic. In other words, electron beams emerge from reconnection zones and propagate into the surrounding medium. We note that this model has been **successfully applied** to the first phase of solar flares (Heyvaerts, 1985 and references therein). It is the time evolution of such beams which may determine the rapid variability in AGN.

3. COHERENT EMISSION OF RELATIVISTIC ELECTRON BEAMS (REB)

3.1 Stability of REB

An electron beam is unstable and excites Langmuir waves with the frequency $\omega_{\rm pe} \simeq 5.6 \cdot 10^4 \sqrt{n}$ (Kaplan and Tsytovich, 1973 and references therein). The linear growth time of the instability is given by (Lesch and Schlickeiser, 1987):

$$t_{\rm lin} \simeq \frac{n}{n_{\rm b}} \frac{\gamma}{\omega_{\rm pe}} . \tag{4}$$

n_b denotes the number density of the beam electrons. The excited Langmuir waves slow down the beam by wave-particle interactions. During this relaxation of the beam distribution about 25% of the beam energy is transformed into wave energy. The energy density W_L of the excited Langmuir turbulence is (Lesch and Schlickeiser, 1987):

$$W_L \simeq \frac{\gamma n_b m_e c^2}{12} \frac{m_e c^2}{k_B T}. \tag{5}$$

W_L presents the energy reservoir for any kind of coherent radiation process which transfers beam energy into radiative energy.

3.2 Inverse Compton Scattering of Langmuir Waves

In recent laboratory experiments where a REB penetrates a thermal plasma strong broad band emission has been observed. To account for the huge power a model was invoked in which the Langmuir waves with high amplitudes are produced by the beam (Benford, 1991). The wave-particle interactions scatter the electrostatic waves up in frequency by the inverse Compton effect. As will be discussed by Benford (1991), the relativistic effects boost the radiation into a frequency range $\omega_{pe} \rightarrow \gamma^2\omega_{pe}$. The high emitted power is due to the fact that the beam electrons are resonating with the Langmuir waves, which are packed in form of solitary structures. The length of the solitons is the growth length of charge bunching. Thus the electromagnetic emission is the result of *coherent* radiation from a large number of beam electrons oscillating in phase.
The inverse Compton scattering process of the Langmuir waves presents an energy loss mechanism for the relativistic electrons and a damping process for the waves. The corresponding damping time t_{IC} is given by

$$t_{IC} \simeq 10^{-4}\ s \left[\frac{n_b}{4\cdot 10^4\ \mathrm{cm}^{-3}}\right]^{-1} \left[\frac{n}{10^6\ \mathrm{cm}^{-3}}\right]^{3/2} \left[\frac{\gamma}{10^3}\right]^{-3/2}. \tag{6}$$

Here we estimated n_b by using the fact that the electrons are accelerated in reconnection sheets, i.e. the energy density of the electrons should be equal to the energy density of the dissipated magnetic field. For an effiency factor of 0.1 and $B \simeq 100$ Gauss we obtain $n_b \simeq 4\cdot 10^4\ \mathrm{cm}^{-3}$.

The time scale for inverse Compton scattering is much smaller than the energy loss time for synchrotron radiation which is about $5\,s$. To determine the luminosity L^* of the coherent radiation it is necessary to estimate its energy density W_{em} and the effective optical thickness τ_{eff} since we have

$$W_{em} \simeq \tau_{eff} W_L \tag{7}$$

and

$$L^* \simeq 4\pi R^2 W_{em} c, \tag{8}$$

with the length R of the region in which relativistic electrons and Langmuir turbulence coexist. R is related to the relaxation length of the beam. Including weak and strong turbulence effects the relaxation length is given by (Lesch and Schlickeiser, 1987)

$$R \simeq 3 \cdot 10^{15}\,\mathrm{cm} \left[\frac{n_b}{4 \cdot 10^4\,\mathrm{cm}^{-3}}\right]^{-1} \left[\frac{n}{10^6\,\mathrm{cm}^{-3}}\right]^{1/2} \left[\frac{\gamma}{10^3}\right]^{3}, \tag{9}$$

which gives a light travel time of about 1.1 days.
The energy density of the electromagnetic emission is (Lesch and Pohl, 1991) $W_{em} \simeq 10^6\,\mathrm{erg\,cm^{-3}}$ and the corresponding luminosity $L^* \simeq 10^{47}\,\mathrm{erg\,s^{-1}}$.
The time scale and the luminosity we obtain are in good agreement with the observations of rapid variability in the optical in AGN.

4. DISCUSSION

If the emitted power of active galactic nuclei is caused by accretion onto a supermassive black hole an accretion disk provides the mass transport of the magnetized plasma towards the central object. The combined action of convective turbulence and differential rotation results in strong magnetic fields which emerge from the disk into the corona. There, the dissipation of magnetic energy accelerates electrons to relativistic energies via reconnection. Since the electrons are accelerated by an electric field they propagate into the surrounding medium in form of a relativistic beam. As is known from laboratory experiments such beams emit strong coherent radiation with huge power. The spectral and polarization characteristics of the coherent emission (which at highest frequency is equivalent to inverse Compton scattering of Langmuir waves) are the same as those of the incoherent synchrotron emission (Benford, 1984). The time scale of the coherent emission is related to the propagation of the beam and its sudden relaxation towards a stable distribution function which no longer excites the plasma turbulence necessary for the coherent radiation. This model explains qualitatively the features of intraday variability in the radio and/or the optical: time scales and luminosity. Moreover, it resolves the problems of huge brightness temperatures and rapid rising and decaying of the variable emission.

We think that in any kind of AGN coherent processes could account for at least a part of the variability since reconnection and the accompanying acceleration of electrons to relativistic energies should appear in any plasma in which the magnetic field is turbulently mixed. Especially the variability in the X-ray on very small time scales (about a few minutes (Hayakawa, 1991)) may be due to coherent emission. This can be seen by the following example: we increase the magnetic field by a factor of 3, which gives a Lorentz factor of $5 \cdot 10^5$ which then results in an emission which peaks at $2 \cdot 10^{18}$ Hz, i.e. well in the X-ray. Since the higher field strengths are expected to appear at smaller radii, the variability in the X-ray might occur on smaller time scales than the variability at lower frequencies, which indeed has been observed.

However, when considering possible astrophysical applications of coherent inverse Compton scattering of Langmuir waves (CIC), the following point should be kept in mind. Apart from factors of the order of unity the power radiated by an electron due to synchrotron radiation inverse Compton and CIC are in a ratio of the energy densities in the magnetic field, the unscattered photons and the Langmuir turbulence, respectively. It would require exceptional circumstances for the energy density in the Langmuir turbulence to exceed the energy density in the magnetic field in the whole system. It is implausible that this inequality could be satisfied other than in localized regions of an extended source. The mechanism of CIC is suitable to explain the variable but not the steady continuum emission. The coexistence of coherent and incoherent emission processes is probably responsible for the variable emission characteristics of AGN.

REFERENCES

Benford, G.: these proceedings
Benford, G.: in *Physics of Energy Transport in Extragalactic Radio Sources*, Ed. A.H. Bridle and J.A. Eilek, p. 185 (1984)
Begelman, M.C., Shloshman, K.G., Frank, J.: in *Theory of Accretion Disks*, Ed. F. Meyer, W.J. Duschl, J. Frank and E. Meyer-Hofmeister, Kluwer, Dordrecht, p. 225 (1989)
Biskamp, D., Welter, H.: *Phys. Fluids B* **2**, 1787 (1990)
Blandford, R.D.: in *Active Galactic Nuclei*, Saas-Fee Advanced Course 20, Springer Press, p. 161 (1990)
Camenzind, M.: these proceedings
Hayakawa, S.: *Nature* **351**, 214 (1991)
Heyvaerts, J.: in *Unstable Current Systems and Plasma Instabilities in Astrophysics*, IAU 107, Ed. M. Kundu and G.D. Holman, Reidel, Dordrecht, p. 85, (1985)
Kaplan, S.A., Tsytovich, V.N.: *Plasma Astrophysics*, Pergamon Press, Oxford
Lesch, H.: A&A **245**, 48 (1991)
Lesch, H., Schlickeiser, R.: A&A **179**, 93 (1987)
Lesch, H., Pohl, M.: A&A in press (1991)
Stella , L., Campana, S.: in *Iron Line Diagnostics in X-Ray Sources*, Ed. A. Treves, G.C. Perola, L. Stella, Springer Press (1990)
Wagner, S.: these proceedings
Witzel, A.: these proceedings

Collective Radiation From Jets

G. BENFORD

University of California, Irvine

ABSTRACT

Collective emission can occur when plasma oscillations scatter relativistic electrons. For typical jet conditions the spectrum has the same polarization as ordinary synchrotron emission, but far greater power, because the electrons bunch together (antenna mechanism). This allows higher emissivity and relaxes the demand for jet energy. Flickering emerges naturally as coherence varies, without demanding highly relativistic jet motions.

1 INTRODUCTION

Perhaps the most ingrained model in extragalactic radio astronomy is the Synchrotron Assumption – that behind observed emission lies a power-law distribution of electrons radiating incoherently. Further postulating rough equality between relativistic particle and magnetic energy, we fix a lower bound on the total energy of a source. But the source brightness temperature should not exceed about 10^{12}K, and until recently even quasars seemed to satisfy this condition. Recent observations of rapid variations in quasar flux density and polarization, as detailed by Quian *et al.* (1991) and at this meeting, suggest a crisis in this worldview.

I am sure that shock models can probably accommodate higher brightness temperatures, though perhaps the pertinent question is whether any data at all could ever falsify shock models, since so many free parameters festoon them. High brightness temperatures abound in astrophysics, in Jovian decametric emission, pulsars, and Type III solar bursts – not from shocks, but from collective radiation of cooperating electrons. Plasma waves can order the accelerations of relativistic electrons, forcing them to emit waves in phase coherence over some length D. These N cooperating electrons act like an antenna, yielding an N^2 scaling in the radiated power.

Abandoning the Synchrotron Assumption means a net lowering of the energy budget needed in radio sources, since collective emissivities are much higher. This lowers greatly the demands on the supposed culprit black holes. Such attractions are

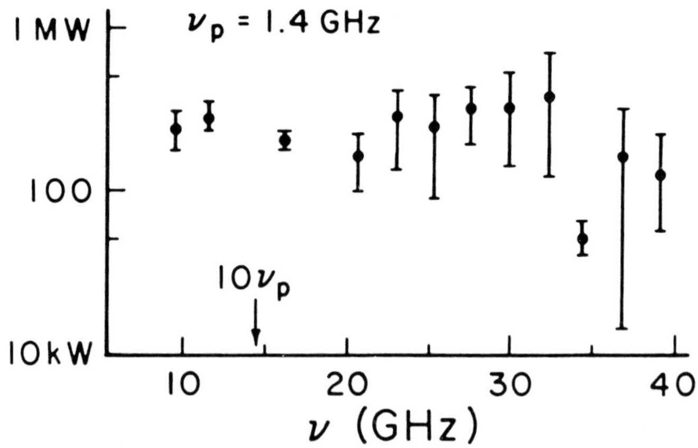

Figure 1

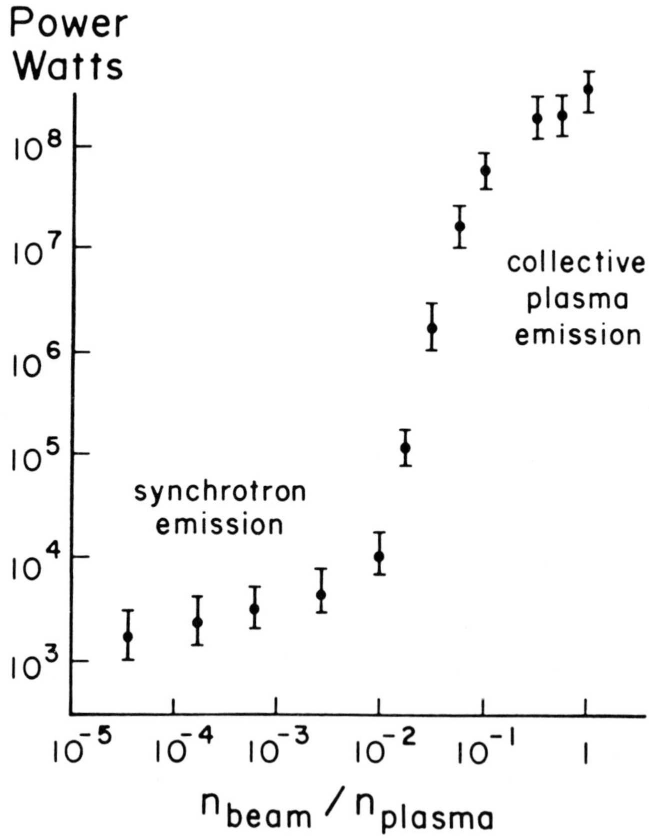

Figure 2

tempting, but the ultimate authority should be experiment.

2 EXPERIMENTS

To this end, in 1976 I began laboratory experiments studying collective emission by relativistic beams penetrating background plasma. We study intense beams propagating a meter through helium or hydrogen plasmas, measuring beam characteristics (voltage, current), emitted rf power, and also the electrostatic plasma wave amplitudes. We have seen beam synchrotron emission, both individual and collective (Tzach and Benford, 1981), plus radiation at the plasma frequency, ν_p, and emission at high harmonics far above ν_p. Here I focus on the beam emission of frequencies in the range $\sim \gamma^2\nu_p$, which resembles physically the synchrotron emission at $\gamma^2\nu_c$, with ν_c the cyclotron frequency. Kato *et al.* (1983) found a clear dipolar pattern Doppler shifted forward in a cone, peaking off the beam axis at the typical beam pitch angle of about 15 degrees. In this pattern is an emission spectrum (Fig. 1) including the plasma frequency at 5 GHz, and much more emission out to about the 6th harmonic. (We could not measure above this frequency). Depending on details of theory, this Compton-boosted radiation should extend to the tenth or even twentieth harmonic, since the beam energy factor was $\gamma = 3$. The observed power spectrum resembles in shape that of synchrotron radiation for a nearly monoenergetic beam, as theory predicts.

Note though that the power is enormous – totalling about 100 megaWatts over all the spectrum, with less than a megaWatt in the ν_p line, whereas we saw incoherent synchrotron emission over the same frequency range at 1 to 10 kW, i.e., enhancement by a factor of up to 10^5. We traced the dependence of this amplification on beam parameters and found that it was primarily due to one simple ratio, beam density to plasma density, or n_b/n_p. Figure 2 shows that when this ratio is below about 10^{-4} single-particle emission dominates, though at very small power levels. This is the realm of weak turbulence theory, in which plasma wave instability surely occurs, but displays little ability to bunch the beam electrons. As the ratio rises, we know from a vast body of plasma turbulence studies that wave coherence increases dramatically. Plasma waves localize into "cavitons" which develop intense fields. Beam electrons bunch, forming the antennas I believe we see radiating her. A model I have developed with James Weatherall views the emission as a scattering of beam electrons by cavitons of size D, leading to a spectrum which extends to about γ^2 c/D. Earlier work treated scattering by individual plasma oscillations, and so found spectra extending to $\gamma^2\nu_p$.

Generally, c/D exceeds ν_p by one or two orders of magnitude, depending on plasma conditions, since D is typically about ten plasma Debye lengths. By the time n_b/n_p

exceeds 0.1, plasma collective emission dominates synchrotron emission by about 10^5. This suggests that beams entering thin plasma regions will emit collectively, and arguments about beam parameters based on the Synchrotron Assumption will be wildly wrong. This work has been confirmed by experiments in Japan and the USA.

3 ASTROPHYSICAL APPLICATIONS

Collective processes can duplicate all features of ordinary, single-particle synchrotron emission, while greatly enhancing the emissivity. This central fact emerges because any accelerating field, whether the static magnetic field of synchrotron radiation or the oscillating electric field of collective processes, makes electrons radiate dipolarity. We compare the two qualitatively:

	synchrotron emission	plasmon emission
spectral peak	$\nu \approx \gamma^2 \nu_c$	$\nu \approx \gamma^2 \nu_p$
base frequency	$\nu_c \approx 280\ B_{-4}$ Hz	$\nu_p \approx 90\sqrt{n_{-4}}$ Hz
polarization	same	
spectrum	same	
power	$\sim NB^2$	$\sim NE^2 + (N^*)^2\ E^2$

Here N is the number of electrons, N^* the number cooperating in bunches of size $\approx c/\nu_p$, and $E = \langle E_K^2 \rangle^{1/2}$ is the rms average electrostatic field strength in plasma oscillations (produced by streaming instabilities). Usually one finds in weak turbulence that $E^2 \ll B^2$. However, intense cases abound where $E^2 \sim 4\pi nT$, the thermal plasma energy. This need not be small compared with B^2 especially when we assume equipartition, $B^2 \sim 4\pi nT$. In any case, we have seen cases where $N^{*2}/N = 10^6$, yielding a measured brightness temperature T_b exceeding 10^{18} (Benford and Smith, 1980). For a discussion of coherently excited synchrotron emission, which has application to pulsars, see Tzach and Benford, 1981. We observed the synchrotron harmonics and measured the beam pitch angle distribution from them. Again, $T_b \geq 10^{18}$.

Astrophysicists have long assumed synchrotron emission as the "default" choice, since no one could make a case for other, more powerful mechanisms. The essential message I bring is that *this is no longer so.* Plasmon emission far above ν_p can occur when $n_b/n_p \geq 0.01$, as experiment shows. *Also, there is no way to distinguish plasmon emission from synchrotron through the polarization.* This arises basically because both processes yield Doppler-shifted dipolar patterns. After averaging over power law distributions for the electrons, there is *no* distinction between the mechanisms (Windsor and Kellogg, 1974).

4 DETECTION

Polarization is no help in distinguishing plasma emission from synchrotron emission, once one makes the usual assumptions of isotropic particles having a power law distribution function. Detection of a high brightness temperature T_b is the obvious test, but if the emission is spotty (i.e., the packing fraction of coherent zones is low) then the average T_b can still ape that of synchrotron emission, $\leq 10^{12}$ K. In such cases the only hope is that there are a relatively small number of coherent zones, N^*, so that

(a) the radiation amplitude has a fluctuation of order $(N^*)^{-1/2}$; or

(b) the signal phase displays an autocorrelation time t_{ac} far greater than the $\sim 10^{-8}$ sec. typical of incoherent emission observed with a linewidth of 100 MHz.

In the laboratory (a) is all we typically see, since in such fast experiments we discard the phase information. In our experiments we often see amplitude fluctuations (lasting $\sim$ 10 ns) in the range 0.01 to 0.1. This implies $N^* \sim 100$ and a packing fraction ~ 0.01 in our volume of $\sim 10^3$ cm^3.

Looking for long t_{ac} can be done using presently digitized data. A clear indication might come from rapid swings in polarization angle. It is difficult to estimate from the current simple theories, how many radiators there are, since there is in addition, the possibility that the collective system acts not merely like an antenna (bunching), but can become a maser, with spatial amplification. Similarly we cannot calculate t_{ac} very well. (One crude standard is that cavitons may last for about 100 ion plasma periods, which yields for our experiments $\sim 10^{-3}$ $n_p^{-1/2}$ sec. Our observed fluctuations last about 30 times this long). Once we make the familiar assumptions of a power law spectrum in electron energy, the spectrum and polarization of collective plasma emission is identical to that of synchrotron radiation, so the customary astrophysical tests cannot distinguish it.

The observed 10 hour flickering of Quian *et al.* probably arises from collective emission waxing and waving, as an energetic beam passes through a region $\sim 10^{15}$ cm where conditions for strong turbulence occur. The collective emission would mimic synchrotron radiation in self-absorption and polarization, but not in rapid variability. Could such high brightness emission be smothered by ordinary synchrotron absorption of surrounding material? Not if the emission has a clear path through a tunnel low in absorbing electrons, as would be probable, since about half of observed sources display this variability.

The principal use of collective emission lies in its ability to pump streaming electron energy into radiation with great efficiency, vs. synchrotron, which inefficiently harnesses only electron energy transverse to the magnetic field. Quantitatively, for a power law distribution of particle energies,

$$\frac{\text{P(collective)}}{\text{P(synchrotron)}} = 2.10^9 \, f \, \frac{E_0^2}{8\pi n_0 kT} \left[\frac{\Delta n_b^2/n_b^2}{2.10^{-4}}\right]\left[\frac{D}{15\lambda_D}\right]\left[\frac{T}{20eV}\right]^{5/2}$$

$$\times \left[\frac{n_p}{10^{-3}\ cm^{-3}}\right]^{1/2}\left[\frac{10^{-6}G}{B_0}\right]^2\left[\frac{n_b/n_p}{10^{-5}}\right]\left[\frac{1n(2\gamma c/D\omega_p)}{16}\right]$$

The numerical parameters corespond to hypothetical conditions in galactic jets. The scaling values used for density, temperature, and magnetic field give $B^2 \approx 4\pi n_p kT$. The turbulent emission has considerable advantage in efficiency over synchrotron emission. $W_0 = [E_0^2/8\pi n_p k_B T]$ is the dimensionless electrostatic energy within soliton structures. Note that the product fW_0 is the normalized field energy density in cavitons in three dimensions. In Langmuir collapse, this is conserved even as cavitons shrink. Thus, the average over a time ensemble of cavitons should reflect a value for this product and the mean caviton volume D^3. Unfortunately, no experiment measures all of these quantities.

Obviously the recently seen, quite exciting quasar flickering is the obvious place to apply these ideas. Brightness temperature is the most intuitive way to compare theory with observation, but it is a somewhat processed quantity, depending on estimates of source size. I strongly point out that the regions suggested by the 10-hour flickerings are very small, yet even there the turbulence necessary for strong collective emission may be rare. The packing fraction of the turbulence, f, remains unknown – though our experiments place it between 0.01 and 0.2, a respectable minority. The true brightness temperature T_b will be diluted to the observed brightness, $T = fT_b$. This means that sources could display flickering but not especially high observed T. Thus many sources may be disguised collective emitters. This was the ground on which I suggested in 1984, (when nobody saw any flickering on months-long time scales), a program of day-to-day observations such as the Bonn group has now carried out.

Studies of polarization swings would imply collective emission, but one cannot clearly predict the dependence on intensity. This is because the electrons encounter cavitons which have varying orientations. In some magnetic field regimes, cavitons are pancake-shaped, in others, spherical. This will lead to differing polarizations of the beam emission, equivalent to differing orientations of the linear dipoles. There should be <u>no</u> circular emission, since there is no circularity enforced by the magnetic field on the cavitons. This differs greatly from the synchrotron assumption.

The observed flickering can tell us much about regions within a solar system radius of the active galactic centers. Collective emission implies a class of dense beams penetrating plasma, and analysis can reveal properties of the region beyond that available through the Synchrotron Assumption.

A deeper question is why collective emission should have $T = fT_b \lesssim 10^{12}K$ skillfully mimicing synchrotron emission. I suspect this arises from a fundamental limit on energy lodged in highly ordered beams with $n_b \sim n_p$. Very high $T \gg 10^{12}K$ will plausibly readjust local pressures by self-absorption, which will move matter to fill such low-n_p regions within times governed by the local sound speed. This facet may tell us new things about the regions near black holes of size $\sim$ 10 A.U..

Work supported by the Air Force Office of Scientific Research.

REFERENCES

Benford, G., 1984, in Physics of Energy Transport in AGNs, NRAO Proceedings, 1984.

Benford, G. and Smith, D., 1982, Phys. Fluids 25 (8), 1450.

Kato, K., Benford, G. and Tzach, D., 1983, Phys. Fluids 26 (12), 3636.

Quian, S.J., Quirrenbach, A., Witzel, A., Krichbaum, T.P., Hummel, C.A. and Zensus, J.A., 1991, Astron. Astrophys. 241, 15–21.

Tzach, D. and Benford, G., 1981, Astr. and Space Sci. 80, 307.

Windsor, R.A. and Kellogg, P.J., 1974, Astrophys. J. 190, 167.

VLBI-Scale Radio Jets

R. T. SCHILIZZI

Netherlands Foundation for Research in Astronomy

1 INTRODUCTION

One of the outstanding problems in extra-galactic astronomy is to understand the nature of the source of energy and its transport in the nuclei of active galaxies. Ground-based VLBI observations have had an important part in forming the standard picture of an AGN, and in the first part of this contribution I will briefly review some of this information. A number of the following papers will elaborate on the details. In the second part, I will review some of the data emerging on the interaction of radio jets with the surrounding medium.

Extensive descriptions of the properties of compact radio sources can be found in Parsec-scale Radio Jets (Eds. J. A. Zensus, T. J. Pearson), and Beams and Jets in Astrophysics (Ed. P. A. Hughes).

2 SYSTEMATIC PROPERTIES

i) All VLBI images of compact radio sources contain unresolved components which are identified with radio emission from the nucleus. Observations on the longest baselines and at the highest observing frequencies (see contribution by L.B. Baath, these Proceedings) provide direct evidence that the central engine is smaller than 10^{16} - 10^{17} cm.

ii) All sources with large-scale (10s of kpc) jets also possess bright pc to 100 pc-scale jets. The reverse is not true; not all bright compact jets have kpc jet counterparts. The fact that the energy-carrying beams in powerful radio sources are observed to be thin jets within 1 pc of the nucleus means that the initial collimation must be caused by processes taking place in close proximity to the central engine rather than by processes in the interstellar medium of the parent galaxy. Compact jets are mostly one-sided and point in the same direction as their larger-scale counterparts; however substantial misalignments do occur. These properties are usually ascribed to the effects of relativistic beaming close to the line of sight to the observer. Misalignments occur in both quasars and galaxies (see 3C273 in Figure 1); there is evidence that the magnitude of the misalignment increases closer to the core component (*e.g.* 3C345, Zensus *et al.*, 1990; 3C273, panel 'e' in Fig 1 and Baath *et al.*, 1991).

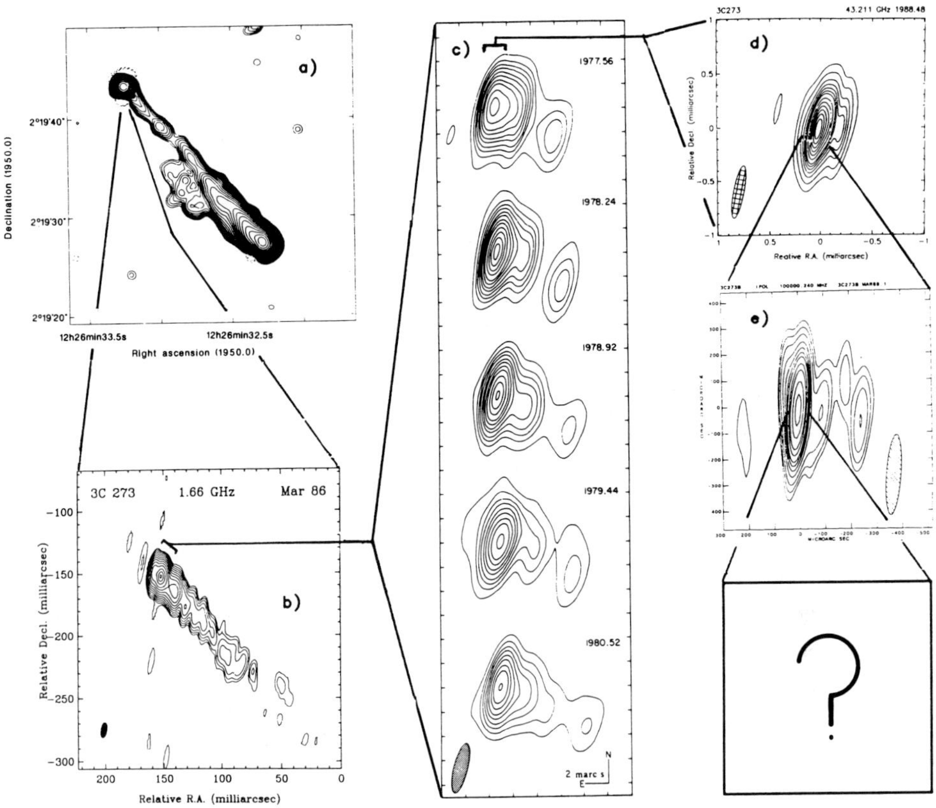

Figure 1: 3C273, the "classical" quasar.

3C273, imaged at different angular resolutions from 408 MHz to 100 GHz. **(a)** MERLIN at 408 MHz showing the core (to the NE) and the 21 arcsec long jet which coincides with the optical jet. **(b)** VLBI at 1.66 GHz on a scale 250 times smaller than panel 'a'. **(c)** a series of VLBI images at 5 GHz on a scale a further factor of 4 smaller than panel 'b'. The changes in angular position of some of the features over the course of several years imply linear velocities in the plane of the sky well in excess of the speed of light. This "superluminal" motion is usually interpreted in terms of relativistic bulk motion close to the line of sight towards the observer. **(d)** the first image obtained of 3C273 at 43 GHz on a scale 10 times smaller than panel 'c'. **(e)** the first image at 100 GHz showing that the "core" at 43 GHz is in fact composed of a number of features. In panels b to e, the synthesised beam of the ground VLBI network used to produce the image is shown.

References: (a) Davis et al. 1985 Nature, **318**,343; (b) Unwin and Davis 1988 Proc. IAU 129, 27; (c) Pearson et al 1981 Nature **290**, 365 ; (d) Krichbaum et al, 1991 A. & A. in press; (e) Baath et al (1991) A. & A. **241**, L1

iii) The widespread occurrence of bright knots of emission in compact jets moving outwards at superluminal speeds shows that matter can be ejected with relativistic bulk velocities from the nucleus. This is clear evidence of the existence of a very deep potential well in at least some AGNs. The measured proper motion decreases as a function of redshift which has implications for cosmology, as discussed by Cohen (these Proceedings). Measurements of the X-ray luminosities of compact radio sources are generally lower than calculated for inverse compton emission from a stationary incoherent synchrotron source. This implies that the radio luminosity is boosted compared to the X-ray and this is taken to be further evidence for the effects of relativistic beaming.

iv) The speeds and trajectories of individual knots can vary as they move outwards (*e.g.* 3C345, Zensus *et al.*, 1990). However some knots appear to be stationary *e.g.* 4C39.25, Marcaide *et al.*, 1990) which may represent stationary shocks in the jet or locations where the jet orientation changes to be closer to the line of sight thus causing boosting of the radio emission. Both these results are suggestive of jets which can "wiggle". Icke (1991) reviews mechanisms for bending jets. Some motion occurs at sub-luminal speeds *e.g.* M87 (Reid *et al.* 1989) and 3C84 (Marr *et al.* 1990).

v) The percentage of linear polarisation found in the cores of quasars and radio galaxies is typically 1 %, while that in the cores of BL Lac objects is 3 – 4 %. The percentage linear polarisation in jets in quasars and radio galaxies is typically 15 % with the magnetic field direction parallel to the jet elongation. In BL Lac jets, the polarisation is typically 10% with the magnetic field perpendicular to the jet. Gabuzda (these Proceedings) reviews polarisation VLBI results in detail.

3 INTERACTION OF JETS WITH THE SURROUNDING MEDIUM

There is both direct and indirect observational evidence for interaction of radio jets with their surroundings. As Miley (1984) pointed out some years ago, for some nearby objects it is clear that interaction does take place and that the effects on jet and gas are mutual. Thermal gas can bend, decollimate and depolarize radio jets and may act as a catalyst for the radio emission, while the jets can affect the ionisation and kinematics of the gas. On the same physical scales but in more distant objects, there is growing circumstantial evidence of interaction from the increasingly detailed VLBI images; equivalent resolution optical images are required to complete the picture. On the even smaller scales which will be probed by space VLBI, Rees (1988) has suggested that very complicated flow patterns are likely to be observed as the jet passes through the multi-phase medium. Entrainment of surrounding gas, bending by transverse pressure gradients, and shocks created by the jet colliding with the dense gas clouds that emit broad emission lines are some of the effects that can be expected.

3.1 Indirect evidence

i) Radio sources associated with Compact Steep Spectrum (CSS) and Seyfert nuclei are about the same size as the optical Narrow Line Region, typically 1-10 kpc. The nuclear radio source in the galaxy 3C236 is a good example (Schilizzi *et al.*, 1988). There one sees radio emission on scales of up to 2 kpc, and thereafter no detectable emission until the extended lobes are reached at distances of $\sim$ 500 kpc to the north-west and 2 Mpc to the south-east.

ii) The radio luminosities of kpc size radio sources are correlated with OIII] luminosities and line widths (de Bruyn and Wilson, 1978).

iii) Gigahertz-peaked-spectrum (GPS) sources appear to be "smothered" (O'Dea, Baum and Stanghellini, 1991). These objects have sizes in the range 10 to 100 pc, and their numbers are too great for them to be young. Optical spectroscopy shows highly reddened narrow line regions, which can be interpreted as due to mergers of galaxies, at least one of which in each case is gas rich. In addition, the percentage polarisation of GPS sources is low and, in some cases, high rotation measures have been observed, all evidence for a gas rich environment which inhibits the growth of the radio emission.

iv) Powerful high redshift radio galaxies have optical and infrared extensions which are usually aligned along the axis of their associated radio sources. The optical and infrared flux could well be emitted by young stars formed by vigorous interaction of the radio jet with the interstellar medium (*e.g.* Chambers, Miley and Joyce 1988). This is supported by the discovery that the spectral energy distribution of the distant galaxies can be well represented by relatively young stellar populations of only 10^8 years old, comparable to the probable age of the radio sources. One should note, however, that an alternative explanation is that the optical continuum is scattered radiation of a quasar beamed preferentially along the radio jet (*e.g.* Fabian, 1988; Scarrott *et al.* 1989).

v) The incidence of distorted radio structures increases with redshift (Barthel and Miley 1988), which has a natural explanation in interaction of the radio emitting plasma jet with inhomogeneities in the ambient medium which at early cosmological epochs is much denser. They also found that the overall sizes of radio sources at redshifts 1.5 are greater than their low redshift counterparts. A similar investigation of distortion in radio sources by Kapahi (1990) did not support the earlier results of Barthel and Miley, but recent work by Bogors and Barthel (in preparation) has again shown the effect to be present.

vi) The last piece of indirect evidence for interaction of radio jets with their environments is that many jets are considerably bent. This was referred to in my remarks on superluminal sources. Leaving aside what Icke (1991) calls "pseudo-bending", that due to ballistic motion of the radio emitting plasma disturbed by nuclear precession for example, a jet can be bent by pressure gradients in an inhomogeneous medium through which it is passing, by bouncing off clouds encountered in its path, or by deflection through a shock. Icke gives a detailed description of the physics involved.

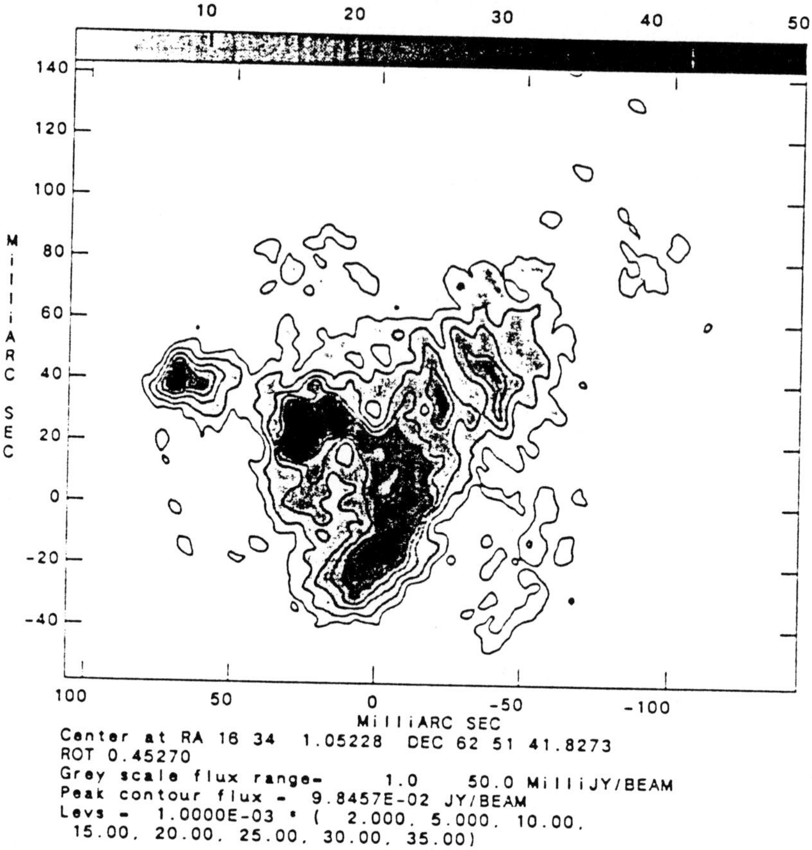

Figure 2: 3C343 imaged at 1662 MHz using a global VLBI + MERLIN array

In 3C48 (Wilkinson *et al.* 1991), and possibly 3C43 (Akujor *et al.* 1991) and 3C343 (Schilizzi *et al.* in preparation), collision of the radio jet and an interstellar cloud appears to have taken place a few hundred pc from the core. In the case of 3C48 severe disruption of the jet occurs after the impact, while in 3C43, the jet seems to be deflected. In 3C343 (Figure 2), it is not clear which component is the core since observations are available at one frequency only, but if, by analogy with 3C48, the eastern component is taken as the core, the jet is disrupted at the bright knot $\sim$ 50 milliarcsec to the west. Standard arguments lead to estimates of the cloud mass of order 10^4 M_0.

3.2 Direct evidence

i) Optical line emitting gas in 3C277.3 coincides in position with knots in the radio emission (*e.g.* Miley 1984). The line/radio associations usually occur just outside bright radio regions and close to bends in the radio source.

ii) There is an anti-correlation between line emitting gas and radio polarisation in 3C305, 3C293, and 4C26.42 (van Breugel *et al.* 1984). The densities and filling factors derived for the line gas are sufficient to depolarise the radio emission by differential Faraday rotation.

iii) The M87 jet seen on parsec scales (Reid *et al.* 1989) is clearly limb brightened, indicating entrainment or other surface interactions have taken place.

4 REFERENCES

Akujor, C. E., Spencer, R. E., Saikia, D. J. (1991) Astron. Astrophys., in press

Baath, L. B. *et al.* (1991) Astron. Astrophys. **241**, L1

Barthel, P. D., Miley G. K. (1988) Nature **333**, 319

van Breugel, W. J. M., Heckman, T. M., Miley, G. K. (1984) Ap. J. **276**, 79

de Bruyn, A.G., Wilson, A. S. (1978) Astron. Astrophys. **64**, 433

Chambers, K. C., Miley, G. K., Joyce R. R. (1988) Ap. J. **329**, L 75

Fabian, A. (1989) MNRAS **238**, 41p

Icke, V. (1991) in Beams and Jets in Astrophysics (Ed. P. A. Hughes), Ch. 5

Kapahi, V. K. (1990) Current Science **59**, 561

Marcaide, J. M. *et al.* (1990) in Parsec-scale Radio Jets (Eds. J. A. Zensus, T. J. Pearson) p. 59

Marr, J. M., Backer, D. C., Wright, M. C. H. (1990) in Parsec-scale Radio Jets (Eds. J. A. Zensus, T. J. Pearson) p. 78

Miley, G. K. (1984) Proc. QUASAT Workshop, ESA SP-213, p. 131

O'Dea, C. P., Baum, S. A., Stanghellini, C. (1991), preprint

Reid, M. J. *et al.* (1989) Ap. J. **336**, 112

Scarrott, S. M. *et al.* (1990) MNRAS **243**, 5p

Schilizzi, R. T. *et al* (1988) in The Impact of VLBI on Astrophysics and Geophysics (Eds. M. J. Reid, J. M. Moran) p. 127

Wilkinson, P. N. *et al.* (1991) Nature **352**, 313

Zensus, J. A. (1990) in Parsec-scale Radio Jets (Eds. J. A. Zensus, T. J. Pearson) p. 28

A Statistical Study of Superluminal Velocities

M.H. COHEN AND R.C. VERMEULEN

California Institute of Technology

ABSTRACT

We show a current (μ, z) diagram for superluminal sources. Core-dominated quasars are faster than lobe-dominated quasars and BL-Lac objects. We compare the cumulative distribution of velocity for core-dominated quasars with a distribution derived from a simple model. The results suggest that the Lorentz factors of the pattern and the beam, γ_p and γ_b, are not equal, in agreement with available evidence indicating that $\gamma_p/\gamma_b > 1$.

1. THE μ-z DIAGRAM

The VLBI observations discussed here were all made at 5 or 10 GHz. It appears that the phenomena may be different at lower frequencies, since one object, CTA102, shows no motion at 5 GHz (Wehrle and Cohen, 1989) but does show motion at 1.6 and 0.9 GHz (Bååth, 1987).

Figure 1 shows the observables, proper motion and redshift, on a (μ, z) diagram (Cohen et al., 1988). This figure contains 44 objects, 39 of which have been reported in the literature. Some objects have more than one moving component, and for these only the fastest one is shown. Five objects have both moving and stationary components, so that there are 49 points in all. The stationary components are shown at their upper limit. The horizontal arrow is for 0735+178 for which only an absorption redshift is known. The quality of the observations and the errors vary widely among the objects, but we have made no attempt to assign weights. A table of these data will appear in a fuller account of this work which now is in preparation.

The objects in Figure 1 may be classified as follows: 20 core-dominated (cd) quasars (filled circles), of which 17 have moving components and 7 have stationary components; 7 lobe-dominated (ld) quasars (filled squares); 6 BL-Lac objects (open circles) with moving components, of which one also has a stationary component; 4 radio galaxies (crosses); and 7 compact doubles (triangles), none with detected motions.

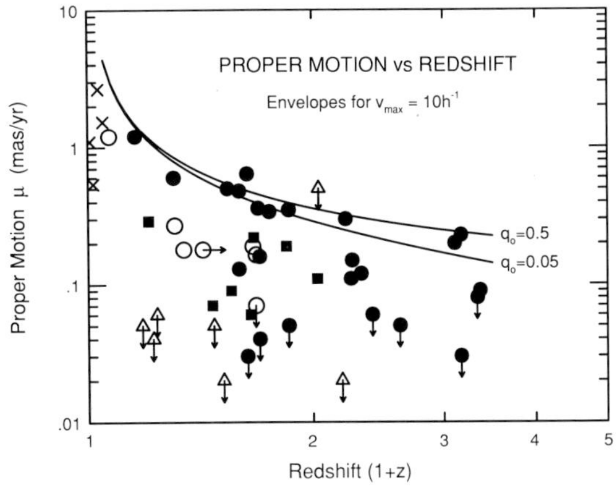

Figure 1. Proper motion-redshift diagram for 44 objects. See the text for a definition of the symbols.

We convert proper motion to apparent transverse velocity with a Friedmann universe with $H_0 = 100h$ km/sec/Mpc and $q_0 = 0.5$ and 0.05. The lines in Figure 1 show $\mu(z)$ for $v = 10h^{-1}$ (we set $c = 1$). Since the maximum apparent transverse velocity for a moving feature is $v_{max} = \beta_p \gamma_p$, the lines in Figure 1 represent the upper envelope of possible proper motions for a distribution of objects at various orientations, all of which have $\gamma_p \sim 10h^{-1}$. Here γ_p is the Lorentz factor of the moving pattern, which we distinguish from that of the relativistic beam, γ_b.

Figures 2a and 2b show the (v, z) diagrams corresponding to Figure 1. The velocity has been set to zero for components with no detected motion, and the error bar shows the upper limit. For $q_0 = 0.5$ it appears that $v = 10h^{-1}$ forms a good envelope, although one object (3C395) is substantially higher. However, we caution against placing too much emphasis on any one velocity in this heterogeneous data base. For $q_0 = 0.05$ the envelope is not as sharp; $v = 15h^{-1}$ might be a reasonable value. This illustrates the danger in discussing these statistics with only one value of q_0, but it also shows that a larger body of data might be able to constrain q_0, along with an evolution of γ_p.

Figure 2 shows that the cd quasars are on average faster than the ld quasars, as would be expected if their axes lie closer to the line-of-sight and the core domination results from relativistic beaming (e.g. Orr and Browne, 1982; Hough and Readhead, 1987). Figure 2 also shows that the BL-Lac objects are slower than the cd quasars, and have speeds comparable to the ld quasars (cf. Mutel, 1991).

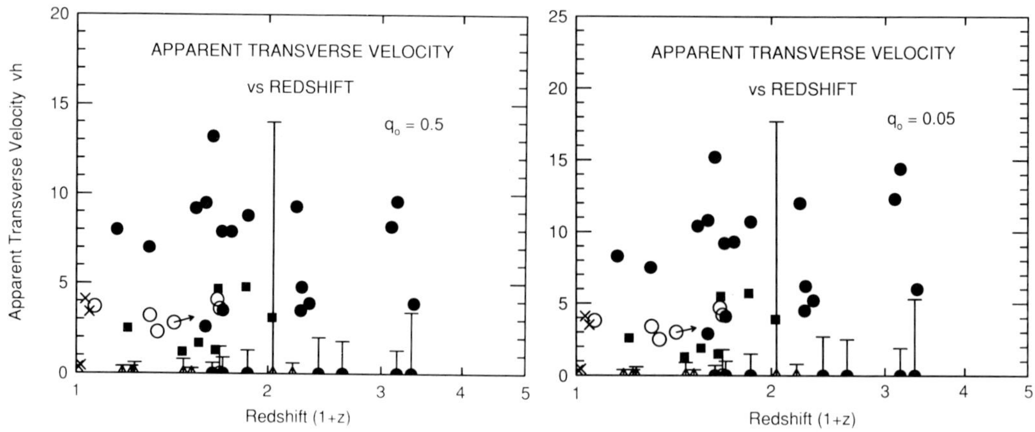

Figure 2. Velocity-redshift diagrams: (a) $q_0 = 0.5$, (b) $q_0 = 0.05$. Note the different scales on the ordinates.

2. BEAMING

There is strong evidence that active galactic nuclei contain relativistic beams. For simplicity we assume that the beam is narrow and characterized by a single bulk Lorentz factor, γ_b. The radio components are of at least two kinds. The core, known to be stationary in one source (Bartel et al., 1986) has a flat spectrum and is relatively stable. It may be the throat ($\tau = 1$) of an expanding relativistic jet. Moving features might be distinct entities, or, more likely, they could be due to shock waves. Stationary components may represent standing shocks, perhaps locations where a beam hits a dense interstellar medium, or places where it is deflected.

Lind and Blandford (1985) have emphasized that in general $\gamma_p \neq \gamma_b$, and that the simple Doppler boosting formula may need modification for an optically thick source. In the following we allow for different values of γ_p and γ_b, and assume a typical boosting index. This may not be correct for the core component, and a deeper study of this point would be useful.

There is evidence to suggest that, typically, $\gamma_b \sim 5$, as follows. (a) Statistical studies of R (the ratio of core to lobe flux density) suggest that $\gamma_b \sim 5$ (Orr and Browne, 1982; Peacock, 1987). (b) Application of the theory of incoherent synchrotron radiation and inverse Compton radiation to compact variable X-ray/radio sources shows that the Doppler factor must be not less than 2–10 in several cases (e.g. Cohen, 1986). Thus the peak values of γ_b must be 5 or greater. This result is essentially independent of H_0. (c) Theoretical estimates of the limiting velocity of a beam advancing into

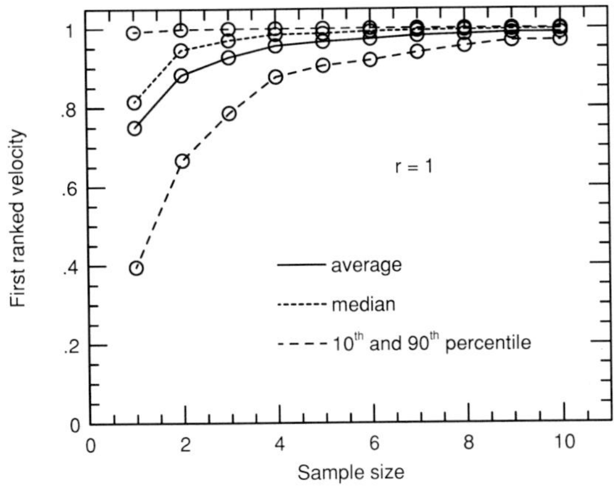

Figure 3. The first-ranked apparent velocity in a randomly-selected sample, relative to the maximum possible value, as a function of sample size.

a black-body radiation field (Compton drag) give $\gamma_b \sim$ 2–10 (Phinney, 1987; Henri, 1991; **Abramowicz**, 1991).

On the other hand, superluminal motions, summarized above, show that γ_p is at least 15–20 for cd quasars, for commonly-adopted values of the Hubble constant. From this it appears that $r \equiv \gamma_p/\gamma_b > 1$ for cd quasars; perhaps $r = 3$ could be typical. The simplest relativistic shock wave model has $r = 1.4$, but higher values can be attained if the jet is carrying a magnetic field (Blandford, 1990).

3. COMPARISON OF OBSERVATIONS WITH A SIMPLE MODEL

We now adopt a simple model for cd quasars, calculate the cumulative probability of v, and compare it with the measured cumulative distribution. The measurement errors are highly non-uniform, and our analysis circumvents some of the inherent difficulties of binning data in this case.

We assume that the sources are picked at random on the basis of their core flux density, and that the parent population is randomly oriented with respect to the line-of-sight ($p(\theta)\,\mathrm{d}\theta = \sin\theta\,\mathrm{d}\theta$). We assume that the luminosity function is a power law, $N(L) \sim L^{-p}$, over a sufficiently broad range. The observed flux density depends on the Doppler boost factor D as $S = S_0 D^n$, and we assume that n is the same in every source. The probability distributions depend on the parameter $a \equiv (np - 1)$ and on γ_b and r, which we have assumed to be the same for every source. In the examples shown below we have used $a = 2.75$. Similar distributions for the special case $r = 1$

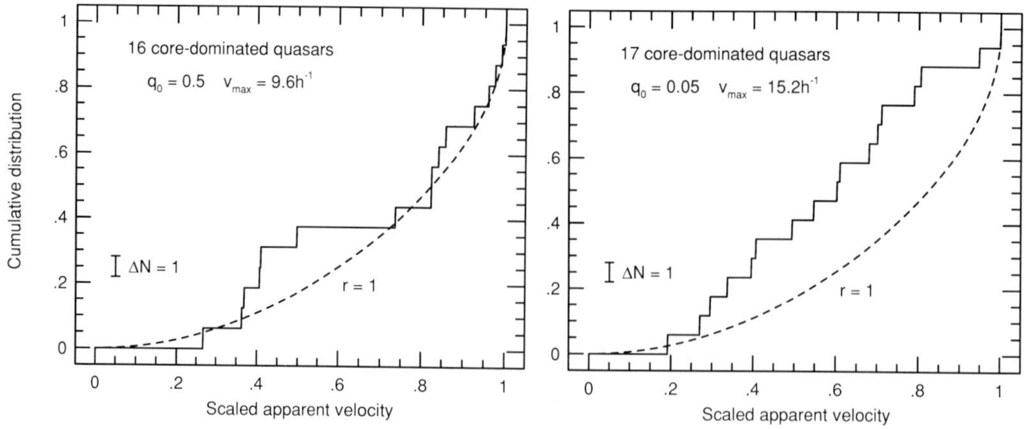

Figure 4. Cumulative distributions of apparent transverse velocity for cd quasars. The solid curve is the measured distribution, the dashed curve is the theoretical distribution for $r \equiv \gamma_p/\gamma_b = 1$.

have also been used by Scheuer and Readhead (1979) and by Cohen (1989).

Figure 3 in Cohen (1989) shows several curves of the probability density for θ, which is independent of r since it does not depend on the pattern speed. The probability peaks at small angles, less than those for which v reaches its maximum. When account is taken of the available solid angle, one finds a high probability that a source randomly selected on core flux will have v near v_{max}. Let v_1 be the first-ranked apparent velocity measured for a set of appropriately selected objects, all having the same γ_b and r. Figure 3 shows the mean values of v_1/v_{max} for $r = 1$ (also the median and the 10th and 90th percentiles) as a function of the sample size. The points were found by a Monte-Carlo calculation. Even for samples as small as 3 the average value of v_1 is within 10% of v_{max}, illustrating that useful estimates of γ_p can be made from small samples of objects.

Figure 4 a,b shows the observed cumulative distribution of velocities for cd quasars. The velocities are scaled to the peak observed value. For $q_0 = 0.5$ the distribution is unstable if all 17 objects are used whereas it becomes stable when the fastest one (3C395) is removed. We show the distribution for the 16 remaining objects. For $q_0 = 0.05$ the distribution is stable against removing any one object.

Figure 4 also shows the theoretical distribution for $r = 1$, with the velocities scaled to the maximum possible, $v_{max} = \beta_p\gamma_p$. Note that this curve is virtually independent

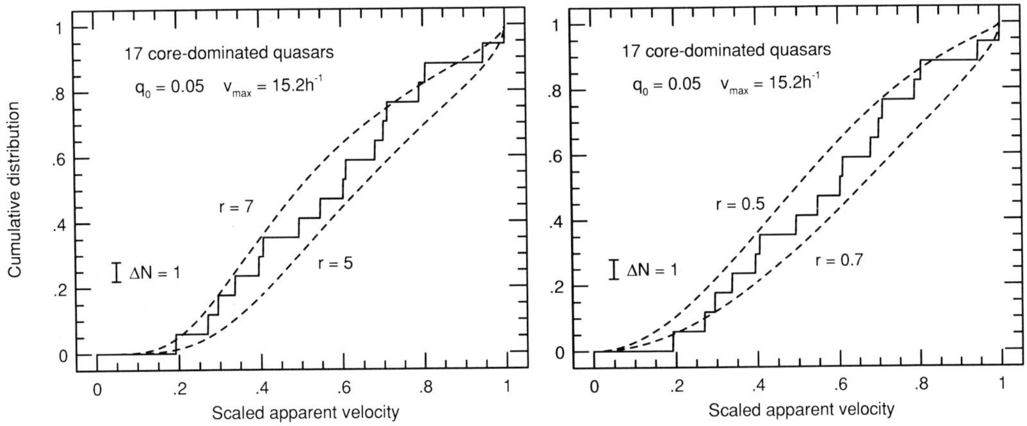

Figure 5. The measured cumulative distribution for $q_0 = 0.05$, with theoretical distributions for: (a) $r = 5$–7, (b) $r = 0.5$–0.7.

of the numerical value of γ_b, for $\gamma_b^2 >> 1$. It clearly fits the data poorly for $q_0 = 0.05$; the Kolmogorov-Smirnov test rejects the hypothesis that the data have been drawn from this distribution at the 99% confidence level. For $q_0 = 0.5$ the fit is not as bad. It is not rejected by the KS test, but it will be interesting to see whether the apparent excess of sources at $v/v_{\text{max}} \sim 0.4$ persists as more sources are added in the future.

For $q_0 = 0.05$ a good fit is obtained with $r = 5$–7, as shown in Figure 5a. On the other hand, a good fit can also be obtained with $r = 0.5$–0.7 as shown in Figure 5b. No similarly good fit can be obtained at any value of r for $q_0 = 0.5$. If we enlarge the parameter space by allowing a distribution of γ_b then, of course, better fits can be obtained.

The objects we are using have been collected from the world's literature, and do not form a complete well-defined set. We can recognize several biases: there will be a deficit of slow objects because they are less interesting and more time-consuming to study, and there may be a deficit of very fast objects because the sampling intervals are too long. Thus statistically valid conclusions cannot be drawn from this study. However, we believe that the objects are representative of their class and that the results from a properly-chosen sample of 17 cd quasars would not be drastically different. Therefore, we find it suggestive that $r = 1$ does not fit the cumulative distribution, whereas $r \sim 6$ does; this matches the expectation outlined above. The numerical combination $\gamma_p \sim 20$, $\gamma_b \sim 4$ would fit the typical core-dominated quasar and also roughly match other estimates of relativistic motion, for $H_0 = 75$. We caution, however, that this is no more than an illustration of how the currently available dataset can be fitted with a simple model containing only a few parameters.

The measurements reported here have been collected by a small band of industrious VLBI workers, over the last 15 years. The work goes slowly because multi-epoch data are needed, and because VLBI is intrinsically laborious. The VLBA should greatly speed up this work, by reducing the labor, providing more observing time, and giving more angular resolution (shorter wavelengths), thereby shortening the required time between epochs. We expect that a decade from now the sample of objects with known internal motions will be perhaps 5 times larger than it is now, and that the dataset will provide powerful constraints on the cosmological parameters.

We are grateful to Drs. D. Gabuzda and D. Hough for giving us data in advance of publication, and to Dr. T.J. Pearson for comments on the manuscript. This work has been supported by NSF grant AST-8814554.

REFERENCES

Abramowicz, M. 1991, this volume.

Baath. L.B. 1987, in Superluminal Radio Sources, ed J.A. Zensus and T.J. Pearson (Cambridge: Cambridge University Press), p. 206.

Bartel, N., et al 1986, Nature, 319, 733.

Blandford, R.D. 1990, in Active Galactic Nuclei by R.D. Blandford, H. Netzer and L. Woltjer (Berlin: Springer-Verlag), p. 239.

Cohen, M.H. 1986, in Highlights of Modern Astrophysics, ed S.L. Shapiro and S.A. Teukolsky (New York: Wiley), p. 299.

Cohen, M.H. 1989, in BL Lac Objects, ed L. Maraschi, T. Maccacaro, and M.-H. Ulrich (Berlin: Springer-Verlag), p. 13.

Cohen, M.H., Barthel, P.D., Pearson, T.J. and Zensus, J.A. 1988, Ap. J., 329, 1.

Henri, G. 1991, this volume.

Hough, D.H. and Readhead, A.C.S. 1987, in Superluminal Radio Sources, ed J.A. Zensus and T.J. Pearson (Cambridge: Cambridge University Press), p. 114.

Lind, K.R. and Blandford, R.D. 1985, Ap. J., 295, 358.

Mutel, R. 1991, this volume.

Orr, M.J.L. and Browne, I.W.A. 1982, MNRAS, 200, 1067.

Peacock, J.A. 1987, in Astrophysical Jets and Their Engines, ed W. Kundt, (Dordrecht: Reidel), p. 185.

Phinney, E.S. 1987, in Superluminal Radio Sources, ed J.A. Zensus and T.J. Pearson (Cambridge: Cambridge University Press), p. 301.

Scheuer, P.A.G. and Readhead, A.C.S. 1979, Nature, 277, 182.

Wehrle, A.E. and Cohen, M.H. 1989, Ap. J. Lett., 346, L69.

Cosmological Manifestation in the VLBI Structure of Active Galactic Nuclei

L.I.GURVITS

Astro Space Center of P.N.Lebedev Physical Institute, 53 Leninsky pr., Moscow, 117924, Russia

1. INTRODUCTION

One of most interesting problem of modern cosmology is connected with evolution and distribution of any discrete sources in the universe. At the moment the deepest range of distances among known sources which are presented by redshift z is covered by quasars, or under more general definition by active galactic nuclei (AGN). This range does already reach the upper limit of about 5. At such redshifts the observed universe is orders of magnitude younger than now and one has the right to expect some differences of properties even for related objects there and here, at $z = 0$. Thus any attempt to search any dependence of AGN properties versus redshift are assumed as quite expedient.

Another aspect of investigation of any possible dependence between any property of AGN and redshift is explained by usual dream of astronomers to find a sort of standard object. Indeed being found such an object could be used for calibration of extragalactic distance scale and measurement of cosmological parameters. Even at that large distances radio astronomical technique, in particular, VLBI can provide as high linear resolution as units of parsecs. High accuracy of VLBI technique together with very deep range of covered cosmological distances give optimistic hope to improve accuracy of such measurements.

So, there are at least two reasons to look carefully on structural patterns of extragalactic sources distributed over deep range of cosmological distances: *(i)* to search for proper qualities of objects, in particular, structural ones at different redshifts, and *(ii)* to search for any their standard property useful for extragalactic scaling.
Are there any known properties of AGN which show dependence of redshift?

2. REDSHIFT-DEPENDENT PROPERTIES OF AGN

First let us restrict ourselves by reviewing only structural or structural-related properties of AGN. More general review of references on AGN in cosmology is given by Shaver (1991).

Assuming declared interest to VLBI structure of AGN one should refer to negative result of investigation for redshift-dependent milliarcsecond structures obtained by Preston & al. (1983). But this result was obtained from the smaller than accessible to the moment source sample and the narrower range of redshifts. The conclusion one could derive from this paper is that if there is a dependence between milliarcsecond structure of extragalactic sources and redshift then it is weak.

But on the other hand there are some indications which can be assumed as a hint at some redshift-dependent properties of AGN structures.

3. CONTINUUM RADIO SPECTRA

As it was shown by Peterson & al. (1982), Wright (1983) it is rather typical for continuum radio spectra of quasars at redshift $z > 3$ to have a "hump" in the range from about 1 to 10 GHz. O'Dea (1990) has shown that percentage of Giga-Hertz peaked sources (GPS) at redshift $z > 3$ is higher than at low redshifts. Both these two statements might be considered as direct indications on distinction of continuum radio emission properties from low and high redshift sources. But significant part of continuum radio flux from discussed sources is formed in the central milliarcsecond region of a structure (see below). Hence, an admission on distinction of structural properties of high and low redshift continuum sources looks quite likely.

4. RADIO AND OPTICAL STRUCTURE

It is known that for close extragalactic sources direction of elongation of a radio structure is near perpendicular to the direction of an optical structure elongation. But there is a trend noted by Chambers & al. (1987, 1988) this alignment relation to be changed at higher redshifts: more distant objects tend to have closer directions of elongation of radio and optical structures. Of course this trend has only marginal character but it should not be forgotten in the list of possible redshift-dependent structural issues. This effect might be connected with another recently discovered effect (Hu & al., 1991): there is significantly higher probability for high redshift ($z > 3$) radio loud quasars to have an optical companion detected by L_α filters specially tuned for given redshift than for the same distant but radio quiet ones.

5. ARCSECOND AND SUB-ARCSECOND SCALE STRUCTURE

This angular scale is filled up by good quality radio maps obtained with VLA. Analysis of few hundreds such maps published in several papers (Barthel, 1986; Miley, 1987; Barthel & al., 1988; Barthel, Miley, 1988) has allowed to conclude that there is increasing of distortion of extended radio structure with increasing of redshift. This distortion is displayed by bending of elongated structures - jets.

Examining an enlarged data base of radio maps of quasars Kapahi (1990) contends for above conclusion. McCarthy & al. (1991) also put in doubt the conclusion on increasing of distortion for a sample of radio galaxies. But even formulating such a negative conclusion Kapahi (1990) supposes an existence of bending dependence on redshift for lobe dominated sources.
So, it seems that some possible dependence between bending angle and redshift on arcsecond and subarcsecond scale (i.e. on kiloparsec linear scale) can not be ruled out.
It is interesting to refer on the analogous investigation on three order of magnitude smaller angular scale. Analyzing VLBI structures of quasars Wehrle & al. (1989, 1990) conclude the greater bending angle on parsec scale to be more typical for the greater redshift. As it is shown below some additional arguments could be done for this statement.

6. VISIBILITY AMPLITUDE AND MILLIARCSECOND STRUCTURE

This value Γ is connected with a brightness distribution in a source $I(x,y)$ by the classical formula of Fourier transformation:

$$\Gamma(u,v)\cdot\exp[i\cdot\phi(u,v)] = \int\int I(x,y)\cdot exp[i\cdot(ux+vy)]dxdy \ ,$$

where x, y are picture plane co-ordinates, u, v are corresponding spatial frequencies. In radio interferometric practice this value measured in flux units is called as correlated flux. Its physical meaning is a part of flux from a source which is convolved with a pattern of a synthesized aperture. Of course, Γ does not contain all structural information of a source (the rest is hidden as a phase $\phi(u,v)$). But it can be used as an indicator of any structure on scales of synthesized beam. In reference to VLBI technique Γ can be assumed as compactness measure. So, if there is any difference in VLBI structures at different redshifts we can be lucky to catch it by analyzing Γ versus z.

The largest homogeneous data array of Γ measurements on VLBI scales for extragalactic sources is provided by Preston & al. (1985). This paper gives values of correlated flux density of 917 extragalactic sources observed with 3-station VLBI array at 2.3 GHz. Typical baseline was $(50 \div 80)\times 10^6$ wavelength, which corresponds to $(4 \div 2.5)\times 10^{-3}$ arcsecond. The initial paper contains data on redshifts for 548 extragalactic sources. I have added redshift values for 83 sources with optical identification published after 1985. The diagram $\Gamma(z)$ for 631 extragalactic sources is shown on Fig. 1. Data are averaged over intervals $\Delta z = 0.2$.

In spite of too generalized character of this diagram, in particular due to mixture of different types of objects (quasars, radio galaxies, etc.) some conclusion could be done. First one may note a clear near-linear increase of Γ in the redshift range from 0 up to about 1. This increase has an obvious explanation: apparent angular sizes of similar objects in quasi-Euclidean region of the universe decreases with increase of a distance to objects. But the shape of the dependence at redshifts higher than 1 looks rather enigmatic. The zero approximation of that part of the diagram might be presented by a constant ≈ 0.30. Such a constancy looks to give some chance for further speculation on epoch (or distant) dependent properties of extragalactic source radio structures.

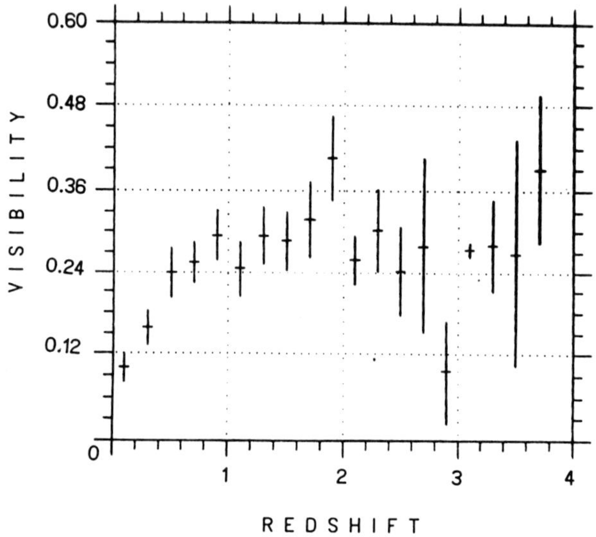

Fig. 1. Visibility amplitude Γ at 2.3 GHz versus redshift z for 631 extragalactic sources averaged over intervals $\Delta z = 0.2$. Visibility data from Preston & al. (1985).

Indeed, as shown in number of papers (Kapahi, 1987 and references therein) apparent angular sizes of extragalactic radio sources θ tend to decrease monotonously with increase of redshifts. But such a tendency can not be presented in terms of visibility amplitude Γ by constant of about 0.3. This tendency might lead to the only constant of 1, and for a given telescope (no difference whether it is a single dish or a synthesized aperture) the only its angular resolution would determine a redshift where this constant will be reached. Of course following this logic one has to remember that refered statements of Kapahi (1987) and others for $\theta(z)$ relation and presented here diagram $\Gamma(z)$ are dealing with angular scales more than three order of magnitude

apart: the former one is composed from data at arcsecond and larger angular scale while the last one is based on the typical VLBI angular scale of milliarcseconds. But it does not matter for cosmological properties of the universe as a whole.
A common explanation of the observed $\theta(z)$ relation is based on the hypothesis on cosmological evolutions of extragalactic sources themselves, in particular their linear sizes. But neither any simple hypothesis on size evolution can explain simultaneously $\theta(z)$ and $\Gamma(z)$ relations. So, one may conclude this hypothesis to have at least different modification for arcsecond (and larger) and milliarcsecond structures of extragalactic radio sources.
The above conclusion is based on very simple analysis of the $\Gamma(z)$ diagram. It is necessary to note that the concrete shape of this relation is exposed by many factors such as evolution of size and structure of sources, observing of different rest frame emitting frequencies due to redshift, possible different behavior of different morphological types of sources, *etc.* Finally, due to its physical nature a visibility amplitude is a sort of presentation of so called isophotal sizes rather than metric ones (Sandage, 1961). But even with all these reservations it seems possible to conclude that milliarcsecond structure of extragalactic sources shows some dependence of redshift.

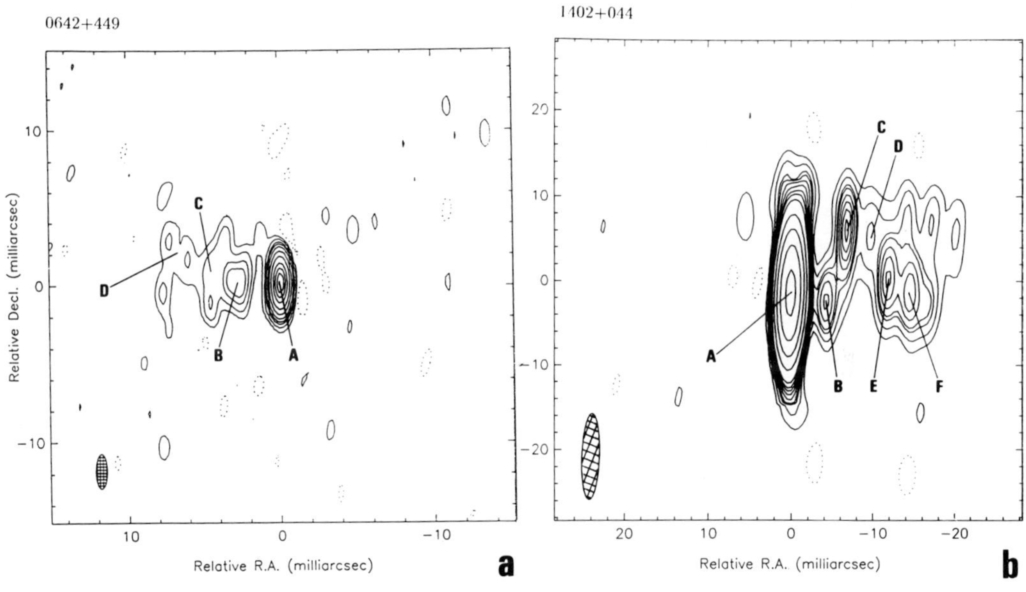

Fig. 2. 5 GHz images of (a) 0642+449 and (b) 1402+044. Contour levels: (a) -1, -0.4, 0.4, 1, 2, 3, 5, 10, 20, 40, 60, 80, 99 % of the peak brightness 888 mJy/beam; (b) -0.5, 0.5, 0.75, 1. 1.5, 2, 2.5, 3.5, 10, 20, 40, 60, 80, 99 % of the peak brightness 619 mJy/beam.

Doing this conclusion it is natural to take an interest in more detail investigation of redshift-dependent behavior of VLBI structures of AGN, specially at extremely high redshifts.

There are very few published VLBI maps of extremely high redshift quasars. In particular, there are only four VLBI maps of quasars at redshift higher than 3: 0014+81 at $z = 3.384$, (Kühr & al. 1986), 0642+449 at $z = 3.406$, 1402+044 at $z = 3.208$, and 1614+051 at $z = 3.212$ (Gurvits & al. 1992). Such a small number does not allow to make any statistically approved conclusions. May be the only mentioned above statement of Wehrle & al. (1989, 1990) on some preliminary trends of more curved milliarcsecond morphologies for sources at very high redshift might be confirmed by the fact that from four listed above high redshift quasars two, 0642+449 and 1402+044, have rather bent jet-like structures (Fig. 2). Keeping in view the title of our meeting it is appropriate to note these jets (or beams?) as the most redshifted ones known to the moment.

7. CONCLUSION

Structures of AGN at cosmological distances on typical VLBI angular scales (milliarcseconds) show some redshift-dependent properties. The nature of these properties needs more attentive investigations and could provide an important observational information for understanding both physics of sources and the universe as the whole.

ACKNOWLEDGEMENTS

It gives me pleasure to thank Drs. N.S. Kardashev, B.V. Komberg, R.T. Schilizzi, V.I. Slysh for stimulating and fruitful discussions of different aspects of the paper. Also I would like to thank Dr. R.A. Preston for providing computer readable version of the VLBI survey at 2.3 GHz. Finally, it is a pleasant debt to thank the LOC and especially Hélène Sol and Jacques Roland for the kind hospitality and excellent organizing of the meeting, and Institut d'Astrophysique de Paris and Observatoire de Meudon for arranging financial support.

REFERENCES

Barthel P.D. 1986, in Quasars, Proc. of the IAU Symp. No. 119, eds. G.Swarup and V.K.Kapahi, Reidel, Dordrecht, p. 181.

Barthel P.D., Miley G.K. 1988, Nature, **333**, 319.

Barthel P.D., Miley G.K., Schilizzi R.T., Lonsdale C.J. 1988, Astron. Astrophys. Suppl. Ser., **73**, 515.

Chambers K.C., Miley G.K., Joyce R.R. 1988, Astrophys. J. Lett., **329**, L75.

Chambers K.C., Miley G.K., van Breugel W. 1987, Nature, **329**, 604.

Gurvits L.I., Kardashev N.S., Popov M.V., Schilizzi R.T., Barthel P.D., Pauliny-Toth

I.I.K., Kellermann K.I. 1992, Astron. Astrophys., in press.

Hu E.M., Songaila A., Cowie L.L., Stockton A. 1991, Astrophys. J., **368**, 28.

Kapahi V.K. 1987, in Observational Cosmology, IAU Symposium No. 124, eds A. Hewitt, G. Burbidge, L.Z. Fang, D. Reidel, Dordrecht, p. 251.

Kapahi V.K. 1990, Current Science, **59**, 561.

Kühr H., Stocke J.T., Strittmatter P.A., Bartel N., Eckart A., Schalinski C., Witzel A., Bierman P. 1986, Astrophys. J., **302**, 52.

Miley G.K. 1987, in Observational Cosmology, IAU Symposium No. 124, eds A. Hewitt, G. Burbidge, L.Z. Fang, D. Reidel, Dordrecht, p. 267.

McCarthy P.J., van Breugel W., Kapahi V.K. 1991, Astrophys. J., **371**, 478.

O'Dea C.P. 1990, Mon. Not. Roy. Astron. Soc., **245**, 20p.

Peterson B.A., Savage A., Jauncey D.L., Wright A.E. 1982, Astrophys. J. Lett., **260**, L27.

Preston R.A., Morabito D.D., Jauncey D.L. 1983, Astrophys. J., **269**, 387.

Preston R.A., Morabito D.D., Williams J.G., Faulkner J., Jauncey D.L., Nicolson G.D. 1985, Astron. J., **90**, 1599.

Sandage A. 1961, Astrophys. J., **133**, 355.

Shaver P.A. 1991, IAU Transactions XXIA, Commission 47.

Wehrle A.E., Cohen M.H., Unwin S.C. 1989, Bull. American Astron. Soc., **21**, 1094.

Wehrle A.E., Cohen M.H., Unwin S.C. 1990, in Parsec-Scale Radio Jets, eds. J.A. Zensus and T.J. Pearson, Cambridge Univ. Press, p. 49.

Wright A.E. 1983, in Quasars and Gravitational Lens, Proceed. of the 24th Liege Astrophys. Col., Universite de Liege, Institut d'Astrophysique, p. 53.

Monitoring of Extragalactic Radio Source Structures with Astrometric/Geodetic VLBI Experiments. The Case of 3C273.

P. CHARLOT

Observatoire de Paris - CNRS/URA 1125

1 INTRODUCTION

Extragalactic radio sources have been observed with the Very Long Baseline Interferometry (VLBI) technique for astrometry and geodesy since 1979. The primary goal of these observations is to build high precision terrestrial and celestial reference frames and to determine variations of the Earth rotation, based on the Band Width Synthesis delay and delay rate VLBI observables. Although the observing mode is different from that used in astrophysics, these data can also be used to produce valuable maps for astrophysics, as shown by Charlot (1990). Such VLBI maps are derived from the amplitude and closure phase VLBI observables with the hybrid-mapping algorithm (Readhead and Wilkinson 1978, Cornwell and Wilkinson 1981).

The VLBI data from the astrometric/geodetic experiments are useful to supplement the current astrophysical programs for monitoring extragalactic radio source structures. For example, they could be used to reduce time intervals between maps for sources with rapid changes like BL Lac objects or like the quasar 3C273. In part 2, we review the characteristics of the astrometric and geodetic VLBI data (sources observed, wavelengths, *u*-*v* coverage, intervals between experiments). In part 3, we present a series of maps of the quasar 3C273 at 6-month intervals produced with such VLBI data, and discuss some points of its structure.

2 CHARACTERISTICS OF THE ASTROMETRIC AND GEODETIC VLBI DATA

Astrometric and geodetic VLBI experiments are carried out simultaneously at two frequencies (2.3 and 8.4 GHz) with the Mark III wide-band data acquisition system. The main observing programs are the IRIS (International Radio Interferometric Survey) and NAVNET (Navy Network) programs, dedicated to Earth orientation measurements, the Crustal Dynamics Program (CDP), dedicated to the measurement of tectonic motions, and the astrometric program of the Deep Space Network (DSN). In all, more than 100 experiments are conducted every year, 30 of which with arrays of five radiotelescopes or more (see Caprette *et al.* 1990). These data are stored in the CDP data base at Goddard Space Flight Center.

A typical astrometric/geodetic VLBI experiment lasts 24 hours, with observations distributed over 10-15 extragalactic radio sources. Each source is observed about once per hour during its mutual visibility by the network, providing a sparse but uniform u-v coverage. For a 6-station array, there are generally 100-200 u-v points for a single source. Correlation amplitudes measured in these experiments are calibrated with the usual scheme combining system temperatures and sensitivities of radiotelescopes to produce correlated flux densities for hybrid mapping (Cohen *et al.* 1975). Sensitivities of radiotelescopes are estimated from the VLBI data acquired on a point-like source or on a source with a simple structure, selected among the other sources observed in the experiment. For this estimation, the total flux density of the calibrator at the epoch of the observations is required; it must be found in the literature since total flux densities are usually not measured in astrometric/geodetic VLBI experiments. System temperatures are generally recorded and are available from the CDP data base.

Since 1979, almost 400 extragalactic radio sources have been observed for astrometry and geodesy (see IERS 1991), about 40 of which are used in ongoing programs. For these 40 sources, VLBI data at intervals of a few months or even a few weeks are available to map their structures. Table 1 lists the 20 sources most observed between 1979 and 1988 and the total number of observations available. It includes superluminal sources (3C345, 4C39.25, 3C454.3, 0212 + 735, 3C273, 0106 + 013) as well as BL Lac objects (OJ287, 1803 + 784, BL Lac). Most of these sources were observed every week by the IRIS program between 1984 and 1988, which explains the huge amount of data.

Table 1: The 20 sources most observed in astrometry and geodesy between 1979 and 1988 (as quoted by Ma *et al.* 1990)

Source	Other name	Nb of obs.	Source	Other name	Nb of obs.
1641 + 399	3C345	22105	0106 + 013		11213
0552 + 398	DA193	20692	0229 + 131		8403
0923 + 392	4C39.25	17116	0528 + 134		8166
0851 + 202	OJ287	16736	2134 + 004		7553
2251 + 158	3C454.3	15217	0234 + 285		5914
0212 + 735		15150	0355 + 508	NRAO150	3982
1803 + 784		15126	2216 − 038		2363
1226 + 023	3C273	13785	2145 + 067		2355
1404 + 286	OQ208	13659	1548 + 056		2281
2200 + 420	BL Lac	13387	1637 + 574		1939

3 MONITORING OF THE STRUCTURE OF THE QUASAR 3C273 WITH CRUSTAL DYNAMICS PROGRAM VLBI DATA

3.1 Observations and data reduction

Our dataset consists of five dual-frequency (2.3 and 8.4 GHz) VLBI experiments conducted by CDP over a North-Pacific network between 1985.8 and 1987.6. Table 2 lists the radiotelescopes, their location, and their size. Such a network provides a significant N-S resolution for low declination sources like 3C273 (see Fig. 1), by contrast with the predominantly E-W combined European-US VLBI network used in astrophysics.

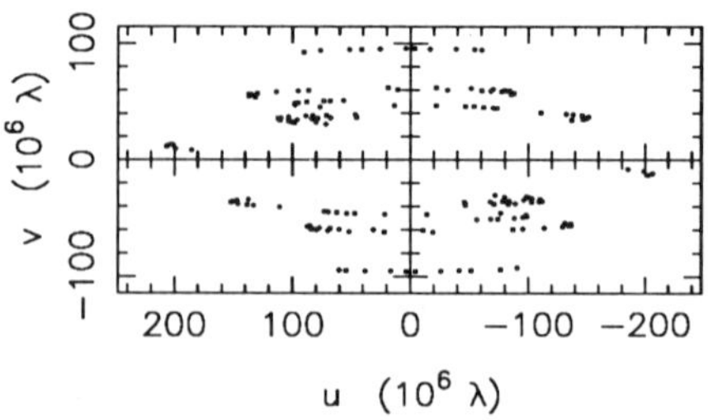

Fig. 1. The *u*-*v* plane of 3C273 at 8.4 GHz with the network listed in Table 2.

The 2.3 GHz data acquired on 3C273 at a single epoch (1986.27) and the 8.4 GHz data acquired on 3C273 in all five experiments have been analysed with the Caltech VLBI package. The single 2.3 GHz map and the complete sequence of 8.4 GHz maps produced with these data are shown rotated counterclockwise by 30° in Fig. 2 and 3. In both figures, the restoring beam is an elliptical Gaussian with FWHM chosen to match the FWHM of the dirty beam produced by the Fourier inversion. The dynamic range of these maps is between 1:100 and 1:40.

Table 2: The North-Pacific network used by CDP

Radiotelescope	Location	Size (m)
Mojave, National Geodetic Survey	Barstow, CA	12
Vandenberg, Crustal Dynamics Program	Vandenberg, CA	9
Hatcreek, Hat Creek Radio Observatory	Hat Creek, CA	26
Gilcreek, Crustal Dynamics Program	Fairbanks, AK	26
Kauai, Crustal Dynamics Program	Kauai, HI	9
Kashima, Kashima Space Research Center	Kashima, Japan	26

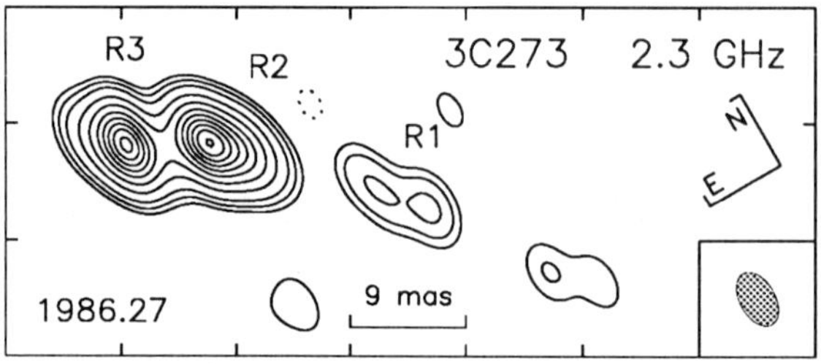

Fig. 2. 2.3 GHz hybrid map of 3C273 at epoch 1986.27, convolved with a beam size of 4.4 × 2.6 mas at PA= 0°, and rotated counterclockwise by 30°. Contour levels are: 0.075, 0.15, 0.30, 0.75, 1.5, 2.5, 3.5, 4.5, 5.5, 6.5 and 7.5 Jy/beam.

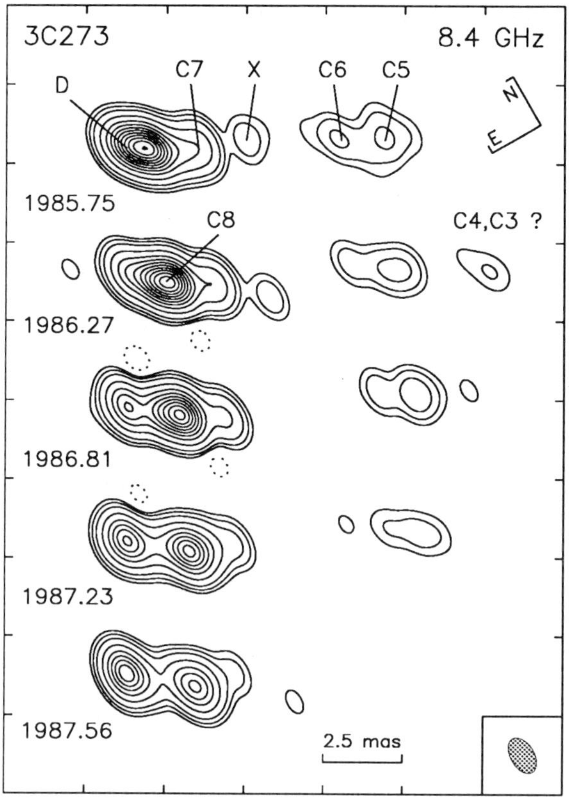

Fig. 3. 8.4 GHz hybrid maps of 3C273 at five epochs, convolved with a beam size of 1.2 × 0.7 mas at PA= 0°, and rotated counter-clockwise by 30°. The maps are displaced vertically in time order and are aligned horizontally on the easternmost component D. Contour levels are: 0.09, 0.18, 0.36, 0.9, 1.8, 2.7, 3.6, 4.5, 5.4, 6.3, 7.2, 8.1 and 9.0 Jy/beam. The peak brightness is 9.1 Jy/beam (epoch 1985.75).

3.2 Morphology of 3C273

The quasar 3C273 is a typical core-jet radio source with the core D located on the eastern side of the structure and jet components moving away from the core at position angles between −100° and −130° (see Fig. 2 and 3). Since 1980, seven superluminal components (C2, C3, C4, C5, C7, C8, C9) with apparent velocities between 0.59 mas/yr and 1.20 mas/yr ($v/c = 4.0h^{-1}$-$8.0h^{-1}$) have been detected with VLBI (Unwin *et al.* 1985, Cohen *et al.* 1987, Zensus *et al.* 1988, Zensus *et al.* 1990, Krichbaum *et al.* 1990). The identification of these components in our maps in Fig. 2 and 3 is based on that proposed by Charlot *et al.* (1988).

At 8.4 GHz, our sequence of maps in Fig. 3 shows the evolution of the jet components C5, C6, C7 and C8. It is to be noted that the identification of components C5 and C6 is uncertain because of a gap of 3 years (1981-1984) in the existing data at 10.7 GHz (see Cohen *et al.* 1987, hereafter Co87). Actually, Co87 identified only the single component C5 in their 1984 and 1985 maps at 10.7 GHz, assuming that C6 has become invisible. Our proposed identification of component C6 with the eastern part of the extended emission located 6-10 mas from the core in Fig. 3, is supported by the apparent motion of this feature over the period 1985-1987 (see Section 3.3). This motion predicts back in time a separation of 1.1 mas at epoch 1981.1, which is the same as that actually measured for component C6 at this epoch (Unwin *et al.* 1985). Indication for the existence of C6 is also found from 5 GHz VLBI polarization observations at epoch 1984.8 (Roberts *et al.* 1990). In the future, analysis of astrometric/geodetic VLBI data acquired on 3C273 between 1981 and 1984 may help to confirm this identification. If it is correct, the weak emission located 10-12 mas from the core at epoch 1986.27 (Fig. 3) may then correspond to the fading components C4 and/or C3. Component X, which is seen in our first two maps, was also detected at epoch 1985.4 (Charlot *et al.* 1988) and at epoch 1985.6 (Co87), but its origin is enigmatic because it was not seen in the earlier 10.7 GHz maps of Co87.

At 2.3 GHz, the structure of 3C273 in Fig. 2 consists of three emission regions labelled R1, R2 and R3, adopting the notation of Zensus *et al.* (1988). Region R1 is identified with component C2, region R2 must contain the components C6, C5, C4 and C3, and region R3 must contain the components D, C8 and C7. Some weak emission, which may correspond to the descendent of component C1, is also detected beyond C2 at 30-40 mas from the core (Fig. 2), but it is close to the dynamic range of our map and cannot be trusted.

3.3 Apparent motions of jet components

Our series of 8.4 GHz maps in Fig. 3 was used to derive the apparent motions of components C5, C6, C7 and C8 (see the expansion diagram in Fig. 4). We have not estimated the apparent motion of component X, because of the limited number of epochs available, and because its position is not well constrained by the data.

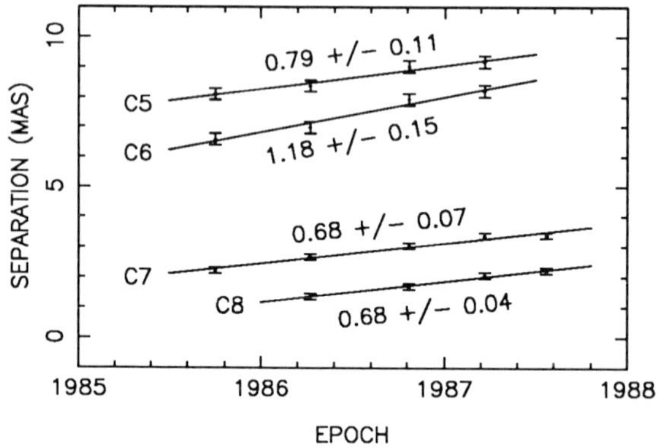

Fig. 4. Apparent motions of components C5, C6, C7 and C8. Positions of the peaks in brightness in maps superresolved by a factor of 2 were used in mesuring separations from the core. Uncertainties in position are 0.1 mas ($\sim$ 15% of the beam size) for C7 and C8, and 0.2 mas ($\sim$ 30% of the beam size) for C5 and C6.

Fig. 4 shows that the apparent motions of the strong jet components C8 and C7 are determined more precisely than those of the weak and resolved components

C6 and C5. For components C7 and C8, our determination (see Fig. 4) is consistent with that of Krichbaum *et al.* (1990), based on data at several frequencies (0.59 ± 0.05 mas/yr for C7 and 0.76 ± 0.05 mas/yr for C8). But for C5, there is a large discrepancy between our determination (0.79 ± 0.11 mas/yr) and that of Co87 at 10.7 GHz (1.20 ± 0.03 mas/yr). This discrepancy may be caused by ambiguity in the identification of C5 (see Section 3.2). It may also be explained by a deceleration of C5, since the period of our observations (1985.8-1987.6) does not overlap that of Co87 (1980.5-1985.6), but deceleration of components has never been observed in 3C273.

3.4 Spectral indices

In Fig. 5, our simultaneous 2.3 and 8.4 GHz hybrid maps of 3C273 at epoch 1986.27 are plotted with the same restoring beam. The 2.3 GHz map is superresolved by a factor of 1.8 along the jet, whereas the 8.4 GHz map is smoothed by a factor 2. Fig. 5 shows that both maps can be separated into five regions corresponding to D, C8/C7, X, C6/C5, and C4/C3. By comparing the integrated fluxes at 2.3 and 8.4 GHz, we have derived the spectral indices between 2.3 and 8.4 GHz of these regions (see Fig. 5). These spectral indices ($S = \nu^\alpha$) indicate that the core is optically thick ($\alpha > 0$) and that the jet becomes optically thin ($\alpha < 0$) at a separation from the core of about 3 mas. This pattern is typical of core-jet radio sources.

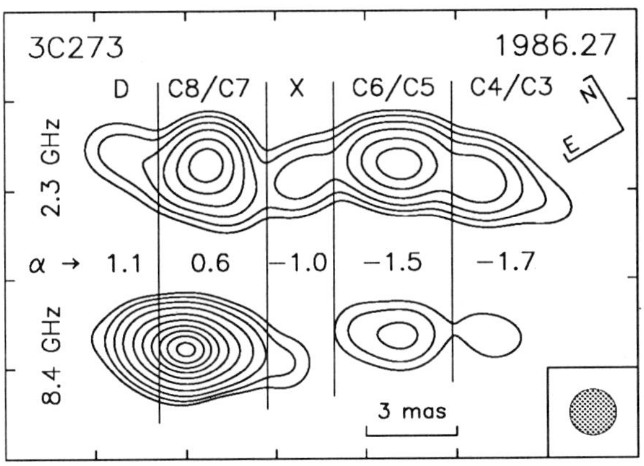

Fig. 5. The 2.3 and 8.4 GHz hybrid maps of 3C273 at epoch 1986.27, convolved with a circular beam of size 1.5 mas, and rotated counterclockwise by 30°. The spectral indices ($S = \nu^\alpha$) of the 5 regions corresponding to D, C8/C7, X, C6/C5 and C4/C3 are printed in between the two maps. At 2.3 GHz, region R1 is not shown. Contour levels are: 0.15, 0.3, 0.6, 1.2, 2.4, 4, 6, 8, 10, 12 Jy/beam.

4. CONCLUSION

The dual-frequency VLBI data on 3C273 analysed in this report illustrate the great potential of astrometric/geodetic VLBI data for monitoring extragalactic radio source structures on short time scales. It is possible to monitor the structure of about 40 sources at intervals of a few months or less with these data. Such a monitoring is useful for sources with rapid changes like BL Lac objects or like the quasar 3C273, to avoid misidentification of components which causes large errors in the determination of superluminal velocities (Blandford 1987). VLBI data acquired in astrometric/geodetic experiments are also useful to study the spectral index distribution in the jets since these experiments are always conducted simultaneously at two frequencies (2.3 and 8.4 GHz). For 3C273, spectral indices have been estimated up to a separation of about 10 mas from the core.

Acknowledgments
We are indebted to the Crustal Dynamics Program for providing the high quality data analysed in this report. The hybrid maps were produced with the Caltech package, and we are grateful to the Caltech VLBI group, especially T. J. Pearson.

REFERENCES

Blandford, R. D.: 1987, in *Superluminal Radio Sources*, Eds J. A. Zensus and T. J. Pearson, Cambridge University Press, Cambridge, p. 310.

Caprette, D. S., Ma, C., Ryan, J. W.: 1990, NASA Technical Memorandum 100765.

Charlot, P.: 1990, *Astron. Astrophys.*, **229**, 51.

Charlot, P., Lestrade, J.-F., Boucher, C.: 1988, in *IAU Symposium 129, The Impact of VLBI on Astrophysics and Geophysics*, Eds. M. J. Reid and J. M. Moran, Kluwer Academic Publishers, Dordrecht, p. 33.

Cohen, M. H., Moffet, A. T., Romney, J. D., Schilizzi, R. T., Shaffer, D. B., Kellermann, K. I., Purcell, G. H., Grove, G., Swenson, G. W., Yen, J. N., Pauliny-Toth, I. I. K., Preuss, E., Witzel, A., Graham, D.: 1975, *Astrophys. J.*, **201**, 249.

Cohen, M. H., Zensus, J. A., Biretta, J. A., Comoretto, G., Kaufmann, P., Abraham, Z.: 1987, *Astrophys. J.*, **315**, L89 (Co87).

Cornwell, T. J., Wilkinson, P. N.: 1981, *M. N. R. A. S.*, **196**, 1067.

International Earth Rotation Service (IERS): 1991, *Annual Report for 1990*, Observatoire de Paris, Paris, p. II-15.

Krichbaum, T. P., Booth, R. S., Kus, A. J., Rönnäng, B. O., Witzel, A., Graham, D. A., Pauliny-Toth, I. I. K., Quirrenbach, A., Hummel, C. A., Alberdi, A., Zensus, J. A., Johnston, K. J., Spencer, J. H., Rogers, A. E. E., Lawrence, C. R., Readhead, A. C. S., Hirabayashi, H., Inoue, M., Morimoto, M., Dhawan, V., Bartel, N., Shapiro, I. I., Burke, B. F., Marcaide, J. M.: 1990, *Astron. Astrophys.*, **237**, 3.

Ma, C., Shaffer, D. B., de Vegt, C., Johnston, K. J., Russell, J. L.: 1990, *Astron. J.*, **99**, 1284.

Readhead, A. C. S., Wilkinson, P. N.: 1978, *Astrophys. J.*, **223**, 25.

Roberts, D. H., Gabuzda, D. C., Wardle, J. F.: 1987, *Astrophys. J.*, **323**, 536.

Unwin, S. C., Cohen, M. H., Biretta, J. A., Pearson, T. J., Seielstad, G. A., Walker, R. C., Simon, R. S., Linfield, R. P.: 1985, *Astrophys. J.*, **289**, 109.

Zensus, J. A., Bååth, L. B., Cohen, M. H., Nicolson, G.: 1988, *Nature*, **334**, 410.

Zensus, J. A., Unwin, S. C., Cohen, M. H., Biretta, J. A.: 1990, *Astron. J.*, **100**, 1777.

MERLIN and VLBI Observations of Two-sided Sources

R. E. SPENCER AND C.E. AKUJOR

University of Manchester, Jodrell Bank, U.K.

ABSTRACT
VLBI observations show that many high luminosity extra-galactic radio sources have compact cores with one-sided jets. The apparent one sidedness is thought to arise from relativistic beaming of intrinsically two-sided jets which are aligned towards the observer. If velocities in the jets are low then we would expect to see a higher fraction of cores with twin jets. VLA observations have shown that kpc scale twin jets are quite common in low luminosity objects but are rare in FRII type sources. In this paper we show that counter-jets on the pc scale are rare but do occur in some sources. In particular some compact steep spectrum sources (CSS) show counter-jets. We compare the rate of occurrence of two-sided jets with that expected on the basis of simple beaming models.

1 INTRODUCTION

The discovery that radio emission from radio loud active galactic nuclei almost always consisted of narrow linear 'jet' on parsec scales was made in the late 1970's when hybrid mapping techniques using closure phase were first employed in VLBI. Beams of fast moving material are needed to supply energy to the outer lobes of extended radio sources (Longair *et al.* 1973, Blandford and Rees 1974) and most of the observed properties of extra-galactic radio sources (cores, jets, twin lobes, bridges and hot-spots) can be explained on this particularly successful model (see *e.g.* Leahy 1991) where we associate the radio 'jets' with the 'beams'. However one of the most mysterious properties of luminous extra-galactic radio sources (the classical 'double' sources) is the predominantly one-sided nature of their jets.

Bridle and Perley (1984) define jets as being 4 times longer than wide, distinguishable from other extended structure and aligned with a compact core where closest to it. A counter-jet can be similarly defined, with a brightness which is say more than 1/4 that of the main jet. Bridle (1990) has investigated two-sided jets on the kpc scale. They are common in low luminosity sources ($P < 10^{25} W Hz^{-1}$ at 1.4 GHz (emitted)) *i.e.* those having edge-darkened FR I structures, as clearly shown in a recent study by Parma *et al.* (1986), but are much rarer in the more luminous edge-brightened FR II sources. Kpc scale counter-jets, or at least 'pieces' of counter-jet have been found in the radio galaxies 3C218, 3C219, 3C288, 3C341, 3C348, Cygnus A, and 3C438 (Bridle 1990), all of which have lobe luminosities in excess of $10^{26.3} WHz^{-1}$, well above the FR I/II divide, but appear to have structures intermediate between those of the FR I and FR II sources. None of the 12 extended quasars investigated by Bridle (1990) have counter-jets satisfying the above criterion but 6 have pieces of possible counter-jets, including 3C215.

Counter-jets on the kpc scale are difficult to find as they contain a few percent or less of the total extended emission and even more difficult on the pc scale, since VLBI observations are often limited in dynamic range. Nevertheless it is important to search for them if we believe in the twin-beam model even though it might seem a hopeless task: a recently published VLBI survey quotes "No two-sided jets are found". Recent observations however have shown six compact steep spectrum quasars with possible counter-jets, and further examination of published VLBI maps shows more counter-jet candidates.

2 WHY SHOULD PC SCALE JETS BE PREDOMINANTLY ONE-SIDED ?

2.1 Physical Causes There are a number of possible reasons why our observations show jets to be predominantly one-sided and we can separate them into intrinsic physical causes or observational selection:

1) There is no counter-jet, with high luminosity sources having intrinsically one sided jets. The problem here is that many sources have lobes and other forms of extended emission on both sides. Flip-flop models have been proposed to explain the appearance of some sources on the kpc scale (Rudnick 1982) where a jet alternately feeds energy into opposite lobes. This leads to the idea of a re-starting jet and recent numerical simulations have shown that many of the features in 3C219, particularly the 'pieces' of counter-jet, can be explained quite well (Clarke and Burns 1991).

2) There is an active beam of material on the 'dark' side but for some reason its intrinsic synchrotron emissivity is much lower than in the jet, perhaps due to less interaction with the interstellar medium, lack of relativistic electrons or a weaker electric field. The twin-beams eventually interact with the interstellar medium and give rise to the double-lobed extended emission seen on larger scales. Lind and Blandford (1985) have shown that the variations in brightness of a jet caused by changing physical conditions associated with shocks or local flaring of a jet can be stronger than

those caused by Doppler boosting with varying geometry. The absence of a counter-jet suggests that these conditions need to be held over a physical length comparable with that of the jet. However the jets themselves often show brightness changes over small distances compared with their lengths and so we require asymmetries in both magnitude and scale size of the ambient medium on each side of the source so that the whole of the counter-jet can remain hidden while the jet shows brightness changes. An alternative is if jets contain two fluids, a relativistic core inside a non-relativistic jet (Pelletier and Roland 1989). The non-relativistic jet is not normally visible but carries most of the mass and gives rise to the outer lobes and hot-spots seen in FR II sources (Pelletier and Roland 1990).

3) A twin-sided jet exists but we do not see a counter-jet if Doppler beaming of radiation from a fast moving jet results in the boosting of the brightness of the jet coming towards us by a factor greater than the dynamic range of the map. In a simple jet model the brightness is boosted by the factor $D^{2+\alpha}$ where α is the spectral index and D the Doppler factor, $D = (\Gamma(1 - \beta\cos(\theta)))^{-1}$. The fraction of sources with a ratio of jet to counter-jet brightness of less than J is given by $f = \cos(\theta)$ where

$$\beta\cos(\theta) = (k-1)/(k+1) \qquad (1)$$

and $k = J^{1/(2+\alpha)}$. A random, uniform distribution of angles of the jets to the line of sight is assumed. Low values for f are found for high Lorentz factors Γ. A further reduction in the fraction of sources with low J will occur if sources are selected preferentially towards the line of sight.

2.2 Observational Selection

VLBI observations are limited by sensitivity requirements to the investigation of high brightness sources, and therefore a collection of objects suitable for study by VLBI is likely to contain a high proportion of sources which are Doppler beamed. Counter-jets are thus less likely to be found in such a sample, given the usual level of dynamic range available in the maps.

A further selection may occur because the generation of artificial counter-jets in the mapping process is *too easy*. Nearly all VLBI maps are made by the hybrid mapping process (Readhead and Wilkinson 1978) which compares the closure phase given by a model with that observed around each triangle of telescopes in the interferometer array. The model is improved iteratively until good agreement in amplitude and closure phase is acheived. The model is not well constrained by the closure phase when two telescopes in a triangle are much closer together compared with the distance to the third. This results in 'spurious symmetrisation' (Wilkinson 1990) where a spurious counter-jet component can be formed symmetrically around a bright core. Mappers take great care to avoid such features by using sophisticated starting models and using windows in the image deconvolution. It is possible that a counter-jet could be missed when arrays with small numbers of telescopes and not particularly well calibrated data are used. Arrays with larger numbers of telescopes (routinely 12-14) are being used these days, and so the loss of such real features is much less likely in recent observations. However observers are still rather sceptical about counter-jets and so confirmation by independent observations is usually needed.

3 A SEARCH FOR PC SCALE COUNTER-JETS

3.1 The Source Samples

Though there have been a large number of VLBI maps published, there have been relatively few surveys of VLBI structures made on a uniform basis. One of the most comprehensive has been the Caltech survey (Pearson and Readhead 1988, hereafter PR) which has examined the milliarcsecond structure of a complete sample of 65 radio sources having flux density at 5GHz $\geq 1.3 Jy$, declination > 35 degrees and galactic latitude $|b| > 10$ degrees. Another uniform sample is the survey of compact steep spectrum sources taken from the 3CR sample (Fanti *et al.* 1985, Spencer *et al.* 1989, Akujor *et al.* 1991) where all the sources have been mapped with VLBI. These are sources which have a projected linear size of $< 15 kpc$ and high frequency spectral index > 0.5. This sample, though restricted to one type of source, may be expected to show more sources with two-sided jets as projection effects are expected to be small (Fanti *et al.* 1990). Combined MERLIN and VLBI maps have been made of a number of CSS sources (Spencer *et al.* 1991) many of which show jets on the scale of 100's of pc.

Examination of the maps given in PR, Spencer *et al.* (1991) and other recent publications enables us to estimate the jet:counter-jet brightness. We have estimated a value or lower limit for J, the ratio of peak jet brightness ≥ 1 beamwidth away from the core to the peak counter-jet brightness or lowest contour for the PR and CSS samples. This definition is used to allow for possible jet brightening away from the core and to endeavour to measure J where the jet is optically thin. There are 4 sources in common in the two samples. Note that we are not yet able to find J for 12 extended sources with weak cores from the PR sample and for the 13 sources which do not have a clear core-jet structure. There are 16 CSS sources in the 3CR sample for which we can assign a limit or value to J out of the total of 25, 9 sources having an unknown or weak core.

There are very few counter-jets found, only 4 in the PR sample and 6 in the CSS's.

There are 6 PR sources out of the 35 listed (17%) and 3 out of 16 (19%) CSS's with J possibly less than 4 whereas we would expect 27% from equation (1) if $\alpha = -0.6$ and $\beta = 1$ and more if $\beta < 1$. There is thus a dearth of twin-sided sources on the Doppler beaming model if the sources are uniformly distributed in θ. However Doppler beaming increases the core flux density of sources and this causes the resulting distribution of angles to the line of sight to be biased away from the plane of the sky, especially in samples selected at high frequencies where the core dominates. At 5 GHz there are very few sources expected to have $\theta > 20°$ if the Lorentz factor $\Gamma = 5$ (Murphy 1988) but we do not expect the 3C sample, selected at 178 MHz and therefore dominated by steep spectrum extended emission (the lobes), to be biased in this way.

The rarity of PR sources with counter-jets is therefore not unexpected on the beaming model, but the relative lack of counter-jets in the CSS sample is at first sight more surprising, particularly in view of the apparent lack of significant projection effects in the sample. Comparison of the size distribution of CSS sources with that of larger sources from the 3C sample suggests that the CSS are not biased towards the line of sight and so the CSS are intrinsically small (Fanti *et al.* 1990).

Can we reconcile this with the observed rarity of counter-jets in the 3C CSS ? The expected fraction, 27%, results since the sources with an angle greater than the corresponding θ (75°) have $J < 4$. We need 30% less sources at $\theta > 75°$ than if they were distributed uniformly to get agreement with the observations. Sources are reduced in projected size by only 3.4% at this angle and it is not possible to see such a small effect in the source size distribution. The selection of CSS by projected linear size from the 3CR therefore only slightly biases the sample to smaller angles to the line of sight, but nevertheless this is sufficient to reduce the number of twin-sided VLBI jets seen.

3.2 Comments on Individual Sources

0429+415, 3C119, $J = 200$. This probable quasar has a highly distorted spiral structure (Fanti *et al.* 1986, Nan Rendong *et al.* 1991) with a projected linear size of 500 pc if the redshift is 0.41. Elongated emission appears on both sides of a probable flat spectrum core. Further observations are needed to confirm the core.

0518+165, 3C138, $J = 16$, figure (1). VLBI maps of this quasar presented by Fanti *et al.* (1989) clearly show the presence of a compact core and main jet leading to the E. At higher resolution the core has a counter-jet pointing to the W. A MERLIN map at $\lambda 18cm$ shows an extended feature $\sim 600pc$ long as does a $\lambda 6cm$ map. This may be a lobe or a continuation of the counter-jet. The radiative lifetime of electrons in the hot spot at the end of the E jet is less than the light travel time from the core and so a continuing supply of energy is required, making the flip-flop mechanism an unlikely explanation for this source.

0538+498, 3C147, $J = 8$. This quasar has been extensively studied with MERLIN and VLBI. A recent map by Zhang *et al.* (1991) shows radio emission on both sides of the core, the northern plume or jet resembling that in 3C48. This component satisfies the Bridle criteria for a jet (four times longer than wide) but does not possess the high brightness narrow features seen in the southern jet.

0710+439, $J \sim 1$. This compact radio galaxy has a Giga-hertz peaked spectrum. VLBI maps by PR and by Conway *et al.* (1990) show a triple structure. Both maps show that the central component has a central peak with elongated structures on each side. Futher observations are needed to confirm that the peak corresponds to the core of the source and is not just a feature in a one-sided jet.

0923+392, 4C39.25, $J = 10$. Zhang *et al.* 1990 show a $\lambda 6cm$ map of this well studied quasar. They find weak components 'd' and 'e' to the west of the brightest feature and they suggest that component 'd' may be the core, in which case the source is twin-sided with 'e' being the counter-jet. Superluminal motion is found in the bright feature and MERLIN and VLA maps show a twin-sided extended structure. The twin-sided nature of the VLBI source needs confirmation, as the site of the core remains in doubt.

1003+351, 3C236, $J \sim 1$. This giant FR II radio galaxy (Barthel *et al.* 1985) has a steep spectrum core region 2 kpc in extent containing half the luminosity of the source. World array VLBI measurements (Schilizzi *et al.* 1988) showed that the central region has complex structure with a compact core and a two-sided jet. The source has a luminosity of $P_{1.4} = 10^{25.54} W Hz^{-1}$, close to that of the FR I/FR II divide.

1328+307, 3C286, $J = 2.8$, figure (2). This high luminosity quasar has a low brightness jet extending 11 kpc to the SW of the VLBI core and a component on the opposite side of the core 3.4 kpc away (Spencer *et al.* 1989, Hines *et al.* 1989). A joint MERLIN and VLBI map at $\lambda 6cm$ shows a VLBI jet 290 pc long in the same position angle as the kpc jet (Spencer *et al.* 1991) and a possible counter-jet $\sim 100pc$ long. Further observations have been made recently in order to confirm this counter-jet and the data are still being analysed.

1517+204, 3C318, $J = 1.6$, figure (3). Combined MERLIN and VLBI data show this CSS galaxy to have complex structure (Spencer *et al.* 1991). If component K2 is the

Fig. 2 3C286

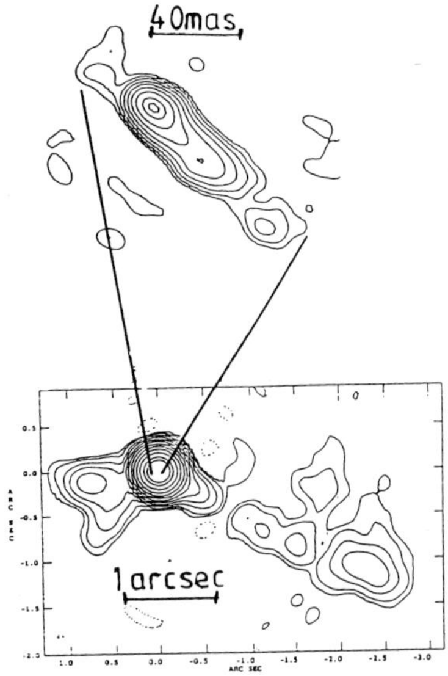

Fig.1 3C138

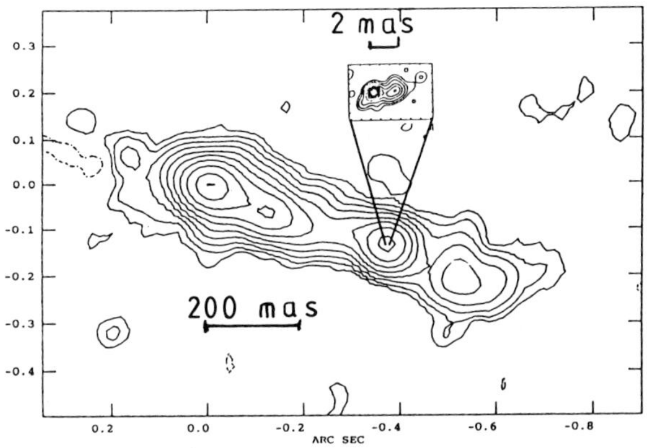

Fig.4 3C454

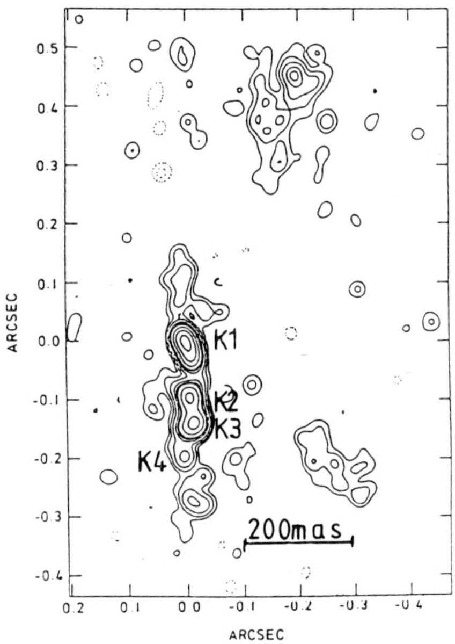

Fig.3 3C318

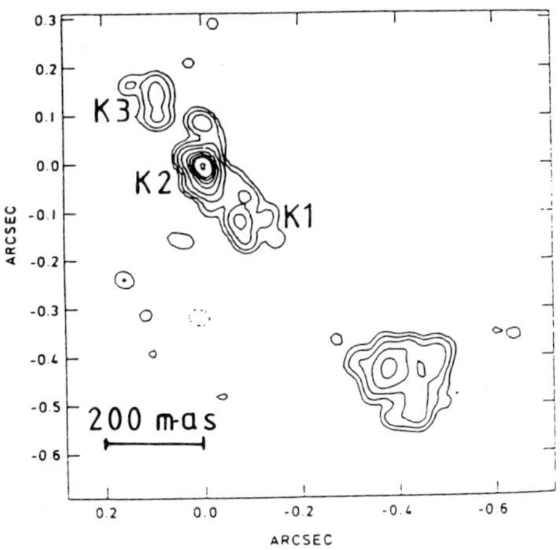

core then the source has a two-sided structure. Further observations are needed to find the spectrum of K2 and to confirm that it is the core.

2249+185, 3C454, $J = 7.7$. This CSS quasar also has a complex structure (Spencer *et al.* 1991). Component K1 may be the core, in which case the source has a two-sided jet on VLBI scales, however the identification of the core remains uncertain and awaits further **data (Figure 4).**

3.3 Other Samples

As seen above orientation bias in some samples may prevent us from finding two-sided sources if they are Doppler beamed. Attempts to remove this bias have been made by studying lobe dominated sources, where we expect that the lobe velocities are small ($< 0.2c$) and their Doppler boosting small. Hough and Readhead (1987,1989) are studying a sample of 28 double lobed quasars and VLBI measurements of a sub-sample of 6 have been made. None of the sources show evidence for two-sided pc scale jets, and two of them (3C245 and 3C263) show superluminal motion suggesting that their cores at least are aligned towards the line of sight. Zensus and Porcas (1987) are studying a similar sample, selected from the Jodrell Bank 966 MHz survey. A sub-sample of 6 sources with sufficiently strong cores to be studied by VLBI again show no two-sided jets.

In the unified scheme of Barthel (1989) and others, quasars are oriented more towards the line of sight than are galaxies, and a degree of selection bias may still be present in the quasar samples. Giovannini *et al.* (1990) are studying a sample of 26 radio galaxies, several of which are low luminosity FR I types and therefore more likely to have two-sided jets. VLBI cores have been found in 15 out of 16 galaxies. Maps of 5 of them have been obtained so far. Only one, 3C338 shows a twin-jetted core with symmetric structure on both kpc and pc scales (Giovannini, private communication).

The sensitivity limitation of VLBI inevitably leads to the selection of sources with bright cores and this may give rise to sufficient orientation bias to prevent us detecting many twin-jetted sources even in lobe-selected samples. This situation will improve as more sensitive VLBI techniques become available.

4 SOURCES WITH TWO-SIDED KPC STRUCTURE AND PC SCALE JETS

Examination of VLA and MERLIN maps of sources from the PR sample reveals 20 core-jet sources with two-sided extended emission (see references in PR). This shows that some form of energy flow must occur on both sides of the source either simultaneously or alternately as in the flip-flop mechanism, and that intrinsically one-sided jets are uncommon if they occur at all.

Two compact sources mentioned by Bridle (1990) as having counter-jets are 3C345 which appears on low dynamic range maps to be one-sided, but a very high dynamic range VLA map (Kollgaard *et al.* 1989) shows a counter-jet at 0.01% of the peak brightness on the arcsec scale leading to a low brightness lobe, and 1928+738 which shows a clear two-sided jet on the kpc scale (Johnston *et al.* 1987) but has no counter-jet visible in VLBI observations. This may be an example of a restarting jet, though superluminal motion has been seen in the VLBI core and Doppler beaming can explain the lack of a VLBI counter-jet.

Recent observations of the OVV source 3C446 (Fejes *et al.* 1991) with MERLIN and VLBI have shown a North pointing extension on the opposite side of the core to a 320 mas long curved jet, again showing that two-sided emission occurs in a source likely to contain a relativistic jet.

5 CONCLUSION

A search of the literature has shown that two-sided VLBI jets do exist, though rare. Only 3 sources (0710+439, 3C236, 3C338) have been found with symmetric jets similar to those on the kpc scale found in low luminosity FR I sources. Eight sources are found to have asymmetric twin-jets, their luminosities at 1.4 GHz are $\geq 10^{26.5} WHz^{-1}$ whereas 3C236 and 3C338 have much lower luminosities ($10^{25.5}$ and $10^{24.6} WHz^{-1}$ respectively). The success rate of finding pc scale counter-jets seems to be higher in low luminosity sources. It is likely that more two-sided jets will be found in the weak cores of extended sources. This requires large VLBI arrays of high sensitivity; a job for the VLBA.

The proportion of sources with counter-jets is consistent with the Doppler beaming model if the source surveys are subject to orientation bias. The 3CR CSS sample is found to possess a small amount of orientation bias even though the observed linear size distribution does not have significant projection effects. There is therefore no compelling reason for believing that intrinsically one-sided jets exist.

The existence of two-sided extended structure around many VLBI jet sources also suggests that jets are two-sided, but we cannot unequivocally distinguish between the Doppler beaming or restarting jet models. Doppler boosting undoubtedly occurs in sources where superluminal motion is seen and it is possible to explain pieces of counter-jets seen on kpc scale as where the jet slows down. Alternatively kpc scale jets appear where electron-positron jets convert into electron-proton jets, as discussed in this conference. More observations which can tell us how jets on the pc scale join up with the kpc scale jets are needed, and this requires high sensitivity combined array measurements with 10 to 100 mas resolution.

REFERENCES

Akujor, C.E., Spencer, R.E., Zhang, F.J., Davis, R.J., Browne, I.W.A., and Fanti, C., 1991, *Mon. Not. R. astr. Soc.*, **250**, 215.

Barthel, P.D., 1989, *Astrophys. J.*, **336**, 606.

Barthel, P.D., Schilizzi, R.T, Miley, G.K., Jägers, W.J. and Strom, R.G.' 1985, *Astron. Astrophys.*, **148**, 243.

Blandford, R.D. and Rees, M.J. 1974, *Mon. Not. R. astr. Soc.* **169**, 395.

Bridle, A.H., 1990, in *"Parsec-scale Radio Jets"* eds Zensus, J.A. and Pearson, T.J., p. 186, Cambridge University Press.

Bridle, A.H. and Perley, R.A. 1984, *Ann. Rev. Astr. Ap.*, **22**, 319.

Clarke, D.A. and Burns, J.O., 1991, *Astrophys. J.*, **369**, 308.

Conway, J.E., Pearson, T.J. Readhead, A.C.S. and Unwin, S.C., 1990 in *"Parsec-Scale Radio Jets"*, eds. Zensus, J.A. and Pearson, T.J. p.167, Cambridge University Press, Cambridge.

Fanti, C., Fanti, R., Parma, P., Schilizzi, R.T. and van Breugel, W.J.M., 1985 *Astr. Astrophys.* **143**, 292.

Fanti, C., Fanti, R., Schilizzi, R.T., Spencer, R.E. and van Breugel, W.J.M. 1986, *Astron. Astrophys.*, **170**, 10.

Fanti, C., Fanti, R., Parma, P., Venturi, T. Schilizzi, R.T., Nan Rendong, Spencer, R.E., Muxlow, T.W.B. and van Breugel, W.J.M., 1989 *Astron. Astrophys.*, **217**, 44.

Fanti, R., Fanti, C.,Schilizzi, R.T., Spencer, R.E., Nan Rendong, Parma, P., van-Breugel, W.J.M and Venturi, T., 1990, *Astr. Astrophys.* **231**, 333.

Fejes, I., Porcas, R.W.P. and Akujor C.E.,1991 *Astron. Astrophys.* in the press.

Giovannini, G., Feretti, L. and Comoretto, G., 1990, *Astrophys. J.*, **358**, 159.

Hines, D.C., Owen, F.N. and Eilek, J.A. *Astrophys. J.* **347**, 713.

Hough, D.H. and Readhead, A.C.S., 1987, in *"Superluminal Radio Sources"*, eds Zensus, J.A. and Pearson, T.J. p. 114, Cambridge University Press, Cambridge.

Hough, D.H. and Readhead, A.C.S., 1989, *Astron. J.*, **98**, 1208.

Johnston, K.J., Simon, R.S., Eckart, A., Biermann, P., Schalinski, C., Witzel, A. and Strom, R.G., 1987, *Astrophys. J.* **313**, 285.

Kollgaard, R.I., Wardle, J.F.C. and Roberts, D.H., 1989, *Astron. J.* **97**, 1550.

Leahy, J.P. 1991, in *"Beams and Jets in Astrophysics"* ed. Hughes, P.A.,p. 100, Cambridge University Press.

Lind K.R. and Blandford R.D. 1985, *Astrophys. J.* **295**, 358.

Longair, M.S., Ryle, M. and Scheuer, P.A.G., 1973, *Mon. Not. R. astr. Soc.* **164**, 243.

Murphy, D.W.,1988, PhD Thesis, University of Manchester.
Parma, P. de Ruiter, H.R, Fanti, C. and Fanti, R., 1986, *Astron. Astrophys.* **64**, 135.
Nan Rendong, Schilizzi, R.T., van Breugel, W.J.M., Fanti, C., Fanti, R., Muxlow, T.W.B. and Spencer, R.E., 1991, *Astron. Astrophys.* **245**, 449.
Pearson T.J. and Readhead, A.C.S., 1988 *Astrophys. J.* **328**, 114 (PR).
Pelletier, G. and Roland, J., 1989, *Ast. Astrophys.* **224**, 24.
Pelletier, G. and Roland, J., 1990, in *"Parsec-scale Radio jets"*, p. 323, Cambridge University Press, Cambridge.
Readhead, A.C.S. and Wilkinson, P.N.W., 1978, *Astrophys. J.* **223**, 25.
Rudnick, L. 1982 in *"Extragalactic Radio Sources"*, IAU Symp. 97, eds. Heeschen, D.S. and Wade, C.M., p. 47, Reidel, Dordrecht.
Schilizzi, R.T., Skillman, E.D., Miley, G.K., Barthel, P.D., Benson, J.M. and Muxlow, T.W.B. 1988, in *"The Impact of VLBI on Astrophysics and Geophysics"*, IAU Symp. 129, eds Reid, M.J and Moran, J.M., p. 127, Kluwer, Dordrecht.
Spencer, R.E., McDowell, J.C., Charlesworth, M., Fanti, C., Parma, P. and Peacock, J.A., 1989, *Mon. Not. R. astr. Soc.* **240**, 657.
Spencer, R.E., Schilizzi, R.T., Fanti, C., Fanti, R., Parma, P. van Breugel, W.J.M. Venturi, T., Muxlow, T.W.B. and Nan Rendong. 1991. *Mon. Not. R. astr. Soc.* **250**, 225.
Wilkinson, P.N., 1990, in *"Very Long Baseline Interferometry, Techniques and Applications"*, eds Felli, M. and Spencer, R.E., p. 69, Kluwer, Dordrecht.
Zensus, J.A. and Porcas, R.W. 1987, in *"Superluminal Radio Sources"*, eds Zensus, J.A. and Pearson, T.J., p. 126, Cambridge University Press, Cambridge.
Zhang, F.J., Akujor, C.E., Chu, H.S., Mutel, R.L., Spencer, R.E., Wilkinson, P.N., Alef, W. Matveyenko, L.I. and Preuss, E., 1991, *Mon. Not. R. astr. Soc.*, **250**, 650.
Zhang, Y.F., Marscher, A.P., Shaffer, D.B., Marcaide, J.M. Alberdi, A. and Elósequi, P., 1990, in *"Parsec-Scale Radio Sources"*, eds Zensus, J.A. and Pearson T.J. p. 66, Cambridge University Press, Cambridge.

VLBI Tests of BL Lac Models

R.L. MUTEL

University of Iowa, Iowa City, IA

1 INTRODUCTION

At a recent meeting on parsec scale radio structure, I presented a summary of all VLB maps of BL Lac objects known to me as of late 1989 (Mutel 1990). In this paper I consider how VLB observations can be used to test various models for BL Lac objects, such as the identity of an unbeamed parent population of radio sources, the importance of gravitational micro-lensing, and intraday flux variability.

2 WHAT IS THE PARENT POPULATION OF BL LACS?

Many of the observed properties of BL Lac objects (high variability, high apparent luminosity, high optical and radio polarization, superluminal radio components) are most easily explained in terms of opposed relativistic jets aligned close to the observer's line of sight. This implies the existence of a 'parent' population of misaligned objects which are intrinsically identical to BL Lac objects but misdirected with respect to the observer. Several such parent populations have been suggested, including radio quiet quasars (Komberg 1991), radio loud quasars (Vagnetti & al. 1991), and Fanaroff-Riley I (FR-I) galaxies (Wardle, Moore, and Angel 1984; Browne 1989; Ulrich 1989).

Padovani, Urry and colleagues have recently published a series of papers which careful analyzes the FR-I/BL Lac unification hypothesis (Padovani & Urry 1990, 1991; Urry, Padovani, & Stickel 1991), including the effects of beaming on number density in flux-limited samples. They conclude that FR-I galaxies could be the parent population if the beaming angle (or equivalently the Lorentz factor) of the relativistic jet is frequency dependent. They find that the x-ray emission is characterized by a relatively low Lorentz factor ($\gamma_X \sim 3$) while the optical and radio Lorentz factors are much higher ($8 < \gamma_{opt} < 20$, $\gamma_{rad} \sim 7$). Unfortunately, there is as yet no physical model of relativistic jets which predicts such wavelength dependent Lorentz factors. In addition, jet models based on direct VLBI measurements of BL Lacs find radio Lorentz factors much lower, usually in the range $2 < \gamma_{rad} < 4$ (e.g. Hughes & al. 1989, 1991; Mutel & al. 1990).

The FR-I/BL Lac unification scheme could be tested by measuring the parsec scale structure and apparent speeds of jet components for a sample of FRI sources. One expects that FR-I radio cores to have somewhat larger jets but smaller average component speeds. Unfortunately, FR-I radio cores are very weak (typically $S < 100$ mJy at 5 GHz) and so VLB mapping is difficult with present arrays. Giovannini & al. (1990) have obtained snapshot VLB images of 16 FR-I galaxies, but with inadequate sensitivity to make high dynamic range maps.

3 ARE BL LACS MICRO-LENSED?

Ostriker and Vietri (1985, 1990) have proposed that the low redshift ($z \leq 0.2$) BL Lacs found in the center of nearby elliptical galaxies may be distant quasars which have been gravitationally lensed by a star in the nucleus of the intervening galaxy. Alternatively, Nottale (1986) suggests that high redshift ($z \geq 0.5$) BL Lacs may be non-variable quasars which are being micro-lensed by stellar components in intervening galaxies which are detected by absorption redshifts in the continuum emission from the background quasar. Recent deep CCD imaging (Stickel 1991) has produced evidence for candidate lensing galaxies along the lines of sight toward four BL Lac objects (0235+164, 0537-441, 0846+513, 1308+326), all of which are high redshift and support the Nottale scenario. There is less compelling evidence for lensing in low redshift BL Lacs, but Falomo & al. (1990) find close companions near the BL Lacs 0323+02 ($z = 0.147$) and 2351-31 ($z = 0.17$) which they suggest may be lensing galaxies.

There are several arguments against microlensing which have appeared in the past couple of years. Probably the strongest is the systematic difference in polarization position angle observed in the jet of BL Lacs and quasars found from polarization VLBI maps (Gabuzda & al. 1989, 1991). They find that for eleven BL Lacs the inferred magnetic field is preferentially perpendicular to the jet axis compared with quasars which have magnetic fields aligned with the jet. Since gravitational lensing cannot systematically rotate the polarization position angle, it implies that the BL Lacs are intrinsically different from quasars. Narayan & Schneider (1990) argue that the absence of multiple images implies an unusually high core density in the lensing galaxies if the absorption redshifts are really that of the lens. Gear (1991) argues that the similarity in broadband spectra between BL Lacs and OVV quasars, along with the rapid time variability of the optical light implies a common region of size $\leq 10^{-3}$ pc. However, the VLBI maps of BL Lacs shown components with widths ≥ 0.1 pc (e.g. Mutel 1990 and references quoted.) which is much too large for microlensing.

A more indirect argument against the microlensing hypothesis is the lack of different v/c distributions for low and high redshift BL Lacs. VLBI images of BL Lacs may be distorted by microlensing so that derived superluminal speeds of jet components are quite different than intrinsic speeds in the parent object. One would expect a large variation in apparent speeds in the lensed objects compared with unlensed objects. However, aside from the extraordinary measurement of $v/c = 45$ for 0235+164 (Bååth 1984), the v/c values for BL Lacs do not seem to depend strongly on redshift (e.g. Table 3, Mutel 1990; also see Gabuzda & al. 1991 for additional v/c values).

4 WHAT IS THE CAUSE OF INTRADAY VARIABILITY?

At centimeter wavelengths a number of active galactic nuclei, including several BL Lac objects, have been shown to vary significantly (up to 25%) on a timescale of $\sim$1 day, implying spectacular brightness temperatures up to $T_B \sim 10^{19}$ K (e.g. Heeschen, & al. 1987; Quirrenbach & al. 1989; Quirrenbach & al. 1991). It is not yet clear whether the variability is intrinsic or a propagation effect. The latter could result from refractive scintillation (RISS) caused by large lens-like density structures perhaps similar to those deduced to explain the extreme scattering events (ESE) reported by Fiedler & al. (1987). Wambsganss & al. (1991) has recently modelled the rapid variability of the quasar 0917+624 using a thin screen and gaussian electron density profiles with a maximum density $n \sim 300$ cm^{-3}. A VLBI test of the RISS hypothesis would involve looking for correlated variations in apparent source size with flux as predicted by Romani & al. (1986) or angular wander of the centroid of the scattering disk (e.g. Cordes & al. 1986). However, there is some evidence that the intraday variations may be correlated between radio and optical wavelengths (Quirrenbach & al. 1991), which probably rules out RISS since interstellar plasma turbulence will not affect optical radiation.

Alternatively, gravitational microlensing (q.v.) would produce wavelength independent intensity variations as the lensing object (star?) passes the line of sight. However, the short variability timescale and multifrequency radio light curves do not appear to be consistent, since the transverse velocity of the lensing star would be $v \sim c$ (Wagner 1991).

5 REFERENCES

Bååth, L., 1984, IAU Symposium 110, VLBI and Compact Radio Sources, ed. R. Fanti, K. Kellermann & G. Setti, Reidel, p 127

Browne, I., 1989, Lecture Notes in Physics, Vol. 334: BL Lac Objects, ed. L. Maraschi, T. Maccacaro & M.H. Ulrich, Springer–Verlag, p 401

Cordes, J., Pidwerbetsky, A. & Lovelace, R., 1986, Astrophys. J. **310**, 737

Falomo, R., Melnick, J. & Tanzi, G., 1990, Nature, **345**, 692

Fiedler, R., Dennison, B., Johnston, K. & Hewish, A., 1987, Nature, **326**, 675

Gabuzda, D., Cawthorne, T., Roberts, D. & Wardle, J., 1989, Lecture Notes in

Physics, Vol. 334: BL Lac Objects, ed. L. Maraschi, T. Maccacaro & M.H. Ulrich, Springer–Verlag, p 22

Gabuzda, D., Cawthorne, T., Roberts, D. & Wardle, J., 1991, *submitted to Ap.J.*

Gear, W., 1991, Nature, **349**, 676

Giovannini, G., Feretti, L, & Comoretto, G., 1990, Astrophys. J. **358**, 159

Heeschen, D.S., Krichbaum, T., Schalinski, C., & Witzel, A., 1987, Astron. J. **94**, 1493

Hughes, P., Aller, H., & Aller, M., 1989, Astrophys. J. **341**, 68

Hughes, P., Aller, H., & Aller, M., 1991, Astrophys. J. **374**, 57

Komberg, B., 1991, *preprint*

Mutel, R.L., 1990, Parsec Scale Radio Jets, ed. J. Zensus & T. Pearson, Cambridge University Press, p 98

Mutel, R.L., Phillips, R.B., Su, B., & Bucciferro, R.B., 1990, Astrophys. J. **352**, 81

Narayan, R. & Schneider, P., 1990, Monthly Notices Roy. Astron. Soc., **243**, 192

Nottale,. L., 1986, Astron. Astrophys, **157**, 383

Ostriker, J. & Vietri, 1985, Nature, **318**, 446

Ostriker, J. & Vietri, 1990, Nature, **344**, 45

Padovani, P. & Urry, C.M., 1990, Astrophys. J. **356**, 75

Padovani, P. & Urry, C.M., 1991, Astrophys. J. **368**, 373

Quirrenbach, A. & al., 1989, Nature, **337**, 442

Quirrenbach, A. & al., 1991, Astrophys. J. **372**, L71

Romani, R., Narayan, R., & Blandford, R., 1986, Monthly Notices Roy. Astron. Soc. **220**, 19

Stickel, M., 1991, Lecture Notes in Physics, Vol. 377: Variability of Active Galaxies, ed. W. Duschl, S. Wagner & M. Camenzind, Springer–Verlag, p 303

Ulrich M.-H., 1989, Lecture Notes in Physics, Vol. 334: BL Lac Objects, ed. L. Maraschi, T. Maccacaro & M.H. Ulrich, Springer–Verlag, p 45

Urry, C.M., Padovani, P., & Stickel, M., 1991, Astrophys. J. **382**, 501

Vagnetti, F., Giallongo, E., & Cavaliere, A., 1991, Astrophys. J. **368**, 366

Wagner, S., 1991, Lecture Notes in Physics, Vol. 377: Variability of Active Galaxies, ed. W. Duschl, S. Wagner & M. Camenzind, Springer–Verlag, p 163

Wambsganss, J., Schneider, P., Quirrenbach, A., & Witzel, A., 1991, *submitted to Astron. Astrophys.*

Wardle, J. Moore, R. & Angel, J., 1984, Astrophys. J. **279**, 93

High Resolution Images of Optically Quiet Quasars

CHIDI E. AKUJOR[1,2,3] & RICHARD W. PORCAS[4]

[1] University of Manchester, Nuffield Radio Astronomy Laboratories, Jodrell Bank, Macclesfield, Cheshire SK11 9DL, England

[2] Astrophysics Group, Department of Physics and Astronomy, University of Nigeria, Nsukka, Nigeria

[3] Onsala Space Observatory, S–439 00 Onsala, Sweden

[4] Max Planck Institut fur Radioastronomie, Auf dem Hugel 69, Bonn, F.R.G.

ABSTRACT
We have embarked on a study of 'optically quiet quasars' (OQQs), in order to clarify the nature of these sources and their relationship with other types of EGRSs. VLBI observations of four OQQs were made in 1984 and followed with recent observations of five sources. These sources have either mas core–jet or compact double structures. An earlier suggestion of a link between the absence of strong optical activity, in *bonafide* OQQs and existence of compact double radio structure remains an attractive possibility.

1 INTRODUCTION

The nuclei of galaxies are classified as active for various reasons: they may contain compact variable radio sources, they may be strong IR/optical/X–ray continuum sources, or they may have strong emission lines. Often, but not always, these properties go together as in radio loud quasars. Much attention has been paid to radio quiet quasars and optically bright radio sources, but very little attention has been paid to the **complementary** objects which have strong flat spectrum radio cores but very little or no optical emission. We refer to such objects with 'empty field' identifications on PSS prints as 'optically quiet quasars' (OQQs) because they resemble quasars in radio structure and spectrum but are faint optically.

In order to investigate the nature of OQQs and the relationship (or lack of it) between this apparent lack of nuclear activity and compact radio structure, pilot VLBI observations were made in 1984 (Zensus & Porcas, 1985) of a set of four OQQs. This was followed with another set of observations in 1989 (Akujor & Porcas, 1991, in prep.), in which five sources were observed with the global network at 5 GHz (1 mas beam).

2 RESULTS

0010+408 was thought to be an empty field object, but a more recent identification suggests an 18mag QSO ? (Patnaik *et al.*, 1991). A VLA map at 8.4 GHz (Patnaik *et al.*, 1991) shows that it has a secondary component in PA 45°. Our map (fig 1a) shows a core jet structure. The mas jet points to the secondary component in the VLA map.
0102+480 is in an empty field (EF) and has no large–scale structure. The radio spectrum suggests a possible GPS source. The mas structure (fig. 1b) shows a core and an apparently curving jet.
0406+121 is in an EF. There is an indication of a jet pointing northwards (fig. 1c).
0646+600 was thought to be an EF object but later identified with a 19mag BSO (Meisenheimer & Roser, 1983) and is classified as a GPS source. It is unresolved with the VLA and has no large–scale extensions. Our map (see fig 1d) shows two bright components separated by about 3.0 mas. The Northern component appears to be extended towards the second component. We do not have any redshift for this source, but a $z \sim 1$ would imply a size ~ 10 pc, making this source possibly among the smallest doubles so far observed.
1044+719: A more recent identification with a 19° QSO has been made (Patnaik *et al.*, 1991). Our map (fig. 1e) indicates a core and a jet pointing directly westwards.

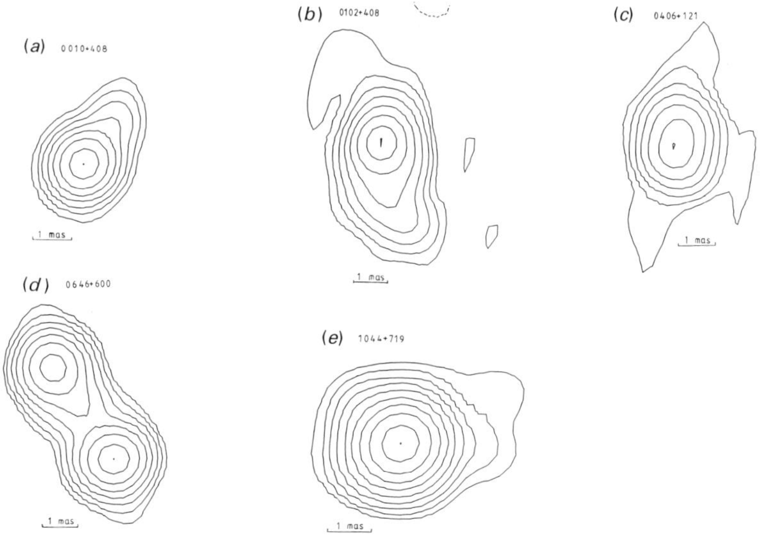

3 DISCUSSION

Four sources in the present list have 'core–jet' structures similar to those of other flat spectrum core–dominated quasars (Pearson & Readhead 1988) and three of these have been identified with faint quasars. One object (0646+600) which is also identified with a faint quasar has a compact double structure similar to the 'compact doubles' (Mutel, Hodges & Phillips 1985) usually found among galaxies. In the first experiment two compact doubles were found among four sources.

The incidence of core–jet structures as in other flat–spectrum sources is consistent with relativistic beaming theories (Orr & Browne 1982). Some OQQs may be 'normal' quasars at very high redshift or those whose identification had earlier been missed. But *bonafide* OQQs may exist and may be related with the incidence of compact double structure. OQQs may be a population of young radio sources in early stages of their evolution and could become stronger in both optical and radio at a future epoch. This scenario is thought to apply to 'compact doubles' due to their compact size and absence of central radio components and large–scale extended emission (Phillips & Mutel, 1982). They may also be active nuclei in which the optical emission is obscured by dust or a new population of radio sources whose structure is determined by such intrinsic factors as low mass of the central engine and hence lower energy transfer in the radio beams related to the absence of relativistic effects.

Indeed, with improved radio positions in the survey of MERLIN phase calibrators (Patnaik *et al.* 1991), it is found that a significant fraction of a representative sample of radio sources (153/853) can be classified as OQQs. We also find that the mean fractional polarisation of the cores of OQQs at 8.4 GHz is 1.8%, compared with that of the quasars (mean 2.4%),(Patnaik & Akujor, in prep). We are now studying a comprehensive list of the 153 OQQs with VLBI, MERLIN and the VLA, in order to clarify the nature of these sources.

References

Meisenheimer, K. & Roser, H. –J. 1983, *Astron. Astrophys. Suppl*, **51**, 41.
Mutel, R.L., Hodges, M.W. & Phillips, R.B. 1985, *Astrophys. J.*, **290**, 86.
Orr, M.J.L & Browne, I.W.A. 1982, *Mon. Not. R.A.S.*, **200**, 1067.
Pearson, T.J. & Readhead, A.C.S. 1988, *Astrophys. J.*, **328**, 114.
Patnaik *et al.* 1991, *To appear in proc. IAU **colloquium 131.***
Phillips, R.B. & Mutel, R.L. 1982, *Astron. & Astrophys.*, **106**, 21
Zensus J.A. & Porcas, R.W. 1985, in Dyson (ed.), *AGN*, p.55.

New millimetric VLBI Results

L.B.BÅÅTH

Onsala Space Observatory, S-439 00 Onsala, Sweden

1. INTRODUCTION

VLBI observations at 100GHz (λ3mm) started as early as 1981, but 1988 was the first epoch aimed for full hybrid mapping. At the same time new data reduction technique was developed to fully handle the new data (Bååth 1991, Bååth et al. 1991a). Since 1988 we have observed three epoches, in March 1988, March 1989, and April 1990. These observations involved telescopes in Sweden, Japan, and USA. The epoch of 1990 also for the first time in VLBI included the SEST telescope in Chile, and the resolution in the north-south direction was drastically improved. The resolution we can now obtain is about 40 μas (microarcsecs.) at λ3mm, resulting in linear resolution, diameter, of 10^{15} - 10^{17} cm depending on the redshift. The resolution we achieve is equivalent to that of a 2km diameter optical telescope which is diffraction limited. We intend to push the development further, to observe at even higher frequencies, e.g. 230GHz and 350GHz. Higher observing frequencies would increase the resolution to 10 μas at 350GHz, but also would enable us to look deeper into the core which may still be optically thick at 100GHz.

Observations with Very Long Baseline Interferometry (VLBI) at mm wavelengths is currently our only possibility to make images of the cores of active galactic nuclei. Such observations can give us direct insight into the core areas on size scales less than the broad line region and close to that of the accretion disc. If we coordinate our VLBI observations with measurements of changes in the continuum spectra we expect to make better models how radio jets are created and sustained and of how shocks propagate through the jet plasma.

The two epoches 1988 and 1989 have been published and discussed in two recent papers (Bååth *et al.* 1991a and Bååth *et al.* 1991b). I refer the reader to these papers for a more detailed discussion on the individual sources. All following linear sizes will be calculated assuming cosmological redshift, H_0 = 100 km sec^{-1} Mpc^{-1} and q_0 = 0.05. This paper will show some of the new results obtained in 1990.

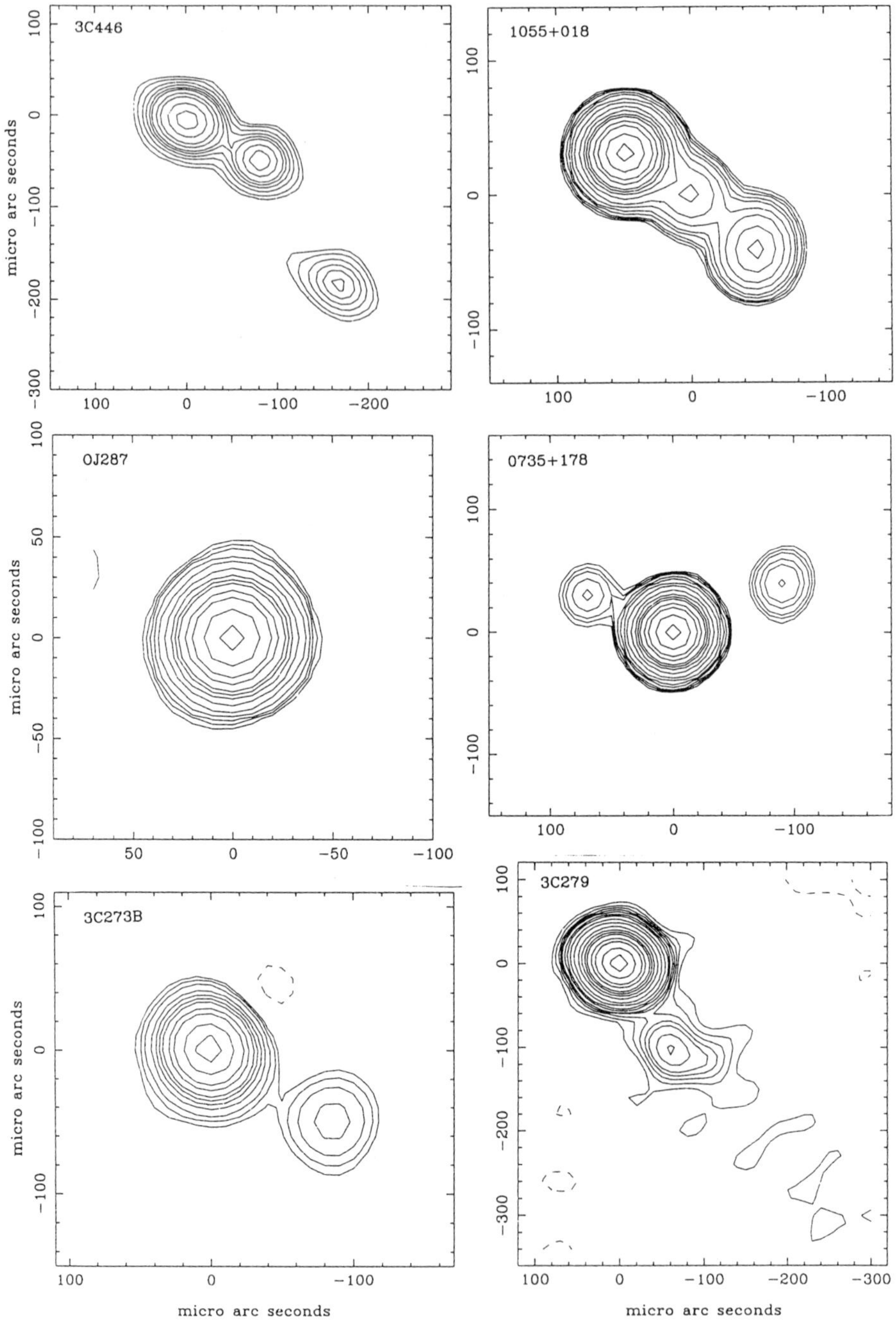

Figure 1. Examples of images from our 100GHz VLBI experiment in 1990.

2. RESULTS

3C446 was not in our original list of objects but was included when the SEST telescope was made available for mmVLBI and low declination sources could be observed. 3C446 is truly a blazar with very rapid temporal variations in continuum as well as line emission (CIII]). Our 100GHz map shows a clear jet which has to bend around by 120° in order to continue as suggested by the VLBI image at 5GHz (Wehrle, Cohen & Unwin 1990). This source was in fact one the most luminous in our sample in 1990 and only 3C273 was more powerful on the longest baselines. Our sparse 230GHz VLBI data also indicate that it could be one of our best candidates for VLBI observations at even higher frequencies.

1055+018 is another example of a jet which has to bend significantly from the inner submilliarcsecond part to the outer arcsecond structure. The VLA structure is mostly north-south while the mas-structure is more east-west (M.Bondi private comm.). Further VLBI observations are necessary in order to tie the structures at different resolutions together.

OJ287 is the most compact of all sources we have observed at 100GHz sofar and was unresolved at all three epochs. The 1990 epoch was close to an outburst, but the source still appears unresolved with a maximum linear size of 10^{16} cm. This size is of the same order as the binary system suggested by Sillanpää *et al.* (1988) It is unlikely that two diverging beams from such kind of system would not have been observed by us at any of our three epochs. OJ287 thus appears to be similar to the other compact radio sources in our sample, with the possible exception that its jets may be observed more end-on than is usual.

0735+178 is a longtime favourite of mine. This source is the original "cosmic conspiracy" in that its radio spectrum was flat from 100MHz to 100GHz for many years. In a recent paper (Bååth, Zhang & Chu 1991) we show that the recent increase in flux density was caused not by a component emerging from the core, but by an older component reaching a bend in the jet. This source also had a more recent outburst, more rapid then the one modelled by us. Our 100GHz VLBI map shows the existence of a very compact component which in time coincides with the new outburst.

3C273 is the brightest of all our objects. We observed it in 1988 just about two months after the start of a major outburst. Our 100GHz map from this epoch has been published in many places and will not be shown here. I refer the reader to Bååth *et al.* (1991) for a more comprehensive discussion. In 1988 we could identify the newly born component as being elongated across the jet and unresolved along the jet. The linear size of this new component was $(56\times5)\cdot10^{16}$ cm at this time, in good agreement with what would be expected from the model by Marscher and Gear (1985). Our map from 1990 shows the core in a more quiescent state and none of the very pronounced wiggles seen in 1988 are visible. A possible expanation for this is that the jet is more lit up in 1988 by a very powerful shock going through it. We will need another event as the outburst in 1988 to verify this. The position angle of the jet in 1990 is very similar to that of the main jet axis in 1988.

3C279 is also exceptional in our sample in that the radio jet is remarkably straight. The position angle deviates from -145° by less than 5° at all scales, going from ours at 100 μas to 10" (dePater & Perley 1983). This suggests that no significant change in viewing angle could have taken place in this source for a very long time. This contradicts the model suggested by Aller (this meeting) and favours an intrinsical change of the Lorentz factor to be responsible for the large initial outburst in this source about 20 years ago.

3C279 was the very first source in which superluminal motion was detected (Whitney *et al.* 1971). The angular expansion speed in this source was measured during the 1970's to be quite large, 500 μas/year (Cotton *et al.* 1979), while observations during the 1980's show a lower expansion rate of 120 μas/year (Unwin *et al.* 1989). Our three epochs at 100GHz indicate an expansion speed in the core of about 150 μas/year, in agreement with the speed measured at this time at cm wavelengths.

The most luminous component at the north-eastern end of the jet has a flux density in our 100 Ghz maps which corresponds to what would be expected from an extrapolation of what is observed at 10 and 22GHz. The size of this "core" component is $\leq10^{17}$ cm.

3C84 is another of our main target sources. Figure 2 shows maps of the most compact structure observed at our three epochs at 100GHz. In addition a deeper image also shows the northern ridge (Marr, Backer & Wright 1990) as more extended emission along the north-south direction. The most compact structure is dominated at all three epochs by two components, with the northern and most intense component being more elongated in 1989 then at the two other epochs. I suggest that we here observe two new components just being ejected from the core. The expansion speed from 1988-1989 is 85 ± 7 μas/year, corresponding to a linear speed of 21000 km/sec. If indeed the southern component in 1990 is a new component seen as an elongation of the core in 1989, then the speed would be the same also for this. Such speed is much lower than has been observed in other compact radio sources, and is of the same order as has been suggested for the shock created when a star moves into the beam.

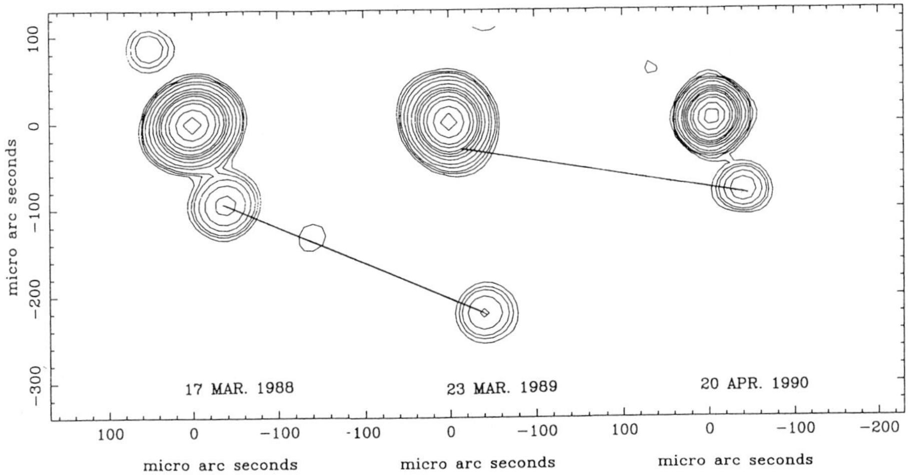

Figure 2. 100 GHz VLBI maps of the core region of 3C84 from 1988, 1989, and 1990. A restoring beam of 50μas was used at all three epochs.

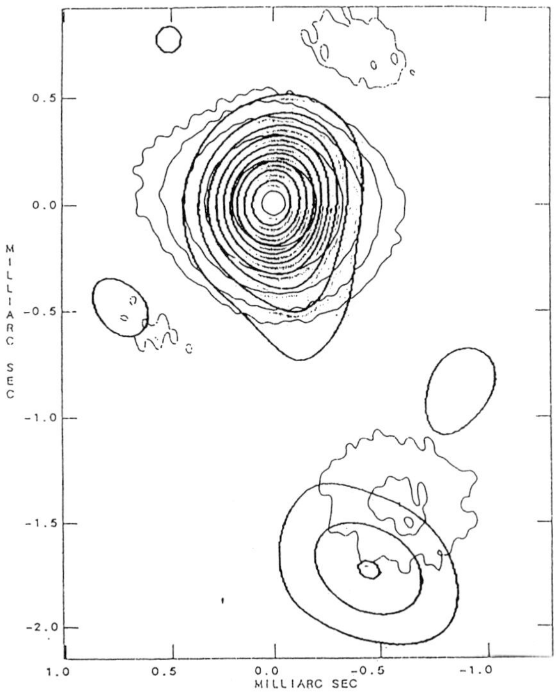

Figure 3. BL Lac observed with global VLBI arrays at 10 GHz (thick contours) and 100GHz (thin contours) plotted on the same scale and with the same restoring beam. The 10GHz map is from Mutel (private communication).

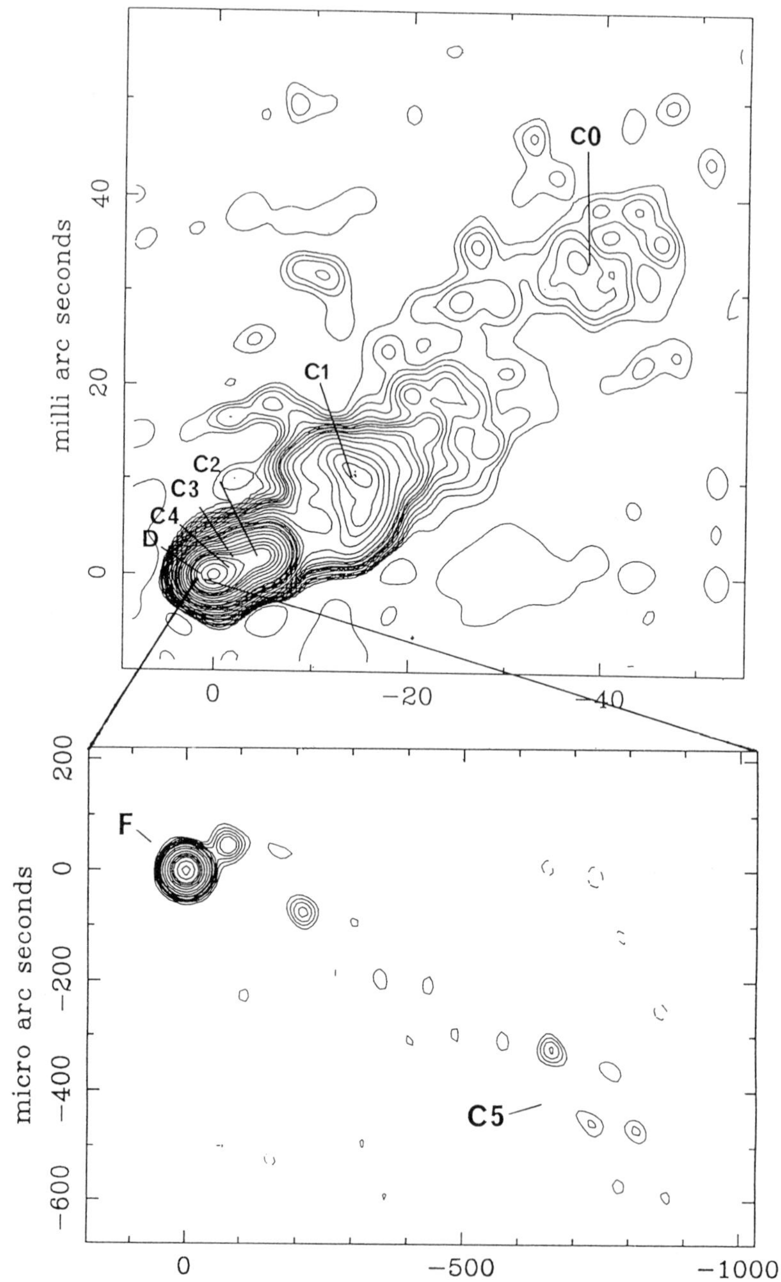

Figure 4. 3C345 observed with global VLBI arrays at 1.6GHz (upper) and 100GHz (lower). The 1.6GHz image is from Rantakyrö *et al.* (1991).

BL Lac is one of the few sources where we have been able to identify a component seen in lower resolution maps. Figure 3 shows our image from 1988 convolved down to the 10 times lower resolution as appropriate to compare with VLBI observations at 10GHz (Mutel, private communications). The outer component is somewhat closer to the core and at slightly different position angle at 100GHz than what is observed at 10GHz. We identify this component as S5, which in our 100 GHz map is at 1486±5 μas at position angle 201.6±0.5° as compared to 1950±160 μas at position angle 195.9±3.2° at 10GHz (Mutel *et al.* 1990). The position at 100GHz is very close to the position of the previous component S4 at the time when this component was stable for a few years. This place may therefore well be where a deflection of the jet, as suggested by Hugh, Aller & Aller (1989), is situated.

3C345 is perhaps the most studied of all superluminals. Figure 4 shows maps obtained with VLBI at 1.6 and 100GHz. Here all components C1-C5 have been marked. The wiggles of the jet are obvious from these images and fits well with the model suggested by Hardee (1990). Thus the jet starts to the north-east, bends to the south-east, and then continues to the north-east.

Our VLBI observations at 1.6 GHz (Rantakyrö *et al.* 1991) shows that the jet must be curved in three dimensions and probably is curling away from the line of sight outside about 5mas from the core. Further from the core, at a linear distance of about 10^{20} cm, there is a stationary component and the jet seems to expand adiabatically at this point. This distance corresponds to the radius of the narrow line region and we suggest the stationary shock to be formed when the jet passes through the narrow line region.

3. DISCUSSION

It is unlikely that we observe the central engine itself at 100GHz, even though the size scale of the most compact components is of the same order as that of the expected accretion disc. We cannot from our data alone tell the true distance between the "engine" and the start of the radio jet which we usually label the "core" for convenience. The radio structure at 100GHz would fit inside the broad line emission region if in fact the high frequency radio core is close to the true nucleus. It is quite possible though, that the radio core is instead formed in the BLR. Only VLBI images at even higher frequency, where the accretion disc also become visible to our instruments, will tell the true geometry.

The images presented here are clear evidence that VLBI observations at 100GHz now are in full operation and can produce results of very good quality. It is my hope that the next few years will see such images produced at 230 and 350GHz as well, and that I in summer 1992 can show you two epochs maps at 230GHz of some of these sources.

REFERENCES

Bååth,L.B., Padin,S., Woody,D., Rogers,A.E.E., Wright,M.C.H., Zensus,A., Kus,A.J., Backer,D.C., Booth,R.S., Carlstrom,J.E., Dickman,R.L., Emerson,D.T., Hirayabashi,H., Hodges,M.W., Inoue,M., Moran,J.M., Morimoto,M., Payne,L., Plambeck,R.L., Predmore,C.R. & Rönnäng,B.: 1991a, *Astron.Astrophys.*, **241**, L1

Bååth,L.B., Rogers,A.E.E., Inoue,M., Padin,S., Wright,M.C.H., Zensus,A., Kus,A.J., Backer,D.C., Booth,R.S., Carlstrom,J.E., Dickman,R.L., Emerson,D.T., Hirayabashi,H., Hodges,M.W., Moran,J.M., Morimoto,M., Payne,L., Plambeck,R.L., Predmore,C.R., Rönnäng,B. & Woody,D.: 1991b, *Astron.Astrophys.*, submitted

Bååth,L.B. Zhang,F.J. & Chu,H.S.: 1991, *Astron.Astrophys.*, submitted

Cotton,W.D.: 1979, *Astrophys.J.(Letters)*, **229**, L115

dePater,I. & Perley,R.A.: 1983, *Astrophys.J.*, **273**, 64

Hardee,P.E.: 1990, *Parsec scale radio jets,* Zensus and Pearson (eds.), Cambridge University Press, p266

Hughes,P.A., Aller,H.D. & Aller,M.F.: 1989, *Astrophys.J.*, **341**, 54

Marr,J.M., Backer,D.C. & Wright,M.C.H.: 1990, *Parsec scale radio jets,* Zensus and Pearson (eds.), Cambridge University Press, p78

Marscher,A. & Gear,W.: 1985, *Astrophys.J.*, **298**, 114

Moore,R.L., Readhead,A.C.S. & Bååth,L.B.: 1983, *Nature*, **306**, 44

Rantakyrö, F.T., Bååth,L.B., Pauliny-Toth,I.I.K., Matveyenko,L.I., Unwin,S.C. & Cohen,M.H.: 1991, *Astron.Astrophys.*, submitted

Sillanpää,A., Haarala,S., Valtonen,M.J., Sundelius,B. & Byrd,G.G.: 1988, *Astrophys.J.*, **325**, 628

Unwin,S.C., Cohen,M.H., Biretta,J.A., Hodges,M.W. & Zensus,J.A.: 1989, *Astrophys.J.*, **340**, 117

Wehrle,A.E., Cohen,M.H. & Unwin,S.C.: 1990, *Parsec scale radio jets,* Zensus and Pearson (eds.), Cambridge University Press, p49

A Polarimetric Perspective on BL Lacertae Objects

D.C. GABUZDA

Canadian Network for Space Research, Department of Physics and Astronomy, University of Calgary

1 INTRODUCTION

The linear polarization of the milliarcsecond-scale radio emission in active galactic nuclei (AGN) yields unique information about conditions in the parsec-scale jets in these sources. The study of BL Lacertae objects has contributed to our understanding both of individual sources (OJ 287: Roberts, Gabuzda, and Wardle 1987; Gabuzda, Wardle, and Roberts 1989a; Cawthorne and Wardle 1988; 0735+178: Gabuzda, Wardle, and Roberts 1989b) and of the relationship between BL Lacertae objects and other AGN (Gabuzda *et al.* 1989, 1991). The characteristic linear polarization structure observed in BL Lacertae objects is quite distinct from that seen in quasars and radio galaxies, suggesting that these different classes contain intrinsically different objects. This provides useful constraints both on unification schemes and on theoretical models of VLBI jets.

2 SUMMARY OF RESULTS

There are now eighteen BL Lacertae objects for which information about the linear polarization of emission on milliarcsecond scales is available (Roberts, Gabuzda, and Wardle 1987; Gabuzda, Roberts, and Wardle 1989b; Gabuzda *et al.* 1989; Gabuzda *et al.* 1991); thus far, all images have been made at $\lambda = 6$ cm. In all but three of these sources, a high fraction of I and P flux detected in the VLA core is present on scales of a few milliarcseconds. In every case in which polarized flux is detected in the VLBI jet, the electric vector is nearly parallel to the jet axis (Figure 1a), implying that the local magnetic field is nearly perpendicular to this axis. The degree of polarization in jet components ranges from $\sim 0-60\%$. Figure 2 shows an image of the BL Lacertae object 1823+568 at epoch 1989.29, which illustrates these features. In contrast, there does not appear to be any preferred orientation to the direction of polarization in the VLBI cores of these sources (Figure 1b); the degree of polarization in these components ($\sim 2-5\%$) is appreciable, but typically smaller than for jet components.

Figure 1. Histograms of the distribution of offset between polarization position angle and VLBI structural axis in a) jet components in BL Lacertae objects, b) core components in BL Lacertae objects, c) jet components in quasars, and d) core components in quasars. In each case, the horizontal axis is divided into bins of ten degrees, and the vertical axis is the number of sources in each bin.

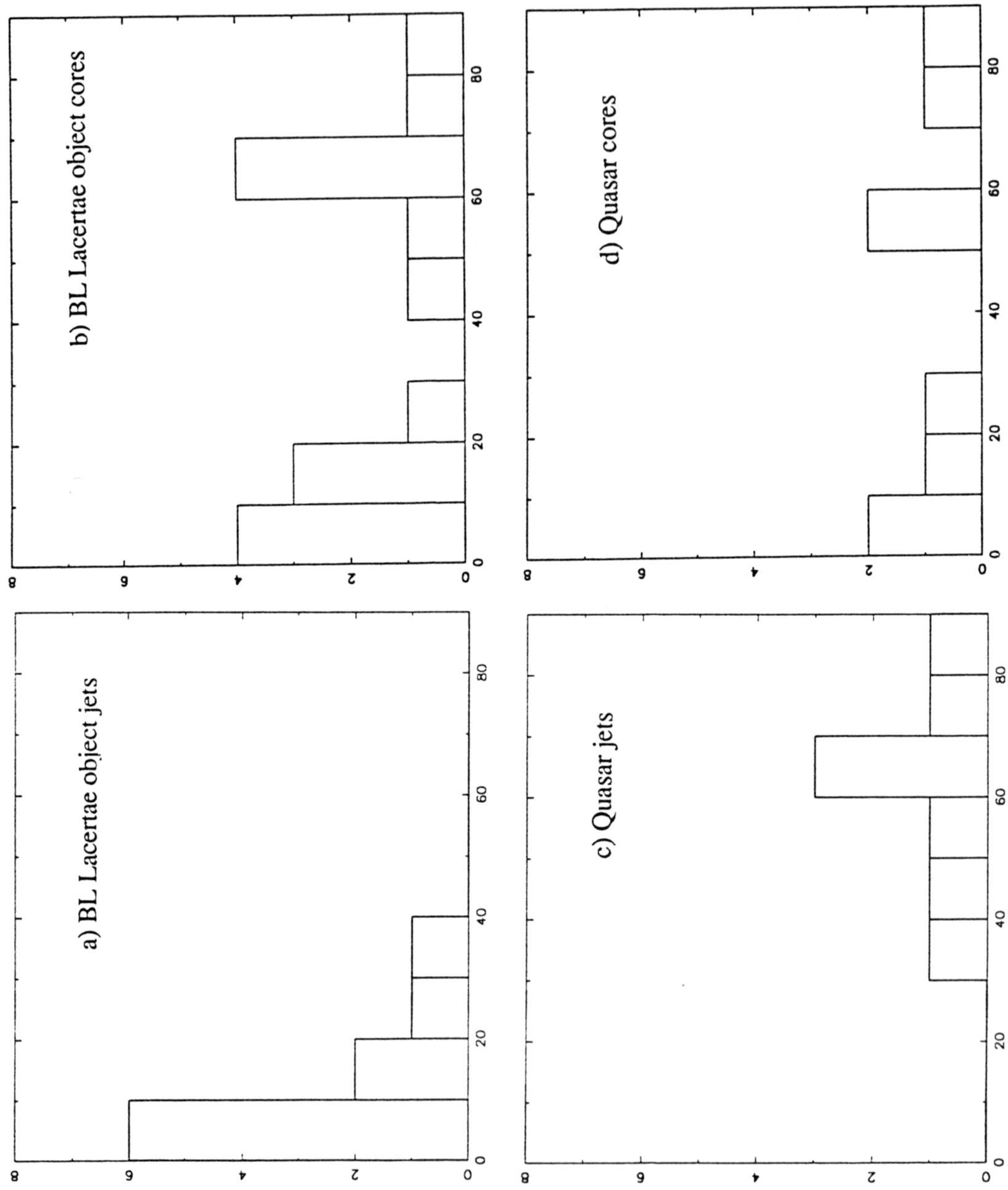

This polarization structure is quite different from that observed in quasars, which show a tendency for the polarization electric vectors to be *perpendicular* to the direction of the VLBI jet, and in which the core polarizations are rather smaller (Figure 1c; Wardle *et al.* 1986; Cawthorne *et al.* 1991; Cawthorne, these proceedings).

3 WHAT DOES THIS TELL US ABOUT BL LACERTAE OBJECTS?

3.1 Conditions in the VLBI Jets

It is striking that in every BL Lacertae object in which polarized flux has been detected in a component in the VLBI jet, the orientation of this polarization implies a magnetic field perpendicular to the direction of the jet. In contrast with this rather uniform distribution of polarization orientation, the range of *degree* of polarization observed in jet components is rather large, and both weakly and highly polarized knots are not uncommon. These features strongly suggest that these components are associated with plane shocks, in which an initially tangled magnetic field is compressed by the shock as it propagates outward from the core (Laing 1980; Hughes, Aller, and Aller 1985, 1989). In the compressed region, the component of the magnetic field perpendicular to the compression is enhanced. For such a shock, the observed orientation of the magnetic field will always be perpendicular to the direction of compression (i.e., perpendicular to the jet), but the observed degree of polarization will depend on the strength of the shock and on the viewing angle.

There is always the possibility that the observed polarization position angles include some contribution due to Faraday rotation along the line of sight to the emission region. De-rotations based on *integrated* rotation measures have been applied to the angles in Figure 1 when these are known. These integrated rotation measures are typically small ($\sim$ 12 degrees or less at $\lambda = 6$ cm (Gabuzda *et al.* 1991)). Although the degree of polarization m for the core is lower than for jet components, the total intensity I is usually much greater for the core than for jet components, so that the polarized flux mI is in most cases greatest for the core. This in turn implies that the integrated rotation measures are usually dominated by the rotation measure along the line of sight to the VLBI core; it is impossible to know if this rotation measure is appropriate for all VLBI components without knowing whether the rotation measure varies across the source. If there were appreciable amounts of local Faraday rotating material in the jets that were not accounted for by the integrated rotation measures, however, then the observed distribution of jet polarization orientation would become "smeared out", and would not be as narrow as that in Figure 1a. We may therefore deduce that the rotation measures in and around the VLBI jets of BL Lacertae objects are typically small.

Figure 2. VLBI hybrid maps of the BL Lacertae object 1823+568 at $\lambda = 6$ cm, epoch 1989.29. *Top:* Total intensity, with contours at −1.0, 1.0, 1.4, 2.0, 2.8, 4.0, 5.6, 8.0, 11.0, 16.0, 22.0, 32.0, 45.0, 64.0, and 91.0% of the peak brightness of 670 mJy/beam. *Bottom:* Linear polarization, with contours of polarized intensity at 7.0, 10.0, 14.0, 20.0, 28.0, 40.0, 56.0, and 80% of the peak brightness of 45 mJy/beam; vectors showing the direction of polarization are superimposed.

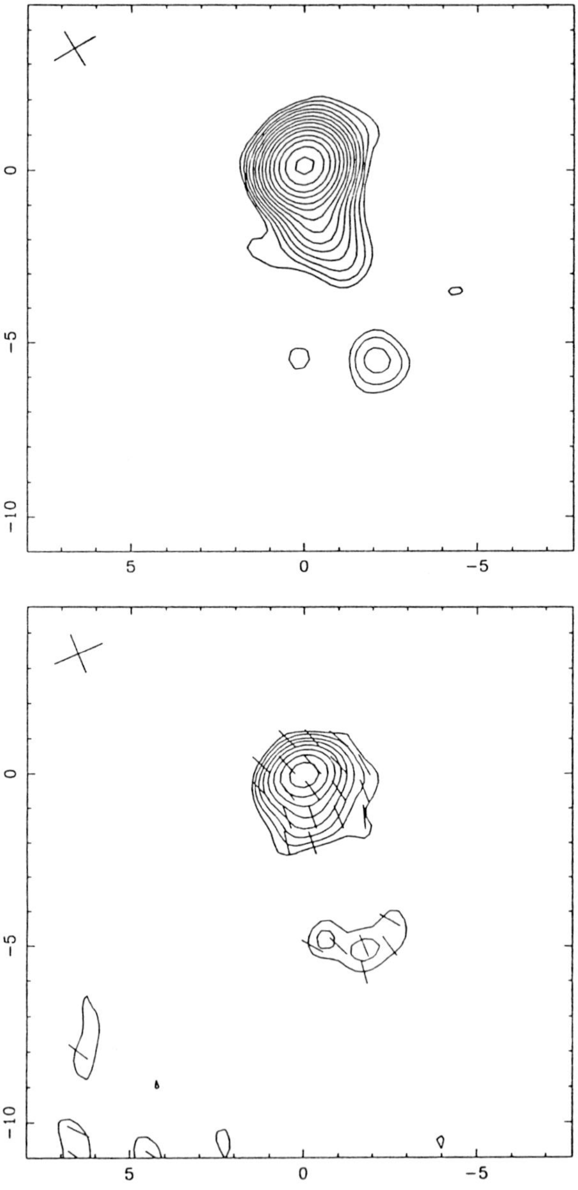

3.2 Conditions in the VLBI Cores

Although the degree of polarization in the VLBI cores of BL Lacertae objects is generally less than that in jet components, they are appreciably polarized ($\sim 2-5\%$). It is therefore somewhat puzzling why the distribution of polarization orientation for these core components should be random (see Figure 1b). One possibility is that the polarization does have some preferred orientation relative to the *local* jet direction, but that this direction is unknown due to bending of the jets on small scales. Large bends between VLBI and VLA scales are common in these objects, and it is possible that this bending starts even within the VLBI core seen at cm wavelengths; bending on such small scales is also indicated by some mm VLBI images (see, for example, Baath, these proceedings). Another possibility is that there is often appreciable local Faraday rotation in the core region which is not accounted for by the integrated rotation measures. A broad distribution of polarization orientation might also be expected if shocks occurring in the VLBI core are oblique rather than planar (as those in the jets seem to be). Lastly, there might simply be some physical reason which is not yet understood why the intrinsic polarization position angle distribution does not show any preferred orientation relative to the VLBI jet.

Observations of eight BL Lacertae objects made at $\lambda = 3.6$ cm in June, 1990 may help discriminate between these possibilities. For instance, if local Faraday rotation in the cores is masking some underlying preferred polarization orientation, then this preferred orientation should begin to emerge in the shorter wavelength observations. The better resolution attained at $\lambda = 3.6$ cm may also give a better indication of whether the jets in these sources begin to bend on small scales.

3.3 The Relationship Between BL Lacertae Objects and Other AGN

The distinct differences in the VLBI polarization structures of BL Lacertae objects and quasars argue strongly against schemes which attempt to unify the two. One such scheme formulated by Ostriker and Vietri (1985, 1990) has suggested that many BL Lacertae objects are images of distant quasars that have been microlensed by stars in intervening galaxies; in this picture, the optical continuum is amplified while the larger line emitting regions are not, giving rise to the weakness of the line emission in these sources. The differences in VLBI polarization structures of BL Lacertae objects and quasars rule out this possibility, however, since the VLBI structure would not be lensed (see Gabuzda *et al.* 1989, 1991, for a more thorough discussion). These differences also make it difficult to construct a unified scheme in which BL Lacertae objects and quasars have different characteristic angles to the line of sight.

It is an interesting question why the characteristic VLBI polarization properties of these two classes of objects are systematically different. The discussion above argues that plane shocks are common in the VLBI jets of BL Lacertae objects; it appears that for some reason, this is not the case in quasars. It is possible, for instance, that shocks do not tend to form in quasars, or that those that do form tend to be oblique rather than planar. Such a systematic difference might in principle be due to differences in the environment, or in the power and variability of the central engines.

There is growing evidence that the superluminal speeds observed in BL Lacertae objects may be systematically lower than those in quasars (see, for example, Mutel, these proceedings). If this proves to be the case, this might be understood if the planar shocks that form in the jets of BL Lacertae objects have characteristically low pattern speeds.

4 THINGS TO COME

Several sets of recent polarization-sensitive VLBI observations of BL Lacertae objects are in various stages of analysis. These include:

• Second epoch $\lambda = 6$ cm observations of ten BL Lacertae objects made in March, 1989. These data should render important information about the variability of VLBI polarization structure in these sources, and help resolve the question of whether component speeds in BL Lacertae objects are indeed systematically slower than those in quasars.

• First epoch $\lambda = 3.6$ cm observations of nine BL Lacertae objects made in June 1990. Information about the frequency dependence of the polarization structure seen at $\lambda = 6$ cm may help increase our understanding of the origin of this structure, particularly as concerns the polarization properties of the core components.

• Observations of several X-ray selected BL Lacertae objects made at $\lambda = 6$ cm in June, 1991. These data should shed some light on the question of how these sources are related to radio selected BL Lacertae objects and to other AGN.

• Observations of three sources (two of which are BL Lacertae objects) known to exhibit intraday cm-wavelength variability (see Witzel, these proceedings) made at $\lambda = 6$ cm in June 1991. Nearly continuous coverage of each source was obtained over 24 hours; by making independent images every few hours we hope to directly study any varying components. Preliminary analysis of data taken at the VLA during the VLBI run indicate that at least one source varied substantially during the experiment.

5 ACKNOWLEDGEMENTS

I would like to thank the US and European VLBI Networks for generous allocations of observing time; and the staff of the Mark III correlator facility for their help, patience, and good humor. Thanks also to my collaborator Tim Cawthorne, who has worked closely on this project. The calibration and mapping of VLBI polarization data have been pioneered by the astrophysics group at Brandeis University; I thank my colleagues Leslie Brown, David Roberts, and John Wardle there for their past and continuing efforts in this research.

6 REFERENCES

Cawthorne, T. V., and Wardle, J. F. C. 1988, *Ap.J.*, **332**, 696.

Cawthorne, T. V., Wardle, J. F. C., Gabuzda, D. C., Roberts, D. H., and Brown, L. F. 1991, in preparation for *Ap.J.*.

Gabuzda, D. C., Cawthorne, T. V., Roberts, D. H., and Wardle, J. F. C. 1989, *Ap.J.*, **347**, 701.

Gabuzda, D. C., Cawthorne, T. V., Roberts, D. H., and Wardle, J. F. C. 1991, submitted to *Ap.J.*.

Gabuzda, D. C., Wardle, J. F. C., and Roberts, D. H. 1989a, *Ap.J. (Letters)*, **336**, L59.

Gabuzda, D. C., Wardle, J. F. C., and Roberts, D. H. 1989b, *Ap.J.*, **338**, 743.

Hughes, P. A., Aller, H. D., and Aller, M. F. 1985, *Ap.J.*, **298**, 301.

Hughes, P. A., Aller, H. D., and Aller, M. F. **1989**, *Ap.J.*, **341**, 68.

Laing, R., 1980, *M.N.R.A.S.*, **193**, 439.

Ostriker, J. P. and Vietri, M. 1985, *Nature*, **318**, 446.

Ostriker, J. P. and Vietri, M. 1990, *Nature*, **344**, 45.

Roberts, D. H., Gabuzda, D. C., and Wardle, J. F. C. 1987, *Ap.J.*, **323**, 536.

Wardle, J. F. C., Roberts, D. H., Potash, R. I., and Rogers, A. E. E. 1986, *Ap.J. (Letters)*, **304**, L1.

The Structure of Linearly Polarized Emission from Quasars on Milliarcsecond Scales.

T.V. CAWTHORNE

Harvard-Smithsonian Centre for Astrophysics.

1 INTRODUCTION.

The results of a survey of the milliarcsecond-scale polarization properties of extragalactic radio sources (Cawthorne *et al.*, in preparation) have demonstrated that Quasars are characterized by properties strikingly different from those of the BL Lacertae objects described by Gabuzda (*these proceedings*). The cores are more weakly polarized in quasars and the jets are more prominent, being longer and brighter than those found in BL Lacertae objects. The polarization usually indicates a longitudinal magnetic field as on larger scales in Quasars (Bridle & Perley, 1982) though there appear to be some sources with field directions that are significantly offset.

2 ALIGNMENT OF THE MAGNETIC FIELD.

The paper by Gabuzda (*these procedings*) shows a histogram illustrating the distribution of offsets between the directions of the jets in quasars from the sample defined by Pearson & Readhead (1988) and their (E-field) polarization. The values are drawn from Cawthorne *et al.* (in preparation). This diagram conceals a wealth of details about rather complicated sources. Rightly, it shows that the angle between the *inferred magnetic field* and jet direction lies between zero (parallel field) and 60 degrees, but is less than 40 degrees for all but one of the sources. What the diagram does not show is that the source with the largest offset (0711+356) is also the least well resolved of all those included. In that source it is possible that bends in the jet are undetected and the estimate of the jet direction at the point from which the polarized flux originates may be greatly in error. Nor does the diagram reveal that the three sources in which the field is best aligned with the jet (3C345, 3C380 and 1928+738) are among the best resolved and have the most prominent jets. In these sources there is little doubt about the local field direction of the jet, and the measurements of the offsets are quite precise. Thus, from the rather small sample of quasars with prominent, polarized parsec-scale jets studied so far, we can conclude that a substantial proportion have magnetic fields that, like those in larger-scale jets, are longitudinal.

Recent observations show that in a number of sources, the magnetic field seems to follow changes in the jet direction. One example is provided by the source 3C454.3, shown in Fig. 1. The four panels illustrate (i) total intensity, (ii) linearly polarized intensity and position angle markers showing the direction of the polarized electric field, (iii) these two superimposed to show registration of the images and (iv) total intensity with the polarization position angle markers showing the direction of the *projected magnetic field.* The source was observed in March 1988 using an 8 antenna global array.

The source is poorly resolved North-South as a result of its low declination. Nevertheless the image shows a bright core component and two outer ("jet") components. The core is 4 percent polarized, an unusually high degree of polarization for a quasar core. The polarization presumably comes from a polarized knot unresolved in these observations. The 2.8cm wavelength observations by Pauliny-Toth (1987) show that the jet runs nearly East-West on smaller scales, and is thus parallel to the direction of the magnetic field inferred from the core polarization. The jet appears to continue West up to the bright knot 3.5 mas from the core, and the magnetic field inferred from the polarization between this knot and the core remains roughly parallel to the jet. At the bright knot the jet appears to turn to the North-West and culminates in these observations in an extended feature. At the knot the polarization reflects the change in direction of the jet as the projected magnetic field there points North-West. In the extended feature beyond, the projected field veers towards the East-West axis, and lies close to the direction of the jet on larger scales detected in Merlin observations by Browne *et al.* (1982). Thus it appears that direction of the magnetic field remains close to the jet axis over a wide range of scales. (As caveat note that the VLBI observations at wavelength 18cm (Pauliny-Toth *et al.* 1987, appear to suggest that the jet may not be straight between the scales present on Fig. 1 and arc second scales of the Merlin map. Beyond 10 mas from the core these observations appear show a weak extension to the North. Observations of higher dynamic range capable following the jet farther downstream would be of great interest in this respect.)

There are, however, sources in which it is difficult to discount substantial misalignments between the field and jet as the result of poor resolution and erroneous estimates of the jet direction. The best example is 4C71.07 (Cawthorne *et al.* (in preparation) - the images were also reproduced in the paper by Roberts *et al.*, 1990). Like many other quasars, this source has a low core polarization, and a prominent, highly polarized jet. Observations in total intensity only that have higher resolution than the published polarization map show that the jet is quite straight for first 8 mas (Krichbaum *et al.*, 1990). Yet in this region the projected magnetic field is misaligned by 25 degrees from the jet direction. The difference

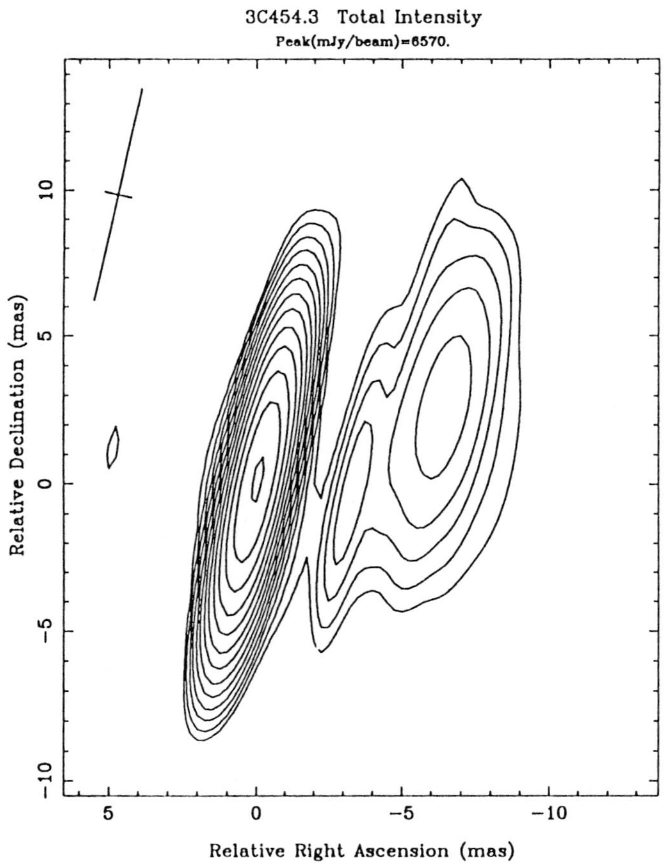

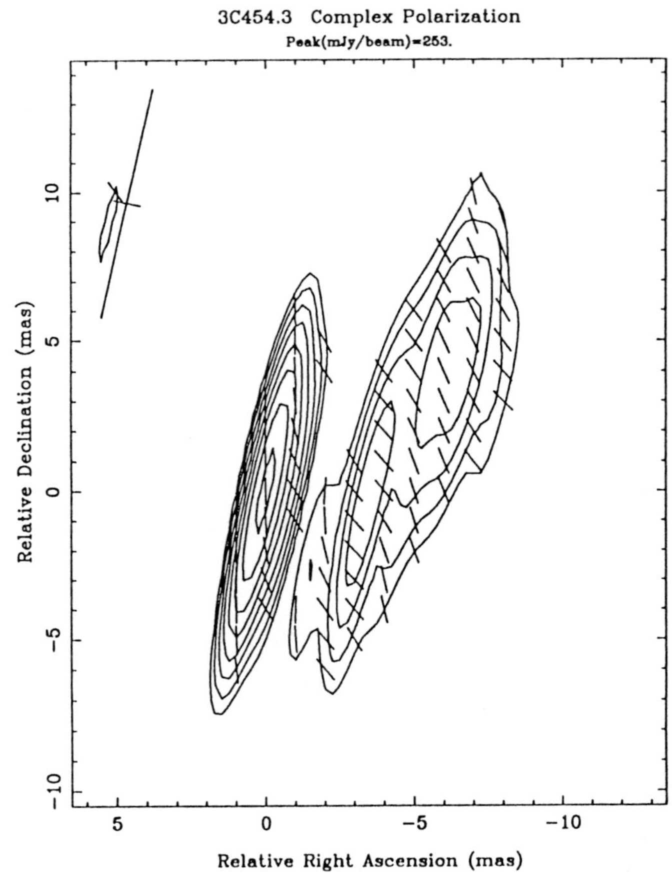

Figs 1a-d. Total Intensity and Polarization maps of 3C454.3 at epoch March 1988.

Fig. 1a (left) Total Intensity, contour levels at -1.4,1.4 2,2.8,4,5.6,8,11.2,16,22.4,32,44.864,89.6 percent of the peak, 6.57Jy/beam. The beam restoring beam is an elliptical gaussian 7.4mas by 0.91mas in PA -12.4°.

Fig. 1b (right) Complex Polarization, contour levels at 8,11.2,16,22.4,32,44.8,64,89.6 percent of the peak, 253mJy/beam. Polarization position angle markers represent the *electric* field direction. The beam restoring beam is an elliptical gaussian 7.9mas by 0.92mas in PA -12.4°.

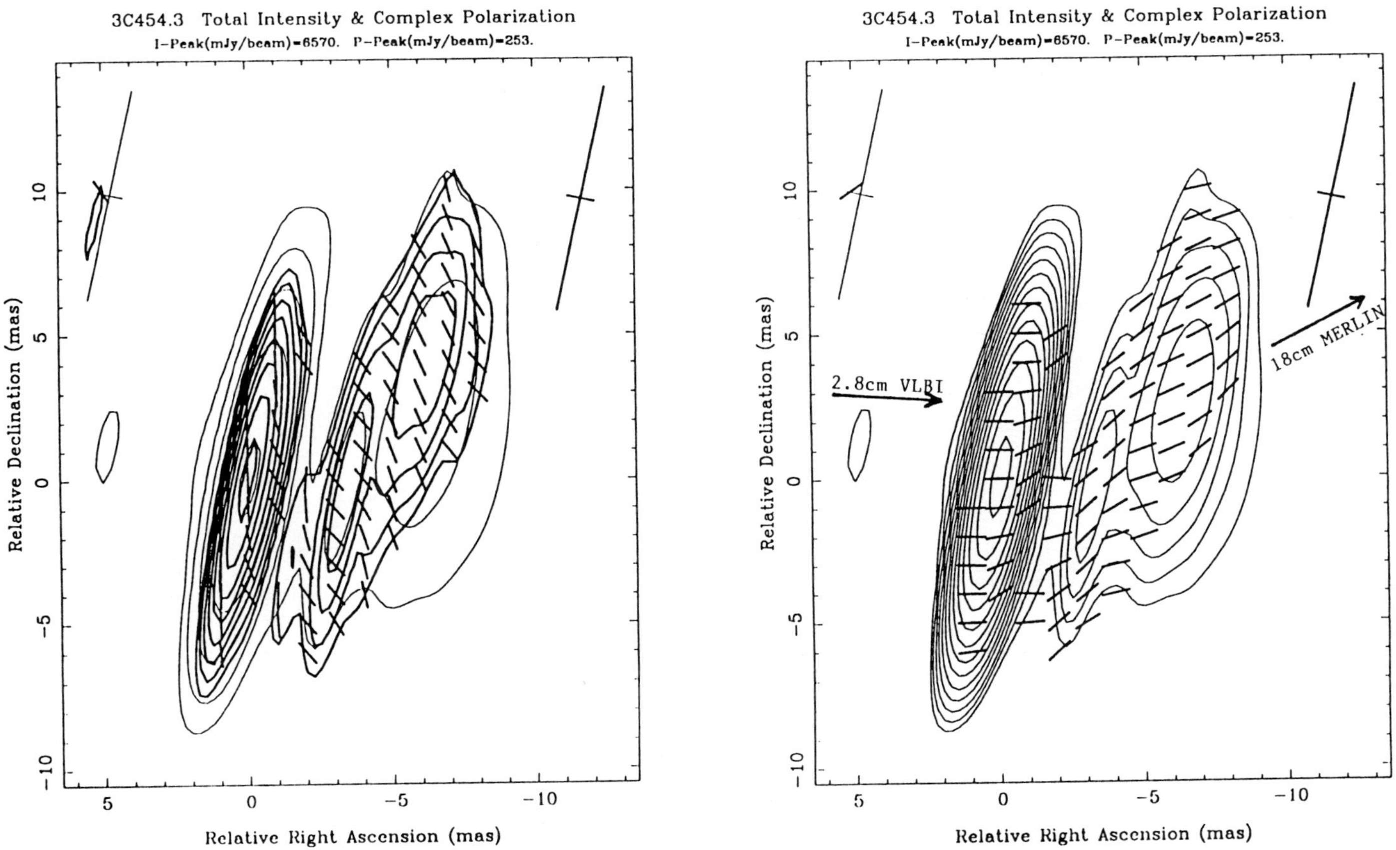

Fig. 1c (left) Complex Polarization map Fig. 1a superimposed on total intensity. map Fig. 1b. Total Intensity contour levels -1.4, 1.4,4,11.2,32,89.6 percent of the peak. Polarized Intensity contour levels at 8,11.2,16,22.4,32,44.8,64,89.6 percent of the peak. Polarization posittion angle markers represent the *electric* field direction.

Fig. 1d (right) Total intensity with polarization position angle markers representing the *magnetic* field direction. The directions of the sub-milliarcsecond structure (from Pauliny-Toth *et al.* 1987) and arc-second scale structure (from Browne *et al.* 1982) are also shown. The total intensity Contour levels are -1.4,1.4 2,2.8,4,5.6,8,11.2,16,22.4,32,44.864,89.6 percent of the peak.

between the polarized flux on the VLBI map and that of simultaneous observations at the VLA indicates that magnetic field becomes more closely aligned with the jet on larger scales. At wavelength 6cm about half the total polarized flux is detected on the VLBI map, but at shorter wavelengths one would normally expect the smaller scale emission to dominate. Thus if there is no Faraday Rotation, the PA of integrated polarization recorded by the VLA should veer towards the PA recorded on the 6cm VLBI map, increasing the misalignment between the inferred magnetic field and the jet direction at shorter wavelengths. In fact the VLA observations of O'Dea (1990) show that the PA changes in the opposite sense from wavelength 6cm to 1.3cm. Thus it is possible that the apparent misalignment between field and jet on millarcsecond scales at wavelength 6cm is due to the existence of about 25 degrees of Faraday Rotation associated with the inner structure. This corresponds to a negligible rotation at the shorter wavelength. However simultaneous polarization sensitive VLBI observations at more than one wavelength are necessary to determine whether this is the correct interpretation. Multifreqency observations of other sources have been attempted, but so far have been only partially successful due to the limitations of current instrumentation.

3 THE STRUCTURE OF THE MAGNETIC FIELD.

The polarization in quasar jets often seems to indicate that they are threaded by a longitudinal magnetic field. Although the degree of polarization of some components can be close to the maximum possible value indicating a highly ordered field, it is usually considerably lower. Hence it seems appropriate to think of the field as predominantly longitudinal with a component of random field superimposed. Assuming there is no Faraday depolarization in or around the *jets* at wavelength 6cm and that they are optically thin, the degree of polarization reflects the order of the magnetic field.

Recent observations have marginally resolved jets in some quasars across their widths. A common feature seems to be that the peaks of polarized intensity in resolved components are displaced from the ridge line of total intensity towards the edge of the jet. For example, note that in the outer component of 3C454.3 (Fig. 1), the peak in polarized flux is displaced North from the peak in total intensity. Such displacements seem to be quite common among quasars. A helical magnetic field of pitch angle close to 45 degrees would produce edge brightened polarized features, but in that case, the polarization should be seen at both edges. Thus it seems likely that processes at the edge of the jet are responsible for these highly polarized patches of emission. Polarization observations cannot directly reveal the dynamical processes responsible for such features, but they can help to infer the field structures within them.

One possibility is that the polarized features are caused by lateral compression due to the ram pressure of an external medium. This leads to a tangled web of field similar to that envisaged in the jets of BL Lacertae objects (eg. Gabuzda, these Proceedings) but here wrapped around the periphery of the jet rather than normal to its axis. This should produce polarized features only at the edge of the jet since, when seen at the centre the polarization will either be very low (the web is seen face on) or have the wrong orientation.

Another possibility is that the plasma is stretched or sheared along the axis in the outer parts of the jet. As a result the tangled field is unwound, and the longitudinal field is enhanced at the expense of the random component. The resulting polarized flux should not depend strongly upon the azimuthal angle of the feature about the jet axis, and so the feature can be seen at any point over the projected width of the jet.

The distribution of the displacements of polarized features from the ridge line of total intensity can thus in principle distinguish between these two mechanisms. An excess of polarized features near the edge of the jet is a signature of lateral compression, while convincing evidence of features centrally located on the total intensity ridge line of the jet would favour shearing or stretching as the mechanism that orders the field. At present polarization observations exist for only a small number of quasars (about 6) that have prominent milliarcsecond scale jets. Of these, several are insufficiently well resolved to detect offsets between total intensity and polarized features. Therefore it is hazardous at present to make any statistical inferences on the nature of the field from such observations. However as more observations of higher quality and resolution accumulate these arguements should prove more useful.

The results of all this work will be discussed more fully in forthcoming papers by Cawthorne *et al.*

4 CONCLUSIONS.

The magnetic field that threads the milliarcsecond scale jets of a significant proportion of quasars is predominantly longitudinal. In cases where the jet is bent, the field often follows the change in direction. Significant misalignments between the inferred magnetic field and jet direction have been detected in several sources. These may be attributed to erroneous estimates of the field direction that result from Faraday Rotation on small scales that is not reflected in VLA measurements of rotation measure, or from erroneous estimates of the jet direction due to poor resolution. Multi-frequency observations and observations of higher resolution are required to determine whether such offsets can be so attributed, or whether they reflect a more complicated magnetic field structure within the jet.

Polarized features in quasar jets often seem to be displaced from the ridge line of total intensity towards the edge of the jet. The distribution of such displacements can help to distinguish between uniform longitudinal fields and compressed random fields as the origin of the polarized features.

ACKNOWLEDGEMENT. The VLBI polarization programme is a collaborative venture; it is a pleasure to acknowledge the many contributions of my colleagues, L.F. Brown, D.C. Gabuzda, M.A. Holdaway, M.F. Ochs, D.H. Roberts and J.F.C. Wardle. Dr. Gabuzda kindly reviewed this manuscript. We are indebted to the staff of the network observatories and the Haystack MkIII correlator for their support.

REFERENCES.

Browne, I.W.A., Clark, R.R., Moore, P.K., Muxlow, T.W.B., Wilkinson, P.N., Cohen, M.H. and Porcas, R.W., 1982 *Nature*, **299**,788.

Krichbaum, T.P., Hummel, C.A., Quirrenbach, A., Schalinski, C.J., Witzel, A., Johnson, K., Muxlow, T.W.B. and Qian, S.J., 1990 *Astron. Atrophys.* **230**, 271.

O'Dea, C.P. 1990 *Astron. Atrophys.* **210**, 35.

Pearson, T.J. and Readhead, A.C.S., 1988 *Ap.J.*, **328**, 114.

Pauliny-Toth, I.K.K., Porcas, R.W., Zensus, J.A., Kellermann, K.I., Wu, S.Y., Nicolson, G.D. and Mantovani, F, 1987 *Nature*, **328**, 778.

Roberts, D.H., Wardle, J.F.C., Brown, L.F., Gabuzda, D.C. and Cawthorne, T.V. 1990, in *Parsec Scale Jets* ed. Zensus, J.A. and Pearson, T.J. (CUP, Cambridge, England).

Physical Parameters of VLBI Structures

T. W. JONES

University of Minnesota

1 INTRODUCTION

Milliarcsecond, compact extragalactic radio sources represent the smallest elements of the active galaxy phenomenon whose spatial structures are presently available to direct observation. Their variability on timescales ranging from days to years offers us the opportunity to study first hand and in some detail a rich variety of gas dynamical and radiative phenomena. The earliest, single or double component Gaussian VLBI 'maps' of compact sources gave us crude estimates of source size and brightness temperature (e.g., Kellermann, et al 1971), and led in turn to spherical models of structure and evolution based on the incoherent electron synchrotron emission process (e.g., van der Laan 1966). From these simple pictures we could derive estimates of the magnetic field strengths, the electron particle energies and total energy contents in the source regions (e.g., Jones, O'Dell and Stein 1974b, Burbidge, Jones and O'Dell 1974). Variability timescales that seemed too small for the observed source sizes coupled with constraints imposed on secondary Compton emission of higher energy photons led fairly early on to the suggestion that both observed rates of variability and brightness temperatures might be 'Doppler' enhanced from motions by at least some portions of the emitting regions at relativistic velocities fairly near the observer's line of sight (e.g., Jones and Burbidge 1973). It was clear from early attempts to analyze source properties that derived source physical parameters would depend rather sensitively on our ability to obtain accurate measures of those effects (e.g., Jones, O'Dell and Stein 1974b).

At the same time that improved VLBI maps began to show elongated and sometimes complex structures which varied and often exhibited apparent projected speeds exceeding the velocity of light (superluminal motions) (see, e.g., Porcas 1990), the ubiquity of highly collimated flows in other astrophysical environments led naturally to relativistic 'jet' models for these compact radio sources, as well (Blandford and Königl 1979, Marscher 1980). Although the origin of these relativistically moving jets remains under discussion (e.g., Lovelace, Wang and Sulkanen 1987, Phinney 1987) they have become the paradigm of compact radio sources. Studies of the behavior of supersonic fluid jets, and especially numerical simulations (of nonrelativistic flows) (e.g., Norman, Smarr, Winkler and Smith 1982) emphasized the likely presence of shocks within supersonic jet flows. That in turn led to the suggestion that shocks propagating through the relativistic jets in compact radio sources might account for such observed phenomena as flux variability (Jones 1983, Marscher and Gear 1985), and polarization variability (Hughes, Aller and Aller 1989). Superluminal motions are also incidentally expected. Such shocks have recently been proposed as the cause of intraday, small amplitude variability (Qian, et al 1991), as well.

In order to properly evaluate all of these interpretations, to realistically simulate the behaviors of relevant jet flows, and certainly in order to understand their origins, we must be able to establish the physical properties of the jets and their environments. In what follows I will briefly outline the derivation of important internal physical parameters for compact sources within the jet paradigm, with special attention paid to the importance of kinematic corrections.

2 BASIC RELATIONS

Most of what we can learn about the internal state of a compact synchrotron source derives from the surface brightness of the emergent synchrotron emission. I will concentrate here on the characteristics of the unpolarized emission, but will include reference to the polarization where it is pertinent. Considerable evidence points towards a significant degree of self-absorption in the emission from these objects. That means that the source brightness temperature, T_b, can be fairly directly related to the energy of the radiating particles; thence we can calculate the magnetic field strength. Consequently, in contrast to extended sources, one is not forced to work with indirect, 'equipartition' magnetic field estimates, for example.

If the radiating electron population has a distribution, $n_\gamma = n_o \gamma^{-(2\alpha+1)}$ (assumed nearly isotropic in a suitable comoving, 'flow' frame) for $\gamma_i < \gamma < \gamma_j$, where γ is the electron Lorentz factor, then in the appropriate frequency range the synchrotron source function and opacity are, respectively

$$J_* = 2m i_{\alpha o}\, \nu_*^{5/2} \nu_{B\perp}^{-1/2}, \tag{1}$$

and

$$\kappa_* = \frac{1}{2} k_{\alpha o}\, r_e c n_o \left(\frac{\nu_{B\perp}}{\nu_*}\right)^{\alpha+3/2} \nu_*^{-1}, \tag{2}$$

where the subscript $_*$ indicates that a quantity is measured in the 'flow' frame, r_e is the classical electron radius, and $i_{\alpha o}$ and $k_{\alpha o}$ are dimensionless constants of order unity (Jones, O'Dell and Stein 1974a). The frequency $\nu_{B\perp} = eB\sin\theta_*/(2\pi mc)$, so that $\nu_* = \gamma^2 \nu_{B\perp}$. θ_* measures the angle between the line of sight and the magnetic field. To transform these to the observer's frame of reference, one uses the fact that J/ν^3 and $\kappa\nu$ are Lorentz invariant quantities (e.g., Rybicki and Lightman 1979), with the observed frequency, $\nu = \delta\nu_*/(1 + z)$, and $\delta = 1/(\Gamma(1 - \beta\cos\mu))$, where β is the speed of the emitting region in the parent object rest frame and μ is the angle between the motion and the line of sight in that frame. Γ is the Lorentz factor for this motion, and z is the cosmological redshift. Then, for example, one can relate the observed brightness temperature, T_b, to the intrinsic brightness temperature, T_{b*}, by

$$T_{b*} = \frac{1 + z}{\delta} T_b .$$

For 'core-jet' source structures that are most commonly associated with variability and superluminal motion, the peak observed brightness temperatures at cm wavelengths are $T_b \sim a\ few\ times\ 10^{11} K$. In what follows it will be more convenient, except where noted, to express T_b in units of $mc^2/k = 5.9\ 10^9\ K$, so that observationally, $T_b \sim a\ few\ times\ 20$. Using equation (1) and standard expressions from radiative transfer, one can then relate T_b to the characteristic Lorentz factor of radiating electrons by the relation

$$\gamma = \left(\frac{1 + z}{\delta}\right) \frac{T_b}{i_{ao}\ \Theta}, \tag{3}$$

where $\Theta \leq 1$ accounts for the fact that J_* is only the limiting surface brightness for infinite optical depth, τ. For a source with uniform source function, $\Theta = 1 - \exp -\tau$. Of course, knowing γ and ν_* is also equivalent to knowing the source magnetic field component $B_\perp$.

The magnetic energy density that follows from this, $u_B = B^2/8\pi$, is

$$u_B = \frac{\pi}{2} \frac{mc^2}{r_e \lambda^2 \sin \theta_*^{\ 2}} \left[\frac{i_{ao}\Theta}{T_b}\right]^4 \left(\frac{\delta}{1 + z}\right)^2, \tag{5}$$

where $\lambda = c/\nu$ is the observed wavelength. Figure 1a illustrates characteristic 'core' values for u_B, assuming $\lambda = 6\ cm$ and $\delta/(1 + z) = 1,\ 7$. Note the strong inverse dependence of u_B on the brightness temperature. As one looks into the transparent portions of a jet, the ratio T_b/Θ depends, of course, on the source structure. It is about as difficult to establish physical conditions here as in extended sources, except through dynamical or kinematic connection to the opaque core. Recalling also the classic result that E_{ssc}, the ratio of power radiated by an electron through Compton scattering of the synchrotron radiation field to the synchrotron power itself, is proportional to $\nu_* J_*/(cu_B)$ (Hoyle, Burbidge and Sargent 1966), we recover the fact that the expected Compton luminosity from a compact source of fixed synchrotron luminosity varies as

$$L_{ssc} \sim E_{ssc} \propto T_b^5 \lambda^{-1} \left(\frac{1 + z}{\delta}\right)^6 . \tag{6}$$

The sensitivity of E_{ssc} to T_b, z and δ makes clear why, on the one hand, one might expect an upper limit to the brightness temperatures in compact sources, but, on the other hand, why it is next to impossible to reliably predict the actual flux density in Compton scattered photons (Jones, O'Dell and Stein 1974b, Bloom and Marscher 1991).

Another quantity that we can compute from the same information and that we need to know to understand the energetics of the compact sources is the energy density in relativistic electrons. (Here we need not distinguish between electrons and positrons.) Using equation 2, estimating that the path through the source along the line of sight is $\sim 2r/\sin\mu$, where r is the transverse radius of the jet, and that the optical depth is τ, we have

$$u_e \approx \frac{mc^2 \sin\mu}{k_{\alpha o} r_e \lambda r}\left(\frac{T_b}{i_{\alpha o}\Theta}\right)^{3+2\alpha} \tau \left(\frac{1 + z}{\delta}\right)^{3+2\alpha} G, \tag{7}$$

where

$$G = [\frac{1}{\gamma_i^{2\alpha-1}} - \frac{1}{\gamma_j^{2\alpha-1}}]/(2\alpha - 1) \quad (\alpha \neq 0.5)$$

$$= \ln\frac{\gamma_j}{\gamma_i} \quad (\alpha = 0.5).$$

u_e increases sharply with T_b, of course, because higher brightness temperatures mean higher energy electrons. Higher energy electrons are less efficient absorbers, so to produce a given τ one requires more of them. In the bright 'core' regions of compact sources we expect that $\tau \sim 1$, although it will drop fairly rapidly according to most models as one views further out in a jet (e.g., Königl 1981, Reynolds 1982, Jones 1988).

It is convenient to rewrite the length r in terms of the observed flux density, S, from a region of that size and the brightness temperature; namely,

$$r/\lambda \approx \left(\frac{S}{2\pi mc^2 T_b}\right)^{1/2} \frac{D_l}{(1+z)^2};$$

$$\approx 4.3\ 10^{18}\left(\frac{S}{T_b}\right)^{1/2} \frac{D_l}{(1+z)^2},$$

where D_l is the luminosity distance to the source, and in the second expression this is expressed in Gigaparsecs, while the flux density, S is in Janskys.

Perhaps the most informative way to examine the electron energy density, u_e, is to form the ratio $\beta_p = u_e/u_B$, which is like the standard quantity used in plasma physics to evaluate the dynamical importance of magnetic fields. Then, large β_p indicates a passive role for the magnetic field, while small β_p means that magnetic fields cannot be ignored. From equations (5) and (7) we find

$$\beta_p \approx \frac{2\lambda\tau}{\pi k_{\alpha o} r} \sin\mu \sin^2\theta_* \left(\frac{T_b}{i_{\alpha o}\Theta}\right)^{7+2\alpha} \left(\frac{1+z}{\delta}\right)^{5+2\alpha} G. \tag{8}$$

This is very sensitive to the observed brightness temperature, but also to the kinematic influence of the Doppler factors. Figure 1b shows $\beta_p S^{1/2} D_l (1 + z)^2$ for two values $T_b = 17$ ($10^{11} K$) and $T_b = 85$ ($5\ 10^{11} K$) as it depends on the redshifted Doppler factor, $\delta/(1 + z)$ when $\alpha = 0.5$. The flux density is in Jy and the distance in Gpc to provide characteristic numbers. I have taken the factor G to equal 10, $\tau = 1$, $\delta \sin\mu = 1$ (maximizing superluminal motion) and $\sin\theta_* = 1$ in this case. Clearly, for very high brightness temperatures the 'plasma' energy is likely to dominate the magnetic field. But, even a modest kinematic Doppler correction alters that situation dramatically. In fact, there is little direct evidence at present for brightness temperatures much above the $10^{11} K$ level. If one includes the observations of superluminal motion that often suggest Doppler factors from 2 to more than 10 (e.g., Porcas 1990), it seems quite likely that *magnetic fields are dynamically important in the compact synchrotron sources exhibiting superluminal motion.*

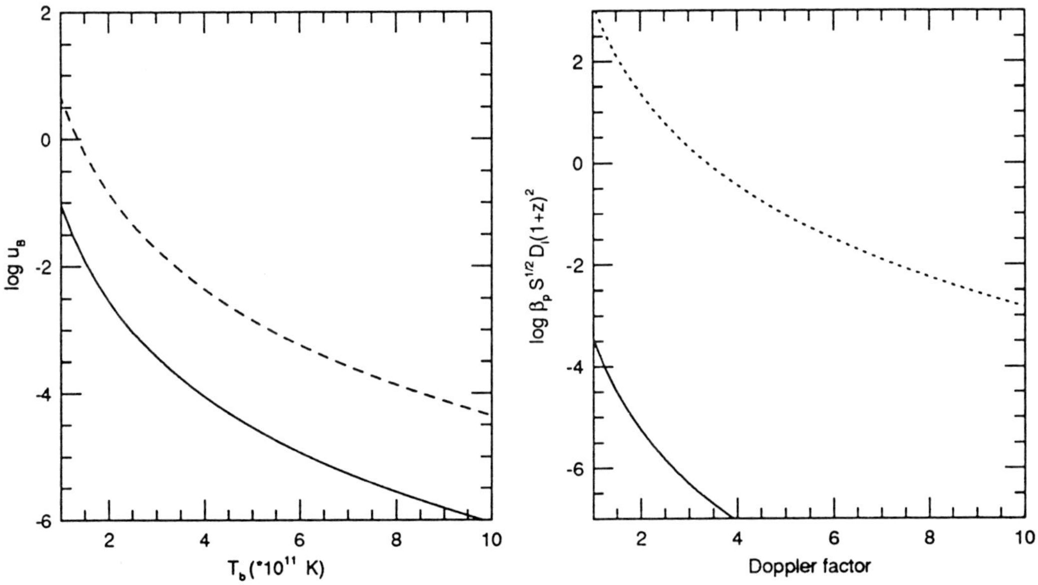

Figure 1 a)Magnetic field energy density derived from surface brightness; Doppler factors $\delta = 1$ (solid line) and $\delta = 7$ (dashed line) are shown. b)Ratio of electron to magnetic energy density in a self-absorbed source with $T_b = 10^{11}$K (solid line) and $T_b = 6\ 10^{11}$K (dotted line).

Among the possible ways to escape this conclusion would be through the presence of large pressures from an unseen energetic ion component or from nonrelativistic plasma, for example. (In the absence of these components one would expect that the relativistic jet maintained charge balance by consisting largely of an electron-positron plasma, which might be produced in several ways (e.g., Zdziarski 1986).) In some sources it is also true that higher surface brightness emission beyond the resolution available to VLBI contributes significantly to the observed flux density (e.g., Pearson and Readhead 1988), leading to weaker field estimates in those regions. It is not clear how significant that uncertainty may be, however. Large amounts of internal nonrelativistic plasma can probably be ruled out from the polarization properties of the jets (e.g., Jones and O'Dell 1978, Jones, et al 1985, Jones 1988). The arguments can become subtle, but basically reduce to noting the fact that the surface brightness in the sources can be linearly polarized at levels as large as several tens of percent (Roberts, et al 1990) It is difficult to construct models with large amounts of 'cold' plasma that will not fully depolarize these sources. Ruling out large densities of energetic ions is somewhat harder to do, although relative concentrations similar to that in galactic cosmic rays (an energy ratio $\sim 100:1$) would not alter the dynamical situation above when one is observing typical sources with moderate Doppler factors.

The idea that magnetic fields might be important in the behavior of radio jets is, of course, not new. Many theorists have proposed models incorporating magnetic fields (e.g., Königl and Choudhuri 1985, Lovelace, Wang and Sulkanen 1987) Even so, I think that their likely importance has not been generally appreciated. It remains, however, to establish what kind of structure the field has and what dynamical processes generate such strong fields, and to determine if the field plays a major role in confining the jet flows. Unless the jets are inertially confined, then it is hard to see how anything but magnetic fields could be the confining agent. Given the observation that some jets appear to bend (e.g., Zensus 1990), which requires some sort of 'illusion' for inertial flows, confined flows seem more likely. From Figure 1a we see that unless T_b is significantly above 10^{11}K, and there is no Doppler enhancement, $u_B \gtrsim 10^{-3}\ erg\ cm^{-3}$. To confine the flow through thermal pressure would require at least $nT \sim 10^{13}$. Although that may be in the range of values appropriate to the narrow line emission region in AGN, the Faraday depth through such a plasma at any reasonable temperature and magnetic field strength would be very high.

The best observational constraints on the magnetic field structures come from polarization measurements. Although it surely differs from one source to another, it seems clear from the nature of time and frequency variability in the linear polarization that the fields are partially and perhaps mostly disordered. Constraints from circular polarizations may also push in that direction (Jones 1988). One can make a generally consistent picture if the typical jet magnetic field divides very roughly into comparable 'large scale' and 'random, small scale' components (Jones, et al 1985). That also seems consistent with the linearly polarized brightness distributions seen through VLBI observations of jets (Roberts, et al 1990). If we accept that the VLBI jets are relativistically beamed at angles that enhance superluminal motions, then it is very difficult to determine confidently the orientation of the magnetic field in the fluid frame. That is because the condition that maximizes projected, 'superluminal motions', $\delta \sin \mu = 1$ is the same that gives $\mu_* = \pi/2$. Deprojection of the magnetic field, unless it is exactly along the jet velocity vector is then especially sensitive to aberration. But, statistically, there is some evidence that the 'mean fields' tend to be roughly poloidal (Jones, et al 1985, Rusk and Seaquist 1985). A confining field would need to be toroidal, but it could be that the toroidal field lies largely around the region containing the emitting plasma, if, for example, the jet carries a distributed electric current.

The last essential parameter that can be estimated straightforwardly from these same data is the energy flux, or luminosity, carried by the jet. This must constrain the energy input into the central engine. If we assume that the magnetic field is disordered, so that we can simply estimate the Poynting flux, then the luminosity is (including only magnetic field, electrons and positrons) is

$$L \sim \beta c \Gamma^2 (\frac{4}{3} u_e + 2 u_B) \pi r^2$$

$$\sim 8\ 10^{54} \frac{\beta c \Gamma^2}{\delta^2 (1 + z)^2} \frac{S}{T_b} \left(\frac{i_{\alpha o} \Theta \delta}{T_b (1 + z)} \right)^4 (1 + \frac{2}{3} \beta_p)\ D_l^2 \ erg\ sec^{-1}, \qquad (9)$$

where S is again in Jy and D_l in Gpc. L is a minimum when $\beta = 3/(1 + (4\alpha)/5)$. With fairly typical brightness temperatures that could also allow $\beta_p \sim 1$, or $T_b \approx 35$, with $S \sim D_l \sim 1$ we would have $L \gtrsim 10^{45} erg\ sec^{-1}$. By comparison, synchrotron luminosities of large scale emission from radio loud quasars range from below $10^{43} erg\ sec^{-1}$ up to $\sim 10^{46}\ erg\ sec^{-1}$; nuclear optical emissions can be up to an order of magnitude or so higher, still. These are rather similar, in fact. So, it seems clear that the power transmitted in the compact jets is substantial in comparison to other source elements.

I am grateful for support from the US NSF and the Minnesota Supercomputer Institute, and for additional travel funds coming from the organizers of this workshop and from the University of Minnesota International Programs Office.

REFERENCES

Blandford, R. D. and Königl, A. 1979, *Astrophys. J.*, **232**, 34.
Bloom, S. D. and Marscher, A. P. 1991, *Astrophys. J.*, **366**, 16.

Burbidge, G. R., Jones, T. W. and O'Dell, S. L. 1974, *Astrophys. J.*, **193**, 43.
Hoyle, F., Burbidge, G. R. and Sargent, W. L. W. 1966, *Nature*, **209**, 751.
Hughes, P. A., Aller,H. D. and Aller, M. F. 1989, *Astrophys. J.*, **341**, 54.
Jones, T. W. 1983, *B. A. A. S.*, **14**, 963.
Jones, T. W. 1988, *Astrophys. J.*, **332**, 678.
Jones, T. W., et al 1985, *Astrophys. J.*, **290**, 627.
Jones, T. W. and Burbidge, G. R. 1973, *Astrophys. J.*, **186**, 791.
Jones, T. W. and O'Dell, S. L. 1977, *Astr. and Ap.*, **61**, 291.
Jones, T. W., O'Dell, S. L. and Stein, W. A. 1974a, *Astrophys. J.*, **188**, 353.
Jones, T. W., O'Dell, S. L. and Stein, W. A. 1974b, *Astrophys. J.*, **192**, 261.
Kellermann, K. I., et al 1971, *Astrophys. J.*, **169**, 1.
Königl, A. 1981, *Astrophys. J.*, **243**, 700.
Königl, A. and Choudhuri, A. R. 1985, *Astrophys. J.*, **289**, 173.
Lovelace, R. V. E., Wang, J. C. L. and Sulkanen, M. E. 1987, *Astrophys. J.*, **315**, 504.
Marscher, A. P. 1980, *Astrophys. J.*, **235**, 386.
Marscher, A. P. and Gear, W. K. 1985, *Astrophys. J.*, **298**, 114.
Norman, M. L., Smarr, L., Winkler, K.-H. and Smith, M. D. 1982, *Astr. and Ap.*, **113**, 285.
Phinney, E. S. 1987, in *Superluminal Radio Sources*, ed. J. Zensus and T. Pearson (Cambridge: Cambridge University Press), p.301.
Porcas, R. W. 1987, in *Superluminal Radio Sources*, ed. J. Zensus and T. Pearson (Cambridge: Cambridge University Press), p.12.
Qian, et al 1991, *Astr. and Ap.*, **241**, 15.
Reynolds 1982, *Astrophys. J.*, **256**, 13.
Roberts, D. H., et al 1990, in *Parsec-scale Radio Jets*, ed. J. Zensus and T. Pearson (Cambridge: Cambridge University Press), p.110.
Rusk, R. E. and Seaquist, E. R. 1985, *J. R. A. S. Canada*, **79**, 246.
Rybicki, G. B. Lightman, A. P. 1979, *Radiative Processes in Astrophysics*, **New York**, Wiley Interscience.
Zensus, J. A. 1990, in *Parsec-scale Radio Jets*, ed. J. Zensus and T. Pearson (Cambridge: Cambridge University Press), p.28.
van der Laan, H. 1966, *Nature*, **211**, 1131.
Zdziarski, A. A. 1986, *Astrophys. J.*, **305**, 45.

Flux and Polarization variability: constraints on AGN modeling

M. F. ALLER, H. D. ALLER, and P. A. HUGHES

Department of Astronomy, University of Michigan

ABSTRACT

Analysis of extensive University of Michigan cm-wavelength observations of total flux and linear polarization from active galactic nuclei shows that variability is common in extragalactic objects; and the polarization frequently exhibits position angle swings and large changes in percentage polarization. The observed behavior can be explained by shocks in the flows of relativistic jets. Detailed modeling of several sources, based on a shock-in-jet scenario, has placed constraints on the flow parameters and shock strengths, and the modeling, combined with VLBI observations of proper motion, have been used to derive an independent estimate of source distance. The long-term polarization data for several sources show evidence for systematic changes in jet orientation with time and also provide evidence for temporal changes in the properties of the underlying flows in the sources.

CHARACTERISTICS OF THE VARIABILITY

The University of Michigan 26-meter telescope has been devoted to the study of variable sources for over 25 years. This large database provides a wealth of information for studying the physical conditions in active objects. About 60 well-known very active objects are observed at least weekly at 14.5, 8.0 and 4.8 GHz, and, additionally, several special classes of objects including a flux-limited survey of BL Lac objects, low frequency variable sources, and the Pearson-Readhead survey sources (Pearson and Readhead 1988) are observed systematically but less frequently. Statistical analysis of these flux limited samples show that significant variability is a *common* feature in extragalactic objects; for the Pearson-Readhead sample, for example, at 14.5 GHz we find that all of the BL Lac objects and all but one of the QSOs exhibit variability above about the 10 percent level. Variability is clearly a phenomenon which must be accounted for in any adequate theoretical interpretation of these objects. While variability occurs in most BL Lac objects and QSOs and in many galaxies, the light curves from individual objects exhibit a broad range of behavior from source to source and even from event to event within the same object: the amplitudes of the fluctuations range from small (on the order of 10 percent) to factors of 5 in flux, the time scales vary from rapid (intraday) to slow (years) and the outburst shape and spectral evolution in a given source vary from

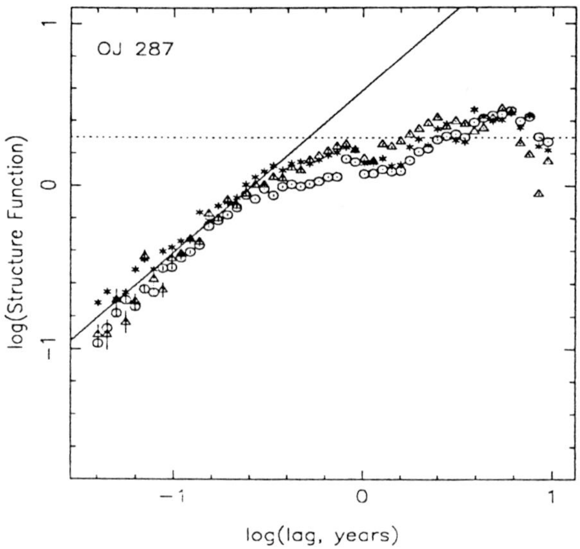

Figure 1. Structure function of the total flux density for OJ 287. The data at 14.5 GHz, 8.0 GHz and 4.8 GHz are represented by crosses, open circles and triangles respectively; this symbol convention is adopted throughout.

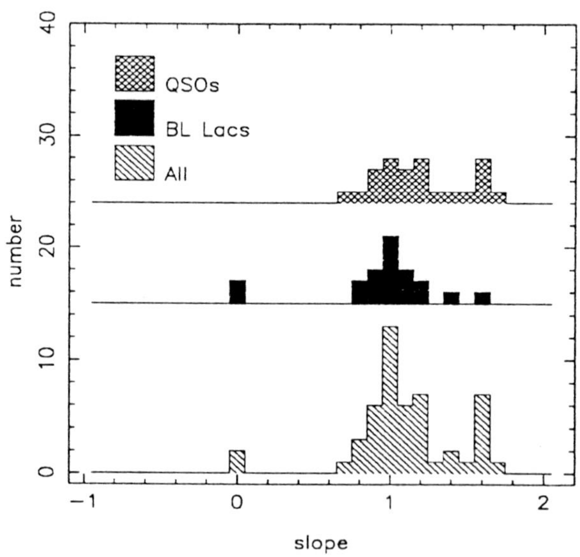

Figure 2. Histogram of the slopes of the structure functions.

event to event (*e.g.* Aller, Aller, Hughes, and Latimer 1991).

In order to search for underlying common features of the process or processes responsible for the variability and to determine characteristic time scales, a structure function analysis of 51 particularly well-observed objects has been carried out. The details of this analysis are given in Hughes, Aller, and Aller (1991b). In Figure 1 we show an example of a structure function. The first order structure function is given by $D^1(\tau) = \langle (S(t) - S(t+\tau))^2 \rangle$. The slope of the rise portion of a plot of log (structure function) versus log (timelag), marked by the full line in the figure, gives information about the statistical characteristics of the process responsible for the variability; a value of 1.0 corresponds to 'shot' noise while 0 corresponds to 'flicker' noise. The log (timelag) turnover gives a characteristic time scale. In Figure 2 we show the distribution of slopes obtained from the structure function analysis of this large group of objects. The analysis shows that the slopes are independent of both observing frequency and optical classification of the object; the narrow range of slope values found suggests that a common mechanism is responsible for the majority of the flux variability. The success of the shock-in-jet model (discussed below) in reproducing the temporal behavior of both the flux and polarization in some of the sources (*e.g.* Hughes, Aller, and Aller 1989) combined with the narrow range of slopes near 1.0 suggests that a single mechanism, shocks, can account for the majority of the observed variability and provides evidence for the ubiquity of shocks in parsec scale flows in extragalactic objects.

THEORETICAL MODELING

BL Lac is unusual among the sources in our database in showing distinct unblended outbursts in both flux and polarization. One feature commonly apparent in our polarization data, as shown in Figure 3 for this source, is the long-term stability in the polarization; this direction is often aligned either parallel to or perpendicular to the structural (VLBI) axis. At times we see swings through 90 degrees accompanied by large changes in percentage polarization, as occurred for example in 1981 in BL Lac. We have interpreted these changes as evidence for an axial compression in the relativistic flow of the jet associated with the passage of a transverse shock. A series of well-sampled VLBI maps are available for this object which allows us to follow the development of the individual components with time (*e.g.* Mutel *et al.* 1990). At least in this source there appears to be good evidence for the temporal association of VLBI hotspots with the cm-wavelength outbursts, as shown in Figure 3, which provides circumstantial evidence that these VLBI features are the result of shocks.

Detailed modeling of the temporal evolution of flux and polarization in individual outbursts based on a shock-in-jet scenario has now been carried out for three sources (Hughes, Aller, and Aller 1989; Hughes, Aller, and Aller 1991a). The modeling gives information on the flow parameters and on the strengths of the shocks. The shock strengths found from the modeling have been low, with Mach numbers typically in the range of 1.1 to 1.4; this result suggests that outbursts represent only minor perturbations to the flow. The modeling also places constraints on the orientation of the flow relative to the line of sight. The strength of the shock (and hence the associated compression along the axis of the shock) determines *both* the degree of

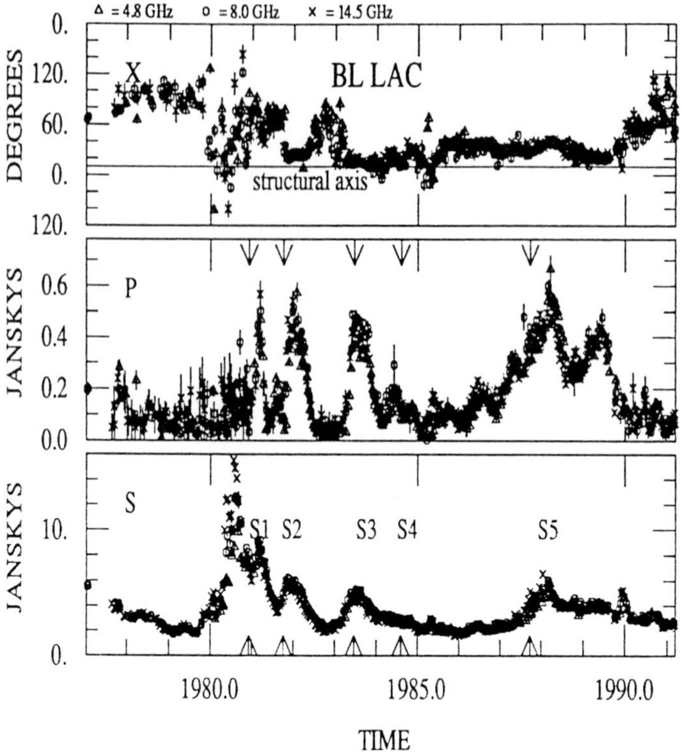

Figure 3. Monthly averages of the long-term flux and polarization for BL Lac. The times when the VLBI components were first seen are marked by the vertical arrows. The horizontal line marks the VLBI structural axis. The data have been corrected for Faraday rotation using a rotation measure of -205 rad m^{-2}.

enhancement of the total flux density and the degree of order of the magnetic field (and hence the degree of polarization in the emitting region). The modeling also places constraints on the orientation of the flow relative to the line-of-sight because of the angle-dependent effects of the relativistic motion on the flux enhancement, the outburst profile, the observed degree of polarization and the emitted versus observed frequency. In Figure 4 we illustrate the sensitivity of some of the fitting parameters in the models to the angle which the flow makes to the line of sight. Here k is the compression, γ_s is the Lorentz factor of the shock transition in the observer's frame, δ is the exponent of the power law energy spectrum for the radiating particles, and γ_{flow} is the Lorentz factor of the shocked flow. The predicted degree of polarization is particularly sensitive to the angle to the line of sight and peaks at an angle $1/\gamma_{flow}$, not at zero as does the boost in total flux. For the case of BL Lac the low observed degree of polarization during the burst (compared with that expected for an axial compression along the flow) and the burst profiles require that the angle of

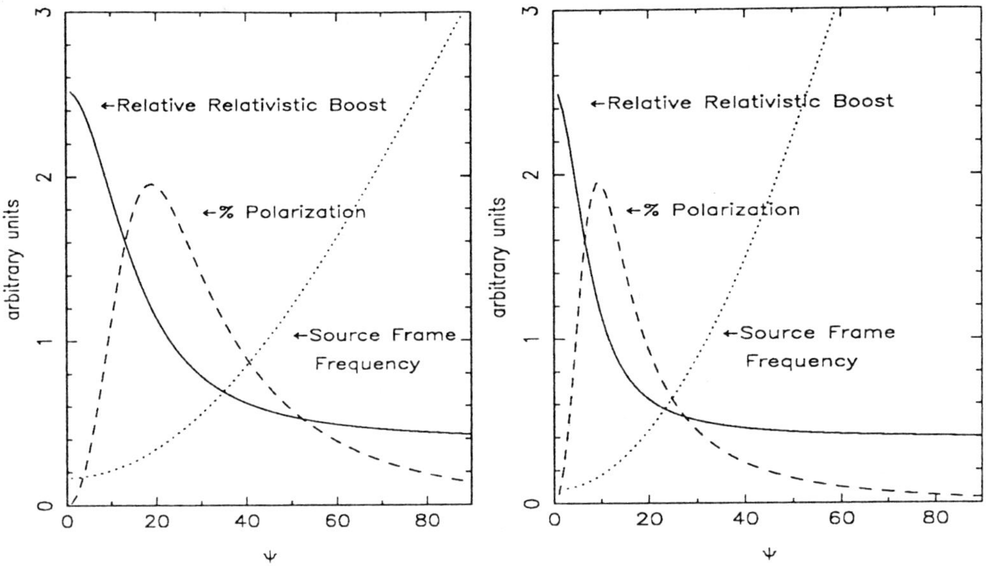

Figure 4. Sensitivity of the modeling to viewing angle Ψ. Left figure: $k = 0.5$, $\gamma_s = 5.0$, $\delta = 1.5$, $\gamma_{flow} \sim 3.1$; Right figure: $k = 0.5$, $\gamma_s = 10.0$, $\delta = 1.5$, $\gamma_{flow} \sim 6$.

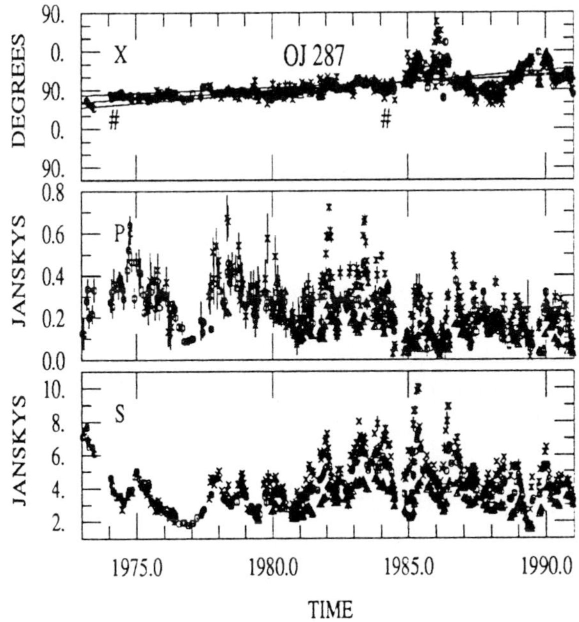

Figure 5. Two-week averages of the long-term flux and polarization for OJ 287. The full lines show the linear change with time of the position angle at each frequency from least-squares fits in the time period delineated by #.

the flow to the line of sight be large: 38 degrees. This angle, when combined with the measured VLBI angular expansion rate (Mutel *et al.* 1990) yields a value of H_0 on the order of 100 km/sec/Mpc. Thus, the modeling of the flux and polarization combined with VLBI observations yields an independent estimate of the distance to the source, and is a new, powerful tool which we hope to apply to other sources with resolved outbursts and well-sampled VLBI observations to determine both H_0 and q_0.

The bursts which we have modeled in 3C 279 and BL Lac all exhibited very stable polarization position angles that are close to the measured VLBI structure: consistent with the hypothesis that the bursts result from axial shocks; and the general correlation that we and others find between polarization and structural orientation (there is a strong preference for them to be either parallel or perpendicular), suggests that the two directions are frequently associated with each other. If this association is true, the long-term polarization observations for several sources provide evidence for changes in jet orientation with time. We see evidence of temporal changes in both the orientation of the source on the plane of the sky, and in its orientation with respect to our line of sight. In Figure 5 we show the flux and polarization data for OJ 287. In this source we find a long term systematic change in position angle which persists for more than a decade, and extends over many distinct outbursts in both the total and polarized flux. Since there is good agreement between our polarization position angle and the orientation of the VLBI structure (Gabuzda *et al.* 1989) at the same epoch, this slow change in the magnetic field orientation may indicate a systematic, steady change in the jet orientation; this might be associated with precession of a jet seen very close to the line of sight. In the case of 3C 279, the slight reorientation of a twisted flow has the potential to explain both the reduction in internal proper motion noted by Unwin *et al.* (1989) and a steady decline in total flux outburst amplitudes evident in the UMRAO data. The outbursts in BL Lac during 1988 and 1989 (Figure 3) exhibited a variation in polarization at a considerable angle to the 'established' jet axis (see Aller, Hughes and Aller 1991) in dramatic contrast to the transverse shock events in the early 1980s. In this case it is not clear whether the position angle change is the result of a bend or twist in the jet or due to the formation of an oblique shock in a stable jet structure. We note, however, that the sequence of VLBI maps in Mutel *et al.* (1990) suggest that a change occurred in the projected jet structure at the last two epochs of mapping. In attempting to model the VLBI structural changes, spectral evolution and polarization of 4C 39.25, Marscher *et al.* (1991) needed to resort to a bend in the jet misdirected slightly from the line-of-sight. Frequent VLBI maps (with sufficient resolution to resolve the jet structures in the transverse direction) and more detailed models are required to unravel these effects.

CONCLUSIONS

The comparison of multifrequency polarization data with models of polarized outbursts has shown that these temporal events, and the associated components observed by VLBI, *can* be accounted for by transverse shocks in a relativistic flow directed at a angle to the line of sight. This fitting process leads to estimates of important parameters of the flow (flow speed, shock strength, angle to the line of sight) and, with the spatial information provided by VLBI, to a direct measure of the distance to the objects. More detailed, multi-epoch examination of these objects, however, are revealing complex levels of behavior (bends, temporal changes in jet orientation, and oblique shock structures) which we are only beginning to understand.

This research was supported in part by NSF grant AST 88-15678.

REFERENCES

Aller, H. D., Hughes, P. A., and Aller, M. F. 1991, in *Variability of Active Galactic Nuclei*, ed. H. R. Miller and P. J. Wiita (Cambridge University Press: Cambridge, England), p. 172.

Aller, M. F., Aller, H. D., Hughes, P. A., and Latimer, G. E. 1991, in *Variability of Blazars*, ed. E. Valtaoja (Cambridge University Press: Cambridge, England), in press.

Gabuzda, D. C., Cawthorne, T. V., Roberts, D. H., and Wardle, J. F. C. 1989, *Astrophys. J.*, **347**, 701.

Hughes, P. A., Aller, H. D., and Aller, M. F. 1989, *Astrophys. J.*, **341**, 68.

Hughes, P. A., Aller, H. D., and Aller, M. F. 1991a, *Astrophys. J.*, **374**, 57.

Hughes, P. A., Aller, H. D., and Aller, M. F. 1991b, *Astrophys. J.*, submitted.

Marscher, A. P., Zhang, Y. F., Shaffer, D. B., Aller, H. D., and Aller, M. F. 1991, *Astrophys. J.*, **371**, 491.

Mutel, R. L., Phillips, R. B., Su, B. and Bucciferro, R. R. 1990, *Astrophys. J.*, **352**, 81.

Pearson T . J. and Readhead, A. C. S. 1988, *Astrophys. J.*, **328**, 114.

Unwin, S. C., Cohen, M. H., Biretta, J. A., Hodges, M. W. and Zensus, J. A. 1989, *Astrophys. J.*, **340**, 117.

Rapid 55 Degree Swing in the Position Angle of the Polarization in OJ 287: A Flare in the Accretion Disk?

A. SILLANPÄÄ, K. NILSSON and L.O. TAKALO

University of Turku, Tuorla Observatory
SF- 21500 Piikkiö, Finland

During our five nights observing campaign in January 1991 at the 2.56m Nordic Optical Telescope on La Palma, we made simultaneous UBVRI - photopolarimetric observations of blazar OJ 287. OJ 287 was monitored about five hours each night. The observing procedure and the used instrument are described in Takalo et al. (1991). During the first night OJ 287 was quite stable at its "normal" quiescent level; V= 15.75, P = 11% and PA = 100 degrees. On the second night we detected some truly amazing behaviour, shown in figure 1: The polarization position angle rotated linearly 55 degrees in five hours, in all five colours. This is the fastest ever observed position angle swing at optical regions in OJ 287 or any blazar. At the same time the degree of polarization was totally stable in the U-band, at a low 5% level. In the I-band the polarization level was decreasing during the night, indicating that the wavelength dependence of the polarization and position angle were both changing during the observations. The flux increased 40% during these five hours, and the spectrum flattened dramatically. The U-R colour index changed from 0.28 to 0.50, while B- magnitude changed from 15.90 to 16.22. There exists a clear linear correlation between the PA and the flux, especially in the U-band.

The "normal" explanation for any polarimetric variability in blazars is the shocks in the jet model (e.g. Königl and Choudhuri 1985). We find it difficult to explain the observed position angle swing using this model. Instead, we prefer a model, consisting of a mildly polarized flare in the accretion disk surrounding the central black hole, (e.g. Pineault 1977). The position angle rotation would then be caused by this flare rotating the black hole, with our line of sight changing accordingly. Assuming the flare occurred a few Schwarzschild radii from the center, we can estimate the black hole mass to be $10^9 M_\circ$. Also the brightening will naturally follow from a flare in the accretion disk, with the observed spectral changes. This follows because at a few Schwarzschild radii the accretion disk effects are stronger at shorter wavelengths.

These observations have clearly demonstrated that if we want to understand what is happening in blazars, we need accurate, long term photopolarimetric monitoring simultaneously in several wavelengths, with a good time coverage.

References

Königl, A., and Choudhuri, A.: 1985, Astrophys.J. 289, 188
Pineault, S.: 1977, Mon.Not.R.Ast.Soc. 179, 691
Takalo, L.O., et al.: 1991, Astron.Astrophys.Suppl., submitted

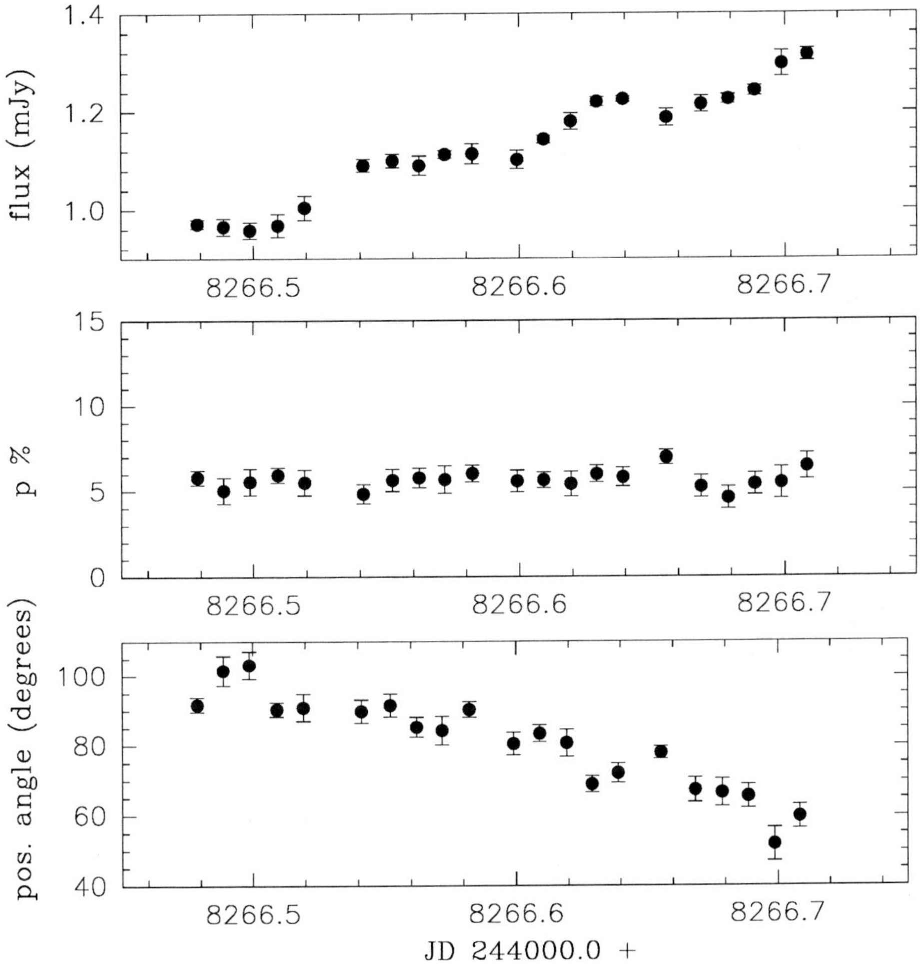

Figure 1. The observed U–band flux, polarization and position angle for OJ 287 on the 9.1.1991.

Particle Acceleration at Relativistic Shocks

J.G. KIRK

Max-Planck-Institut für Kernphysik W-6900 Heidelberg 1 F.R. Germany

ABSTRACT A brief review is given of the status of the theory of particle acceleration at relativistic shocks, indicating the importance of the effects which arise in oblique shocks. In the case of ultra-relativistic shocks, it is suggested that the main acceleration mechanism is shock-drift acceleration, because even a slight obliquity of the magnetic field suffices to inhibit the ability of a particle to cross and recross the shock front. However, for mildly relativistic flows in which the magnetic field is almost parallel to the shock normal the first order Fermi mechanism can be effective. In this case it is possible to estimate the efficiency of acceleration by considering the back reaction of particles on the fluid. The results of an investigation of the stationary solutions of this nonlinear problem show interesting deviations from the nonrelativistic results even at modest shock speeds ($\lesssim c/3$).

1. INTRODUCTION

The motivation for the study of particle acceleration in connection with beams and jets is probably clear to every participant at this conference. However, there are many mechanisms by which particles can be accelerated. At this conference we have heard about magnetic reconnection (contributions by Lesch, Romanova), acceleration close to a black hole (Beskin), acceleration by MHD turbulence (Henri, Schlickeiser) as well as diffusive acceleration by nonrelativistic shocks (Fedorenko, Lehoucq,Fraix-Burnet, Roland, Jones) and simple adiabatic compression (Aller and Aller). Given the variety of observed jets, it is quite probable that each of these mechanisms is important somewhere. Nevertheless, I feel that relativistic shocks must be the favourite candidate in situations where:

i) relativistic bulk motion is inferred more or less directly from observation

ii) emission is seen in compact hot-spots or knots.

and, especially,

iii) there are reasons for believing the radiation pattern moves with a relativistic speed different from that of the bulk motion (Cohen, these proceedings).

There is an additional reason for the study of acceleration at relativistic shocks. It is that one can hope to make some progress by exploiting and generalizing the large body of work already performed on acceleration at the more sedate shocks to be found in the solar system and in supernova remnants. With this in mind, I shall mention in this paper two processes which occur in nonrelativistic shocks and try to show under which conditions they could be of importance in relativistic beams and jets.

2. ACCELERATION AT NONRELATIVISTIC SHOCK FRONTS

At a fast mode shock (i.e., one in which the speed of the upstream flow exceeds that of the fast magnetosonic mode) the magnetic field is in general compressed – only for exact alignment of field and shock normal is there no change in the field. As a result, particles incident from upstream can be reflected by a mechanism familiar as the principle of the magnetic bottle. During reflection, the particle trajectory is dragged across field lines and can gain (or, in fact, lose) energy in the $\mathbf{v} \wedge \mathbf{B}$ electric field. This process goes under the name of 'shock-drift' acceleration. It can also occur during transmission of a particle through the shock, but is usually more powerful at reflections. It is a fully deterministic acceleration mechanism, in the sense that no scattering or other stochasticity is involved. The particle merely has to see the shock as a discontinuity whilst following its unperturbed path according to the Lorentz equation. In the case of nonrelativistic shocks, a useful approximation can be made when calculating the trajectory: that of conservation of the first adiabatic invariant or magnetic moment (Webb, Axford and Terasawa 1983). The second mechanism I will discuss is that of first order Fermi acceleration at shock fronts, better known in its nonrelativistic guise as 'diffusive acceleration' (Axford, Leer and Skadron 1977, Krymsky 1977, Bell 1978, Blandford and Ostriker 1978). This is a stochastic mechanism which depends on particles being scattered in both the upstream and downstream regions around a shock front in such a manner that some fraction of them returns to cross the front repeatedly. Each shock crossing results in an energy gain, even if the individual scatterings conserve particle energy in the local fluid frame. In the nonrelativistic case, particle transport can then be described as spatial diffusion superimposed upon advection at the fluid speed. The scattering itself is thought to occur off Alfvén waves which the particles upstream of the shock can themselves excite via the cosmic ray streaming instability (Wentzel 1969). The result of the nonrelativistic calculation neglecting the backreaction of particles on the flow is the well-known formula $s = 3r/(r-1)$ for the power-law index s of accelerated particles (i.e., the phase-space density is $f \propto p^{-s}$) in terms of the compression ratio r of the shock.

3. RELATIVISTIC GENERALIZATION

At first sight, it seems that the first order Fermi mechanism might be rather ineffective at relativistic shocks. Particles must swim against the flow if they are to return to the shock front from downstream in order to be accelerated, and this clearly gets more difficult when the fluid is moving nearly as fast as the particles themselves

(Webb 1987, Blandford and Eichler 1987, Fedorenko – these proceedings). One can also think of this problem in terms of an effective 'sound speed' for the accelerated particles. If the upstream speed exceeds this sound speed, particles ought not to be able to penetrate through the shock from downstream. For a sound speed of $c/\sqrt{3}$ appropriate for a relativistic gas a tight limit would then result. However, a more detailed look at the problem shows that the energy gained by a particle on crossing a relativistic shock offsets the extra difficulty it encounters in returning to it (Kirk and Schneider 1987). In terms of the effective sound speed, the explanation is that the accelerated particles have an anisotropic distribution with which one should associate a sound speed greater than $c/\sqrt{3}$. In fact, the anisotropy adjusts itself so that the sound speed just exceeds the speed of the incoming fluid (Kirk and Webb 1988). Although no simple relation exist between the power-law index and compression ratio for a *relativistic* shock, it is possible to calculate the slope expected at high momentum given the parameters of the shock (Heavens and Drury 1988, Kirk 1988). This picture becomes more complicated when the nature of the scattering process is looked at more closely. Alfvén waves move at a slow but finite speed through the fluid and lead predominantly to pitch angle scattering. Particles can then readily diffuse up and down field lines, but find it difficult to move across them. For a parallel shock front, this is unimportant, but if the magnetic field is oblique, then particles have to move along field lines much faster in order to catch up with a retreating shock front (Kirk and Heavens 1989). In fact, if the point of intersection of a field line with the shock front moves superluminally, particles cannot catch up with the shock at all in the absence of a process which scatters them across field lines. If we assume this to be the case, first order Fermi acceleration is forbidden in such 'superluminal' shocks and only the shock drift process remains (Begelman and Kirk 1990). Almost all ultrarelativistic shocks can be expected to fall into this category, since only a slight obliquity of the upstream field suffices to make the point of intersection move superluminally. To put it another way, for an ultrarelativistic shock front to be subluminal and thus capable of first order Fermi acceleration requires fine-tuning of the magnetic field direction to within an angle $\sim 1/\gamma$ of the shock normal (where γ is the Lorentz factor of the shock front in the frame in which the upstream fluid is at rest).

For shocks which are not so finely tuned, it is a relatively simple matter to calculate the downstream distribution produced by shock drift acceleration. Given the incoming distribution, one just has to follow the particle orbits as they traverse the shock. Nonrelativistically, the conservation of magnetic moment makes an analytic treatment possible; in the extreme ultra-relativistic limit an analytic approximation can also be found. However, for Lorentz factors of the order of a few, trajectories must be traced numerically (Begelman and Kirk 1990). Although the mechanism is capable of producing very large increases in the synchrotron emissivity it is essentially an enhanced form of compression which just boosts the incoming distribution of relativistic electrons in density and energy without changing its basic shape. The hypothesis that only shock drift acceleration is important at highly relativistic or very oblique shocks is quite vulnerable to observational refutation, because if a power law spectrum is put in upstream, the same power law emerges downstream in the amplified distribution. Therefore, a significant variation in spectral index along a jet argues against this process and thus against ultrarelativistic or highly oblique shocks as the site of acceleration.

4. THE EFFICIENCY OF PARTICLE ACCELERATION

As has been frequently pointed out, (e.g., Begelman 1988) one of the problems in our understanding of active galaxies is that the *overall* efficiency of conversion of (presumably) accretion energy into the observed relativistic particles and magnetic fields must be very high ($\sim$ 10%). The overall efficiency results from a series of processes, one of which may be particle acceleration at shock fronts. If this theory is to be applied to such objects it must be capable of high efficiency, which means one must consider the way in which the accelerated particles themselves succeed in slowing down the flow around – and especially in front of – a shock front.

A similar problem is encountered when applying the theory of diffusive acceleration to the production of cosmic rays by supernova remnants, where, once again an efficiency of the order of 10% is required. Quite early, investigations were made into the types of stationary solution permitted when the accelerated particles are allowed to push back on the fluid which accelerates them (Drury and Völk 1981, Achterberg, Blandford and Periwal 1984, Heavens 1983, 1984). It emerged that there is only a narrow range of initial conditions which allows high efficiency. If the Mach number of the shock front is low, the efficiency remains low, whereas if the Mach number is too high, acceleration is too effective and the entire shock front is smoothed out by the particles. Although it is not completely clear what would then actually happen, it is quite conceivable that the source of seed particles would dry up. The result would then at best be episodic acceleration (Drury 1983). Subsequent to these early papers, fully time-dependent investigations have been performed (Bell 1987, Falle and Giddings 1987, Duffy 1990, Jones – these proceedings). It has also emerged from simplified models of supernova remnants (Drury, Markiewicz and Völk 1989) that the stationary solutions are irrelevant in practise. The reason is that diffusive acceleration is a slow process with a typical acceleration time scale of $t_{\rm acc} \sim \kappa/u^2$ (where κ is the spatial diffusion coefficient and u the fluid speed). A supernova remnant, which accelerates particles for only a million years or so, simply does not have enough time to approach stationarity. In addition to this, if the mean free path of a cosmic ray particle increases indefinitely with energy, stationary solutions formally tend to a spectrum $f \propto p^{-s}$ with $s < 4$ at high momentum p.

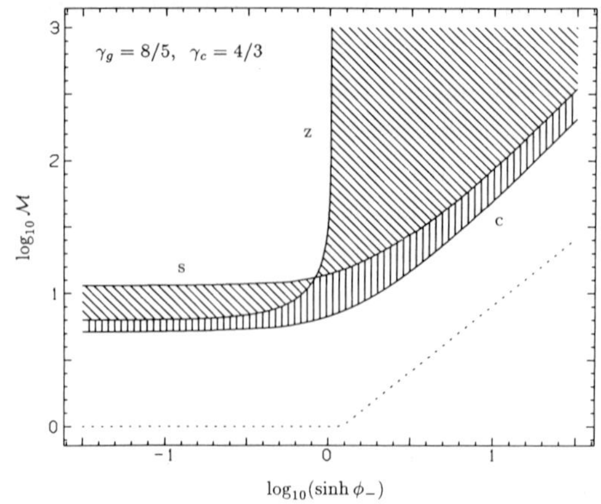

Fig. 1: *The shock solution parameter-space, illustrating the number of permitted stationary shock structures which are modified by the cosmic-ray pressure. Each solution is specified by the spatial component of the four velocity* $\sinh\phi_-$ *of the upstream flow expressed in units of the speed of light and the Mach number* $\mathcal{M}$ *of the upstream flow. The number of solutions for given initial conditions is indicated by the shading: no shading corresponds to no solutions, diagonal shading to one and vertical shading to two.*

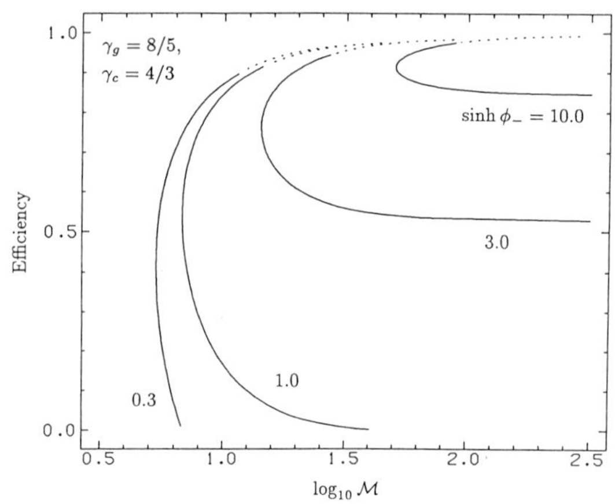

Fig. 2: *The efficiency of particle acceleration by the first order Fermi process operating at a plane parallel shock front as a function of* $\mathcal{M}$ *for various values of* ϕ_- *Solid lines depict stationary solutions which contain a precursor and a gas sub-shock. Dotted lines are solutions in which no sub-shock is present. The ratio of specific heats of the gas* γ_g *is assumed to be* $8/5$*, that of the accelerated particles* $4/3$.

This cannot be realized in nature because the pressure exerted by the particles diverges. Eichler and Ellison (1984, 1985) have suggested introducing a cut-off at high momentum to take account of escaping particles, which changes the character of the stationary solutions available. However, this does not escape the basic difficulty that supernova remnants do not live long enough to approach these solutions.
In the relativistic case stationary solutions are probably more relevant to the actual situation for two reasons. Firstly, relativistic shocks are much more rapid particle accelerators than nonrelativistic ones (Ostrowski 1988, Quenby and Lieu 1989, Ellison, Jones and Reynolds 1990). Secondly, even if the mean free path of a particle increases indefinitely with energy, it is not necessarily the case that the particle pressure diverges. Relativistic shocks produce spectra steeper than $s = 4$ under a wide range of conditions (depending on the equation of state of the fluid). In this case,there is no need to introduce a cut-off, since high energy particles are dynamically unimportant. Figures 1 and 2 present the main results of a recent investigation of stationary, relativistic solutions (Baring and Kirk 1991). In general, it is found that there exists more than one solution for a given set of upstream parameters. This happens too in the nonrelativistic case, and we can adopt a similar classification here into a) gas-dominated solutions and b) particle-dominated ones. If no energetic particles are allowed into the system from far upstream, (i.e., all are injected at the sub shock) there are four possibilities: no solutions, one of type a), one of type b) or one of each. (This way of counting excludes the completely smoothed out shocks.) Figure 1 shows that there is only a small range of Mach numbers for nonrelativistic shocks in which stationary solutions can be found. For relativistic shocks, however, the range is much larger. In particular, there is always a solution of type a) at very high Mach number. Figure 2 shows the efficiency of these solutions for various fixed values of the upstream speed. Whereas at nonrelativistic speeds the range of Mach numbers leading to efficient solutions is rather small, this is not the case when the upstream flow is relativistic. In fact, as well as the highly efficient but smoothed out solution, the type a) solution remains present at high Mach numbers and has an efficiency $\approx 50\%$ for an upstream Lorentz factor of 3.

5. CONCLUSIONS

Relativistic shocks can be efficient particle accelerators by the first order Fermi mechanism, provided they are not too oblique and not too relativistic (i.e., provided they are 'subluminal'). If, on the other hand, they are highly oblique, shock drift acceleration will take over. This mechanism boosts the input particle spectrum without changing its shape, but can produce a potentially large increase in synchrotron emissivity; it does not change the spectral index unless there are specific spectral features such as cut-offs or turnovers. This suggests that it may play a role in the intra-day variability observed in BL-Lacs (Wagner, Witzel – these proceedings), where constant spectral slope is observed.

In contrast to the nonrelativistic case, relativistic shocks can occur in stationary structures in which particles are accelerated with high efficiency, but without smoothing out the sub-shock. This is because they both have a faster acceleration rate – enabling them to reach stationarity – and can produce a steeper spectrum than a strong, nonrelativistic shock – thus eliminating the problem of divergent particle pressure. As a result, it is quite likely that such shocks achieve efficiencies as high as 50%.

REFERENCES

Achterberg, A., Blandford, R.D., Periwal, V.: 1984, *Astron. Astrophys.* **132**, 97

Axford, W.I., Leer, E., Skadron, G.: 1977, *Proc. 15th. ICRC* (Plovdic), **11**, 132

Baring, M.G. and Kirk, J.G.: 1991, *Astron. Astrophys.* **241**, 329

Begelman, M.C.: 1988, *Active Galactic Nuclei* IAU Symposium 134, Eds: D.E. Osterbrock, J.S. Miller (Kluwer, Dordrecht) page 141.

Begelman, M.C., Kirk, J.G.: 1990, *Astrophys. J.* **353**, 66

Bell, A.R.: 1978, *Mon. Not. R. astr. Soc.* **182**, 147

Bell, A.R.: 1987, *Mon. Not. R. astr. Soc.* **225**, 615

Blandford, R.D., Eichler, D.: 1987, *Phys. Reports* **154**, 1.

Blandford, R.D., Ostriker, J.P.: 1978, *Astrophys. J.* **211**, L29

Drury, L.O'C.: 1983, *Rep. Prog. Phys.* **46**, 973

Drury, L.O'C., Markiewicz, W.J., Völk, H.J.: 1989, *Astron. Astrophys.* **225**, 179

Drury, L.O'C., Völk H.-J.: 1981, *Astrophys. J.* **248**, 344

Duffy, P.: 1990, Ph.D. thesis, Trinity College Dublin.

Ellison, D.C., Eichler, D.: 1984, *Astrophys. J.* **286**, 691

Ellison, D.C., Eichler, D.: 1985, *Phys. Rev. Lett.* **55**, 2735.

Ellison, D.C., Jones, F.C. and Reynolds, S.P. 1990, *Astrophys. J.* **360**, 702

Falle, S.A.E.G., Giddings, J.R.: 1987, *Mon. Not. R. astr. Soc.* **225**, 399

Heavens, A.F.: 1983, *Mon. Not. R. astr. Soc.* **204**, 699

Heavens, A.F.: 1984, *Mon. Not. R. astr. Soc.* **210**, 813

Heavens, A.F., Drury, L.O'C.: 1988, *Mon. Not. R. astr. Soc.* **235**, 997

Kirk, J.G.: 1988, *Hot Spots in Extragalactic Radio Sources* Eds: K. Meisenheimer, H.-J. Röser, (Springer, Berlin) page 241

Kirk, J.G., Heavens, A.F.: 1989, *Mon. Not. R. astr. Soc.* **239**, 995

Kirk, J.G., Schneider, P.: 1987, *Astrophys. J.* **315**, 425

Kirk, J.G., Webb, G.M.: 1988, *Astrophys. J.* **331**, 336

Krymsky, G.F.: 1977, *Sov. Phys. Dokl.* **22**, 327

Ostrowski, M.: 1988, *Mon. Not. R. astr. Soc.* **233**, 257

Quenby, J.J. and Lieu, R. 1989, *Nature* **342**, 654.

Webb, G.M.: 1987, *Astrophys. J.* **319**, 215

Webb, G.M., Axford, W.I., Terasawa, T.: 1983, *Astrophys. J.* **270**, 537

Wentzel, D.G.: 1969, *Astrophys. J.* **157**, 545

Time Dependent Numerical Simulations of Diffusive Particle Acceleration in Astrophysical Environments

T.W. JONES AND HYESUNG KANG

University of Minnesota

1 INTRODUCTION

As part of an effort to understand better the relationships between dynamical processes in and electromagnetic radiation from cosmic radio sources we have developed a program to study particle acceleration and propagation through numerical simulations. Shocks are likely to be fairly ubiquitous in cosmic gas flows, including jets (e.g., Clarke, Norman and Burns 1989, Lind,*et al* 1989). Because diffusive acceleration in shocks is a first order Fermi process, so that it happens relatively quickly and possibly with high efficiency, it is particularly important to understand. We simulate the diffusive propagation and evolution of energetic particle populations (more conveniently called cosmic rays, or CR), dynamically coupled to the Euler's equations of ideal gas dynamics in one and two spatial dimensions. We have studied various aspects of diffusive acceleration at shocks, especially including the consequences of *in situ* injection of nonthermal particles from the underlying thermal fluid.

The basic characteristics of diffusive shock acceleration in nonrelativistic flows are now generally well known (see, e.g., Blandford and Eichler 1987, Jones and Ellison 1991). For a plane, simply discontinuous flow with a compression ratio $\rho_2/\rho_1 = r$, the steady test particle CR momentum distribution, $f(p)$, is a power law, $f(p) \propto p^{-q}$ with index $q = 3r/(r - 1)$, independent of details of the diffusion coefficient. It was largely this simple result that initially stimulated interest in diffusive shock acceleration. Especially when the shocks are strong, the expected equilibrium values of the CR pressure, P_c, near and behind a shock can be comparable to or even greatly exceed the gas pressure, P_g. Then, the gradient in P_c in front of the shock can substantially decelerate the incoming gas flow and adiabatically preheat the gas. If that happens, the character of the actual gas shock transition can change dramatically or even be eliminated (Drury and Völk 1981, Kang and Jones 1990 [KJ90]). This nonlinearity can have important observational consequences in several ways that involve the CR themselves (e.g., Kang and Jones 1991 [KJ91]) and the gas, too (Boulares and Cox 1988). In fact, however, none of these characteristics develop instantaneously. The characteristic timescale in a diffusive flow is $t_d \sim \kappa/u_s^2$, where u_s is the shock speed in the upstream gas, and κ is the diffusion coefficient. On this timescale there is a balance between spatial diffusion upstream and convection downstream. Both analytic test particle relations (e.g., Drury 1991) and more complete, nonlinear numerical simulations (Bell 1987, Falle and Giddings 1987, KJ91) show that the evolution in momentum space of the particle distribution also takes place on timescales related to t_d, but typically greater by one or more orders of magnitude, especially for the most energetic particles, if the diffusion coefficient is an increasing function of momentum as predicted in many instances. Dynamical equilibrium between the fluid and the particles can only be established on a time scale comparable to that for energy exchange with the particles. Thus, it tends to be similar to the time scale for particle acceleration. But, it should be remembered that dynamical equilibrium and equilibrium in the particle distribution are separate processes. Spatially, the CR distribution upstream is concentrated in a precursor of length scale $x_d \sim t_d u_s$. Unless that length is small compared to the lateral extent of the shock, local geometry has a significant impact on the acceleration process. Thus for shocks of finite size and/or age, the size of κ, its momentum dependence and spatial dependence can be quite important. Furthermore, κ would not in general be the same for all particle species. For electrons, one must also be concerned about the likelihood of radiative loses. In general, the picture clearly can become complex.

As another complication, there is a dynamical, acoustic instability in shocks dominated by CR pressure (Dorfi and Drury 1985, Drury and Falle 1987). The instability is capable of amplifying sound waves incident on a CR mediated shock until they also steepen into shocks. These shocks once again preheat the incoming gas and could possibly influence the injection of fresh CR and other aspects of particle acceleration in the shock.

2 TIME DEPENDENT PROPERTIES OF CR SHOCKS

2.1 Background

From the previous discussion the necessity should be clear of examining the time dependent evolution of shocks containing CR, including dynamical feedback between the CR and the gas and of considering more realistic, nonplanar flows. We have carried out a number of numerical simulations to explore the basic physics of these time dependent phenomena as well as to understand them in astrophysical situations. Initial results have been reported in KJ90, Jones and Kang 1990 (JK90), KJ91, and Kang, Jones and Ryu (1991). Other writeups are in preparation.

We have three methods of following the behavior of CR depending on the application. Each derives from the conventional diffusion-convection equation (e.g., Skilling 1975), which assumes scattering off 'heavy' centers at rest with respect to the underlying fluid. The scattering is assumed to be strong enough to keep the CR distribution almost isotropic in that frame. We restrict ourselves to nonrelativistic flows. (See the paper in this volume by Kirk for some of the distinctive attributes of relativistic shocks.) Then we have:

$$\frac{df}{dt} = \frac{1}{3}(\vec{\nabla} \cdot \vec{u})p\frac{\partial f}{\partial p} + \vec{\nabla} \cdot (\kappa \vec{\nabla} f) + Q, \tag{1}$$

where dt is the Lagrangian time derivative, Q is a source term allowing for injection (or escape), and $\vec{u}$ is the background fluid velocity. Generally, κ is a tensor and depends upon the direction of the local magnetic field. In most of our simulations we have, however, treated κ as a scalar. Several authors (see Jones and Ellison 1991 and references therein) have demonstrated the importance of injection of 'fresh' nonthermal CR from the ambient fluid. (Represented in the Q term of equation 1). If equation 1 is solved from a given $\vec{u}(\vec{r})$, $f(p)$ represents a distribution of 'test particles'. In some situations that may be adequate (especially for the electron CR component, if it does not produce a dynamically significant pressure, for example). Although we have carried out some simulations using this test particle approach, there is not room here to discuss them. As was pointed out above, however, pressure from the CR (particularly the protons) may in general become significant. The dynamical system is self-consistently described when equation 1 is coupled to the Euler's equations generalized to include CR pressure, P_c (see, for example KJ91). For simplicity, we will refer below to calculations done using equation 1 as 'transport' calculations, to distinguish them from another, more restricted 'two fluid' treatment described below.

We presently assume given properties for κ. A simple model diffusion coefficient that is convenient for comparison, relates the scattering mean free path to the particle Larmor radius, r_L; namely, $\kappa = (1/3)\alpha r_L v = \alpha\kappa_B$, where v is the particle speed, and κ_B is sometimes called the Bohm diffusion coefficient. In quasilinear theory α depends upon the energy density in Alfvén waves (e.g., Bell 1978, Blandford and Eichler 1987).

In many applications, especially for complex 2-D flows, the full transport solution becomes very time consuming to compute (something like a 3 spatial dimension calculation). Therefore, we have adopted for those situations the 'two-fluid' model of Drury and Völk (1981), in which the CR are treated as a massless, diffusive fluid. The needed energy conservation relation for the CR fluid is derived by multiplying equation 1 by the particle kinetic energy and integrating over momentum. The result is (JK90):

$$\frac{dE_c}{dt} = -\gamma_c E_c(\vec{\nabla} \cdot \vec{u}) + \vec{\nabla} \cdot (\langle\kappa\rangle \vec{\nabla} E_c) + S, \tag{2}$$

where E_c and $\gamma_c = 1 + P_c/E_c$ are the CR energy density and the ratio of specific heats, respectively, and S is related to Q. $\langle\kappa\rangle$ is averaged over momentum, weighted by $f(p)$ (e.g., Drury and Völk 1981). The most obvious limitation to this method is that one must assume something about the form of the distribution function $f(p)$, in order to evaluate γ_c and $\langle\kappa\rangle$. If $f(p)$ is a power law, it turns out that $\gamma_c = q/3$ when $4 \leq q \leq 5$. Our approach has been to carry out simulations for a range of γ_c between 5/3 and 4/3. We have confirmed in a number of situations that we get consistently good agreement between the full transport treatment and the two-fluid one. Because we do find such agreement and because they are so much more economical, we have depended heavily on two-fluid calculations to provide insight into diffusive acceleration. This requires caution, however, and we have tried to confirm important conclusions against transport calculations to the extent possible. Numerically, we solve equations (1) and (2) using the implicit, Crank-Nicholson scheme. The gas dynamic equations are solved using the explicit Piecewise Parabolic Method (PPM) hydrodynamic scheme (Colella and Woodward 1984), including the necessary CR pressure terms.

2.2 Results

Given sufficient time many plane shock models will approach something resembling a dynamical equilibrium. The characteristic time for this scales with the diffusion coefficient as described above. It also depends sensitively upon the adiabatic index for the CR, γ_c. For stiffer equations of state (larger γ_c) the equilibrium is achieved much more quickly, because more work is done on the CR in compressing them through the shock structure. In some applications where one expects shocks to be short lived or to change dynamically on short timescales that dependence could be important.

One relatively simple example we have studied in some detail is the spherical blast waves of supernovae. We have examined blasts expanding into uniform media (JK90, KJ91 and Jones and Kang 1991a (in preparation)) (JK91a), and into a medium whose density decreases with distance from the blast center (JK91a). These shocks are clearly of finite extent, but more importantly, continuously decelerating. Supernova blast waves are perhaps the best candidate sites for the acceleration of galactic CR up to energies $\sim 10^5$ GeV. Critical to that interpretation, however, are whether the 'efficiency' of diffusive acceleration in the shocks is high enough that $\sim 10 - 30\%$ of the initial blast energy can be transferred to CR, whether low energy particles have sufficient time to be accelerated to the highest energies, and whether the particle distribution that ultimately escapes from the supernova remnant approximates a power law with $q \approx 4.3$ as determined for galactic CR. A number of previous approximate analytic calculations have explored these issues. But they have either been based on test particle (and sometimes equilibrium plane shock) theory (e.g., Bogdan and Völk 1983) or restricted two-fluid models (Drury, Markiewicz and Völk 1989). The equilibrium test particle calculations are particularly suspect, because in order for a large fraction of the blast energy to be transferred to CR, some shock structure modifications are inevitable.

Using the two-fluid model with mean diffusion coefficients comparable to the mean Bohm value for interstellar magnetic field strengths, and constant values of $\gamma_c = 5/3$ we find acceleration efficiencies of 30% or more. For the same mean diffusion coefficient, but $\gamma_c = 4/3$ the efficiencies tend to be about two orders of magnitude less in the models we have studied (JK90, JK91). If the CR population of interest is supplied entirely *in situ* by injection of nonrelativistic particles one would expect and simulations confirm that the value of γ_c drops initially from values close to 5/3 to something close to 4/3 (~ 1.4 if the CR distribution is a power law with $q = 4.3$). To better understand how these changing values of γ_c influence results we carried out simulations in which γ_c drops with time in a prescribed manner. If γ_c drops slowly enough (not less than a few initial dynamical times) the net efficiency at the end of the simulation is not much smaller than for constant $\gamma_c = 5/3$. That is because the shocks are initially strong and with small κ rapidly approach equilibrium. Then more of the momentum flux is converted to P_c (JK91). Dorfi (e.g., Dorfi 1990) has also used two fluid methods similar to ours to study the evolution of CR mediated supernova remnants prior to the adiabatic blast wave phase.

In KJ91 we concluded from transport calculations for spherical blasts into uniform media that one might reasonably expect the appropriate CR particle distribution to be generated from nonrelativistically injected protons, provided κ is sufficiently small. In particular we concluded that both the slope, q and the 'maximum' energy, 10^5GeV, could result in such simulations if the diffusion coefficient were comparable to the Bohm diffusion coefficient for interstellar magnetic field strengths. Due to numerical convergence constraints, we could not afford to carry out transport simulations using the form for the Bohm coefficient nor for the full duration of the blast with small enough κ, but based those conclusions on extrapolations from simulations using a weaker momentum dependence to κ and the observation that once t_d became sufficiently small the shock structures did indeed reach apparent equilibrium, so that the details of the diffusion coefficient were not essential.

Although it is well known that simple, constant velocity, single fluid shocks are stable, that is not necessarily true when one adds in other features. In particular, CR dominated shocks are unstable to the amplification of incident compressional disturbances with scale lengths less than the scale height of the CR shock precursor. The nature of the instability is not to disrupt the shock itself, but rather to amplify the compressional disturbances within the precursor. We are examining this instability in some detail to see what importance it may have. Physically, the instability is a consequence of the minimal response of P_c to small scale compression, because the CR diffuse too fast. If there is a gradient in P_c this leads to a constant body force that for sufficiently steep gradients amplifies compressions and rarefactions of disturbances that are 'climbing up' the P_c gradient. As predicted by Dorfi and Drury (1985) from linear theory, we confirm through our nonlinear simulations that sound waves propagating through a precursor into a CR dominated shock can grow into 'minishocks' (Kang, Jones and Ryu 1991; see also, Drury and Falle 1987). Depending upon the nature of any postshock gradients in P_c these features can either continue to propagate behind the primary shock or be damped out. Because gas in the precursor is decelerated relative to the shock, incident waves will tend to be refracted so that their propagation is more nearly parallel to the shock normal. The development of minishocks in the precursor primarily increases the preshock entropy, slightly weakening the main gas shock discontinuity. In our plane 1-D simulations there are slight changes in the form of the CR distribution function due to injection of low energy particles in the minishocks and to reduction in the mean time to accelerate high energy particles because the minishocks increase the 'effective' velocity jump seen by higher energy particles with longer mean free paths. There is some reduction in the net efficiency of acceleration because of the weaker main shock, as well. It also seems possible that the existence of this instability could lead to an increase in the intensity of some components of interstellar turbulence in front of interstellar shocks. To fully appreciate the consequences of this instability it would also clearly be useful to extend nonlinear treatments to two and three spatial dimensions and to MHD.

Few, if any, real problems can be entirely understood through one dimensional treatments. As is well established from many other studies 2-D flows tend to be much more complex and more difficult to understand. That is also true for flows including CR, of course. For example, we have some preliminary results of a time dependent two fluid study of the interaction between initially plane shocks and finite, dense circular clouds initially in pressure equilibrium (Jones and Kang 1991b, in preparation). Such structures might resemble features in jets and other astrophysical environments, as well. The usual hydrodynamic treatment of this problem (e.g., Bedogni and Woodward 1990) depends upon such parameters as the Mach number of the incident shock and the density contrast between the cloud and the ambient medium. The only length scale that matters is the cloud radius, so that two characteristic time scales are produced; namely that for the shock to penetrate the cloud and that for the pressure difference across the cloud to accelerate the cloud to move with the flow behind the original plane shock (roughly the time for the cloud to 'sweep' its own mass). The incident shock 'refracts' around the cloud and, especially for a symmetric cloud, produces a shock behind the cloud that propagates into the cloud from behind.

Then there is, of course, a bow shock that moves upwind. Thus, there are at least three separate shock structures where particle acceleration might take place. Clouds of high density will eventually be disrupted by surface instabilities, with the flow at intermediate times becoming rather turbulent behind the cloud.

With the addition of diffusive CR, one must add the diffusion length and time scales to define the nature of the problem. For acceleration to be an important feature of the flow the diffusion length inside the cloud must be very much smaller than the cloud radius (That turns out to guarantee that the diffusion time will also be shorter than the other relevant times.) Furthermore, if one allows for the possibility that the diffusion coefficient depends on gas density (through magnetic field compression, for example) then the relative importance of acceleration in the various regions will depend upon what form that dependence takes. For example, the shocks which penetrate the cloud are 'stronger' than those outside because of the higher density inside. Thus if equilibrium shocks develop, they will tend to accelerate more efficiently than the bow shock, for example. But the shock velocities there are also lower, so that for a constant diffusion coefficient x_d and t_d would be greater, making it more difficult for such structures to develop. On the other hand, if the diffusion coefficient were to scale more strongly that inversely with the square root of gas density, for example, x_d and t_d would actually be smaller inside the cloud than outside.

When considering the emission from relativistic electrons that might be produced in such an environment, it is important not only to include the particle acceleration, but also to take into account what happens to the magnetic field. For this particular example, the development of turbulent flows behind the cloud are especially important. This area acts as a kind of dynamo, amplifying the field. Thus we can expect that synchrotron emission will become enhanced behind the cloud. In our initial simulations, in fact the synchrotron emissivity tends to be greatest in and behind the cloud, rather than in front.

We gratefully acknowledge support from the US NSF and the Minnesota Supercomputer Institute, with additional travel support coming from the organizers of this workshop and from the University of Minnesota International Programs Office.

REFERENCES

Bedogni, R. and Woodward, P. R. 1990, *Astr. Astrophys.*, **231**, 481.
Bell, A. R. 1978, *M. N. R. A. S.*, **182**, 443.
Bell A. R. 1987, *M. N. R. A. S.*, **225**, 615.
Blandford, R. D. and Eichler, D. 1987, *Phys. Rept.*, **154**, 1.
Bogdan, T. J. and Völk, H. J. 1983, *Astr. Astrophys.*, **122**, 129.

Boulares, A. and Cox, D. P. 1988, *Astrophys. J.*, **333**, 198.
Clarke, D. E., Norman, M. L., and Burns, J. O. 1989, *Astrophys. J.*, **342**, 700.
Colella, P. and Woodward, P. R. 1984, *J. Comp. Phys.*, **54**, 174.
Dorfi, E. A. 1990, *Astr. Astrophys.*, **234**, 419.
Dorfi E. A. and Drury, L. O'C. 1985, *Proc. 19th Internat. Cosmic Ray Conference (La Jolla)*, **3**, 121.
Drury, L. O'C. 1991, *M. N. R. A. S.*, , in press.
Drury, L. O'C. and Falle, S. A. E. G. 1985, *M. N. R. A. S.*, **223**, 353.
Drury, L. O'C Markiewicz, W. J. and Völk, H. J. 1989, *Astr. Astrophys.*, **225**, 179.
Drury, L. O'C. and Völk, H. J. 1981, *Astrophys. J.*, **248**, 344.
Falle S. A. E. G. and Giddings J. R. 1987, *M. N. R. A. S.*, **225**, 399.
Jones, F. C. and Ellison, D. C. 1991, *Space Science Reviews*, , in press.
Jones, T. W. and Kang, H. 1990, *Astrophys. J.*, **363**, 499 JK90.
Kang, H. and Jones, T. W. 1990, *Astrophys. J.*, **353**, 149 KJ90.
Kang, H. and Jones, T. W. 1991, *M. N. R. A. S.*, **249**, 439 KJ91.
Kang, H. Jones, T. W. and Ryu, D. 1991, *Astrophys. J.*, , submitted.
Lind, K. R., Payne, D. G., Meier, D. L. and Blandford, R. D. 1989, *Astrophys. J.*, **344**, 89.
Skilling, J. 1975, *M. N. R. A. S.*, **172**, 557.

Particle Acceleration and Flux Outbursts

KLAUS-DIETER FRITZ

Max-Planck-Institut für Kernphysik, Heidelberg

1 ABSTRACT

Flux outbursts of Blazar objects in the infrared to optical regime offer an opportunity to study particle acceleration in a *time dependent* scenario. The outburst spectra which show variability on timescales smaller than a day are interpreted as synchrotron radiation of high relativistic electrons. It is evident that common steady-state first order Fermi theory is not applicable in this case. Intrigued by polarization measurements during outbursts the possibility of a two-component synchrotron source is investigated.

2 TIME DEPENDENT PARTICLE ACCELERATION

Steady-state first-order Fermi acceleration leads in its simplest form to a particle spectrum which has power law shape up to a critical energy and cuts off exponentially above that energy due to dominant synchrotron losses. The spectral index is completely determined by the compression ratio of the fluid at the shock.

This simple model seems to work quite well for the smooth power law component which is observed in the quiescent state of Blazars. The situation is different when during a *flare* event the flux density increases suddenly. Also the degree of polarization (DP) and the polarization angle (PA) show drastic variability. It seems that a new particle source is suddenly turned on and serves as an injection into the acceleration mechanism. Steady-state theory no longer applies for this component. To follow the particle spectrum *in time* we solved the time-dependent transport equation for cosmic ray particles assuming nonrelativistic flow velocities, a simple one dimensional geometry, and a constant spatial diffusion coefficient. Two Laplace transforms with respect to the time and momentum coordinates reduce the transport equation to an ODE which can easily be solved with the appropriate boundary conditions. The back-transformation of the analytic solution has to be done numerically. (For results see Fritz and Webb 1990).

The type of particle injection into the system decides about the resulting spectrum. The simplest particle sources are a continuous injection starting at $\tau = 0$ and an impulsive or δ-injection at $\tau = 0$. The first one leads to a gradual approach of the steady-state case after a characteristic time scale that depends mainly on the spatial diffusion coefficient. Since flux outbursts tend to be fast coming and going events we concentrated on the second type of particle injection: a short pulse of monoenergetic, slightly relativistic particles is injected at $\tau = 0$ at the shock. To simulate a finite telescope resolution we integrated the solutions over space. Integrating to infinity yields a power law spectrum (after a certain rise time) for small momenta and a steep cutoff at high momenta which steepens with time. When integrating over a *finite* volume we obtain a spectrum that decays (after reaching its maximum) at all momenta but fastest at low momenta. This means the spectrum becomes rather flat after a while. In addition, a high momentum bump develops below an exponential cutoff due to synchrotron losses.

3 POLARIZATION PROPERTIES

The DP for a synchrotron spectrum is given by the ratio of the polarized flux to the total flux and turns out to be related to the local spectral index. For straight power laws it is therefore constant. For synchrotron spectra with steepening spectral index towards higher frequencies (the usual case) the DP increases with frequency. Observations of flux outbursts show however that local minima in the DP are seen in certain frequency ranges (e.g. Holmes et al. 1984). A single synchrotron component with a bump in the spectrum (as we obtained) could therefore explain such a behavior. But the polarization angle (PA) during outbursts also shows strong changes with frequency and sometimes even a flip by almost 90°. It is thus rather likely that at least one more component is responsible for that flip. Let us therefore consider the following scenario:

A new shock front develops in the core region of the source and travels outward along a (presumed) jet. The magnetic field will be compressed and changes its direction at the shock according to its angle with the shock normal. The particles of the pre-existing quiescent synchrotron component are freshly accelerated and emit at higher frequencies and amplitudes than before. The DP and the PA are determined by both synchrotron components and crucially depend on the dominant component which as we have seen before might drastically change with frequency as well as with time. Figs.1 and 2 show the results for the superposition of a steady component (power law with high frequency cutoff) with the variable component (impulsive injection) where the position angle of the two components differ by 80°. The different curves correspond to different (normalized) times after starting with the injection. We see a rather complicated behavior of both quantities with frequency. A strong minimum in the DP is obtained as well as flip in PA for certain times, a behavior that was clearly seen to occur during an outburst of BL Lac object OJ287 (Holmes et al. 1984). The requirement for sudden changes in both quantities is the change of dominance of the polarized flux of one component over the other which should occur whenever the

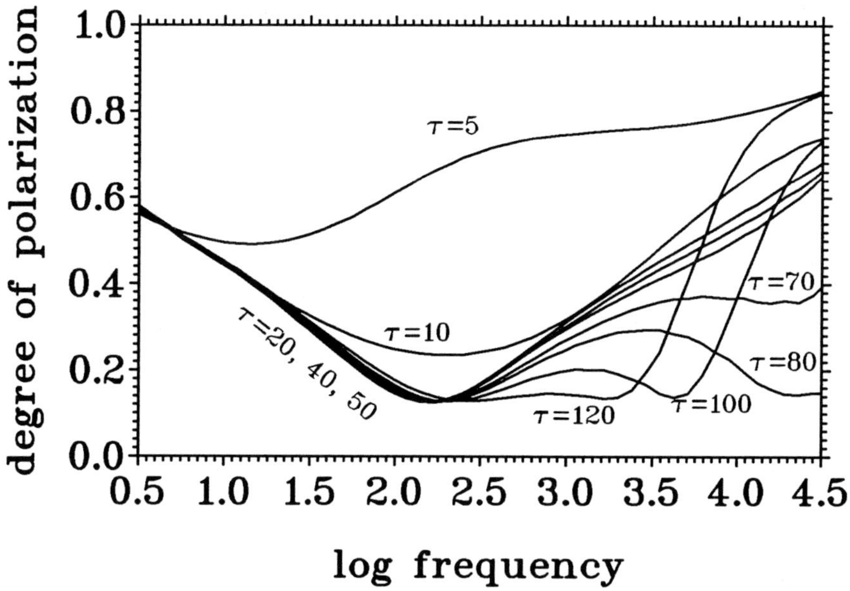

Figure 1. Resulting DP for different times after injection at the shock.

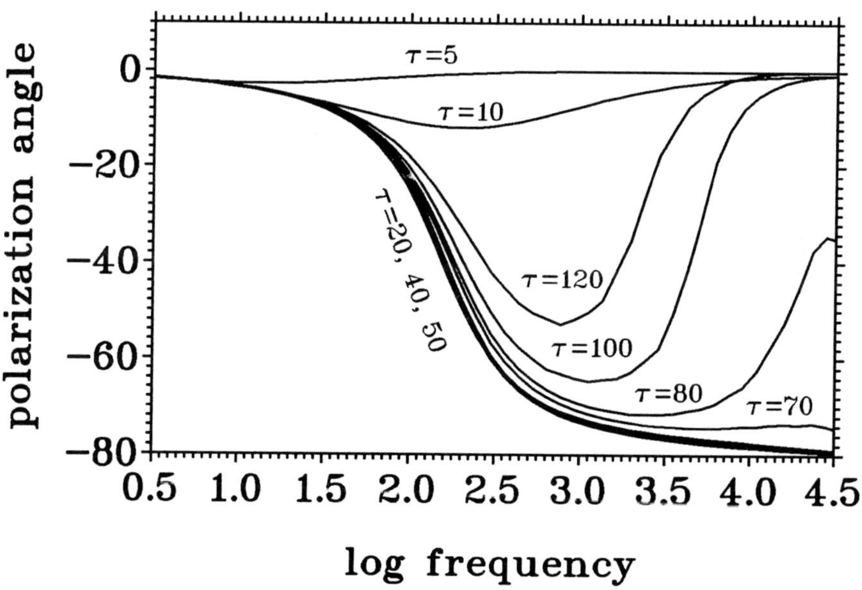

Figure 2. Resulting PA for different times after injection at the shock.

magnetic field of the variable component is strong enough. It would be interesting to have a larger frequency range during an outburst available to be able to follow specific features through the spectrum.

REFERENCES

Fritz, K.-D., Webb, G.M. 1990, Ap.J., **360**, 387

Holmes, P.A., Brand, P.W.J.L., Impey, C.D., Williams, P.M., Smith, P., Elston, R., Balonek, T., Zeilik, M., Burns, J., Heckert, P., Barvainis, R., Kenny, J., Schmidt, G., Puschell, J. 1984, MNRAS, **211**, 497

From Ultrarelativistic Beams to Radio Jets: The Accretion Disk Connection

A. KÖNIGL

Department of Astronomy & Astrophysics, The University of Chicago

1 INTRODUCTION

The origin of the ubiquitous jets in extragalactic radio sources is one of the major puzzles of modern astronomy. The jets have been probed by interferometric radio observations down to a (projected) scale of parsecs, and many of their salient features have been interpreted in terms of collimated fluid outflows. It is believed, however, that the production and initial collimation of the jets occur on much smaller scales still, most likely in the vicinity of the horizon of the central black hole or near the inner edge of the nuclear accretion disk (at radii $\gtrsim 10^{14}$ cm for a $10^8\ M_\odot$ black hole). The jets are thought to originate as relativistic particle beams that tap the rotational kinetic energy of the hole or the disk and that are accelerated by electromagnetic or hydromagnetic means (e.g., Lovelace 1976; Blandford 1976; Phinney 1983; Lovelace, Wang & Sulkanen 1987).

Although the inner beams are not directly observable, one can, nevertheless, place a number of constraints on their transformation into jets from the properties of the large-scale outflows: (1) The jets often retain a relativistically propagating flow component. This is indicated by the apparent superluminal motions measured in many compact radio sources on parsec scales (e.g., Porcas 1987). Such motions may persist also on larger (kiloparsec) scales, as indicated by direct measurements (Walker, Walker & Benson 1988) as well as by the one-sidedness of quasar jets (attributable to Doppler boosting; e. g., Bridle & Perley 1984) and by the morphology of the "hot spots" at the ends of edge-brightened radio sources (Laing 1989). (2) The parsec-scale jets contain very little thermal (electron-proton) plasma. This follows from the low Faraday rotations measured in these objects (Wardle 1977). One attractive explanation of the low rotation measures is in terms of a sizable electron-positron outflow component.

Additional constraints on the beam-jet transition follow from the properties of the radiation field in the vicinity of the galactic nucleus: (1) The inferred radiation field will strongly decelerate highly relativistic outflows by the Compton drag interaction (e.g., Rees 1984). It is, in fact, possible that the characteristic Lorentz factors of superluminal radio sources ($\gamma_\infty \lesssim 10$) are determined by the inverse-Compton deceleration of injected ultrarelativistic beams in the thermal radiation field of a nuclear accretion disk (Phinney 1987). (2) For a sufficiently high beam energy, the interaction with the ambient radiation field will lead to the production of a particle shower. In particular, a charged particle moving in a blackbody radiation field of temperature $T = 10^5 T_5$ K will create e^+e^- pairs if its Lorentz factor exceeds the Klein-Nishina critical value $\gamma_{\rm KN} \equiv m_e c^2/2.7kT \approx 2 \times 10^4 T_5^{-1}$, whereas pion production will be initiated at $\gamma_\pi \approx 3 \times 10^6 T_5^{-1}$.

The production of particle showers in the context of extragalactic jets was first studied by Lovelace, MacAuslan & Burns (1979) and by Burns & Lovelace (1982), who considered the joint effect of inverse-Compton scattering of soft (disk-generated) photons by relativistic electrons and positrons to γ-ray energies and of the annihilation of these γ-rays with ambient low-energy photons that creates e^+e^- pairs. Other relevant interactions have been discussed in subsequent investigations. In particular, it was noted that the cross section for triplet pair production (TPP: $e^\pm + \gamma \rightarrow e^\pm + e^- + e^+$) increases with particle energy as $log\,\gamma$ and comes to exceed the Compton scattering cross section (which *decreases* as $1/\gamma$) for $\gamma \gtrsim 250\gamma_{\rm KN}$ (Mastichiadis, Marscher & Brecher 1986; Mastichiadis 1991). Several authors (e.g., Mastichiadis 1986; Sikora et al. 1987; Begelman, Rudak & Sikora 1990) have studied the development of *proton-initiated* showers. In this case, the dominant interactions are photomeson production, $p + \gamma \rightarrow p + \pi^0$, $\pi^0 \rightarrow \gamma + \gamma$ and $p + \gamma \rightarrow n + \pi^+$, $\pi^+ \rightarrow \mu^+ + \nu_\mu$, $\mu^+ \rightarrow e^+ + \nu_e + \bar{\nu}_\mu$ (with subsequent beta decay of the neutron), as well as proton-photon pair production ($p + \gamma \rightarrow p + e^- + e^+$). A variant on these ideas involving the collision of relativistic protons with ambient thermal protons has also been examined in the literature.

It is clear from the above considerations that, even without specifying the particle acceleration mechanism, one may expect to obtain useful information about the transition from beams to jets from a study of the interaction of a directed, ultrarelativistic particle beam with the radiation field of a nuclear accretion disk. The purpose of the present contribution is to provide an interim progress report on such a study, which I have undertaken in collaboration with P. Coppi and J. Kartje at the University of Chicago. The goal of this investigation is to examine whether one can invoke a radiative interaction of this type to account, in a self-consistent manner, for the production of an electron-positron jet component, for the distribution of terminal Lorentz factors γ_∞ in extragalactic radio sources (as inferred from superluminal motions, rapid flux variations, and source statistics), and for some of the characteristic features of the X-ray and γ-ray spectra of Blazars (BL Lac objects and optically violently variable quasars, sources in which the beamed radiation component appears to dominate the observed emission). In this picture, the high-energy emission represents the upscattered disk radiation produced in the course of the Compton deceleration of the beam (Melia & Königl 1989, hereafter MK89) and thus serves as a direct probe of the innermost regions of the jet.

2 RADIATIVE INTERACTION MODEL

2.1 Particle Injection

In our simple model, we assume that a perfectly collimated beam of relativistic *protons* with initial Lorentz factors $\gamma_{\rm p,i}$ is injected into the thermal radiation field of an accretion disk that has a maximum surface temperature $T_{\rm m}$. As in MK89, we model the disk spectrum as a sum-of-blackbodies and employ a "flat surface plus conical funnel" representation of the disk geometry, which allows us to examine both thin and thick disks of finite radial extent. The height $z_{\rm b}$ of the injection point above the midplane is a free parameter designed to correspond to the characteristic acceleration length of the beam.

As was remarked above, our model focuses on the radiative deceleration of ultrarelativistic beams without specifying the details of the initial acceleration. The proposed acceleration mechanisms include first-order Fermi acceleration in shocks, the unipolar induction dynamo, electric fields induced by magnetic field reconnection, and second-order Fermi acceleration associated with plasma turbulence. As a fiducial value, we adopt $\gamma_{\rm p,i} = 10^8$, which corresponds roughly to the maximum Lorentz factor estimated by balancing first-order Fermi acceleration with radiative cooling in an active galactic nucleus environment (Sikora et al. 1987; **Begelman** et al. 1990).

2.2 Shower Characteristics

A first approximation to the solution of the problem at hand can be obtained by assuming that the injected beam propagates into an isotropic blackbody radiation field of a given temperature T. In this case one can use simple analytic approximations to the reaction cross sections to estimate the characteristic deceleration lengths of the different constituents of the expected particle shower. Table 1 presents the results of such a calculation for the case $T = 10^5$ K, which is a representative value of the maximum disk temperature $T_{\rm m}$. The deceleration of the injected protons has been considered in the limits of a charge-separated and of a fully neutralized beam (electrons comoving with the protons) to illustrate the sensitivity of the deceleration lengths to the presence of neutralizing electrons. The actual lengths should be bracketed by these two limits. The inverse-Compton deceleration distance for highly relativistic pairs in the extreme Klein-Nishina regime was estimated by taking account of the fact that any such particle initiates a chain reaction ($e^{\pm}\gamma \rightarrow e^{\pm}\gamma$,

$\gamma\gamma \to e^+e^-$,...) in which most of its energy is transferred to a newly created e^+e^- pair member at the subsequent stage of the cascade, so that the effective deceleration length involves several generations of pairs.

TABLE 1: PROTON-INITIATED SHOWER

Particle:	Energy†:	Process:	$l_{\rm dec}^{\natural}$:
injected p	10^8	photomeson prod.	7×10^{11}
injected p	$\gamma_\pi = 3 \times 10^6$	i.c. scatt. (neutral) γp pairs (charged)	9×10^{12} 1×10^{14}
injected p	$\gamma_{\rm KN} = 2 \times 10^4$	Thomson (neutral) Thomson (charged)	1×10^{11} 3×10^{17}
primary γ (from π^0 decay)	3×10^{10}‡	$\gamma\gamma$ pair prod.	9×10^{12}
primary $e^\pm$ (from primary γ)	3×10^{10}	TPP, i.c. scatt.	2×10^{13}
TPP $e^\pm$ (1st generation)	2×10^6	i.c. scatt.	3×10^9
$e^\pm$	$\gamma_{\rm KN}$	Thomson scatt.	6×10^7

†In mc^2.

♮Deceleration length in cm, assuming an isotropic 10^5 K blackbody radiation field.

‡In m_ec^2.

The results presented in Table 1 indicate that the pair deceleration lengths in an isotropic blackbody radiation field are much smaller than the characteristic scale ($\gtrsim 10^{14}$ cm) on which the radiation field is expected to vary, with the same also holding true for the injected protons provided that they can be effectively neutralized. In reality, however, the radiation field affecting a particle beam that is injected at a finite height above the disk midplane will be anisotropic and diluted, resulting in significantly larger deceleration lengths. For example, using the disk parameters and the approximate solution of the equation of motion given in §IV of MK89, one finds that the deceleration length in the Thomson regime for an e^+e^- jet with $z_b = 4 \times 10^{14}$ cm is a factor $\sim 10^3$ higher than that calculated under the isotropic blackbody assumption. If a similar factor multiplies the other entries in Table 1, then the radiation field might be nonuniform on the deceleration scale of the injected proton beam. An immediate corollary of this possibility is that successive generations of pairs would then be subject to different radiative deceleration forces and could have correspondingly different values of γ_∞. I will consider some of the implications of this possibility to relativistic jet sources in the following sections.

3 MODEL CALCULATIONS

We have been investigating the dynamical interaction of the injected beam and the development of a particle-photon cascade with the help of a numerical code originally constructed by Coppi (1991) to study the time evolution of e^+e^- plasmas. The code, which self-consistently solves the kinetic equations of particles and photons, has been modified to follow the *spatial* evolution of the beam and to incorporate an *anisotropic* ambient radiation field as well as additional high-energy interactions. To illustrate the code's capabilities, I now present a calculation of a pair-photon cascade corresponding to one stage of the shower considered in Table 1. Specifically, this example details the evolution of an e^+e^- beam that is injected with an initial Lorentz factor $\gamma_{e,i} = 10^8$ into an isotropic blackbody radiation field of temperature $T = 10^5$ K. The processes included in this calculation are TPP, inverse-Compton scattering, and photon annihilation. (As discussed in §4, pair annihilation is insignificant on the scales considered in this example.)

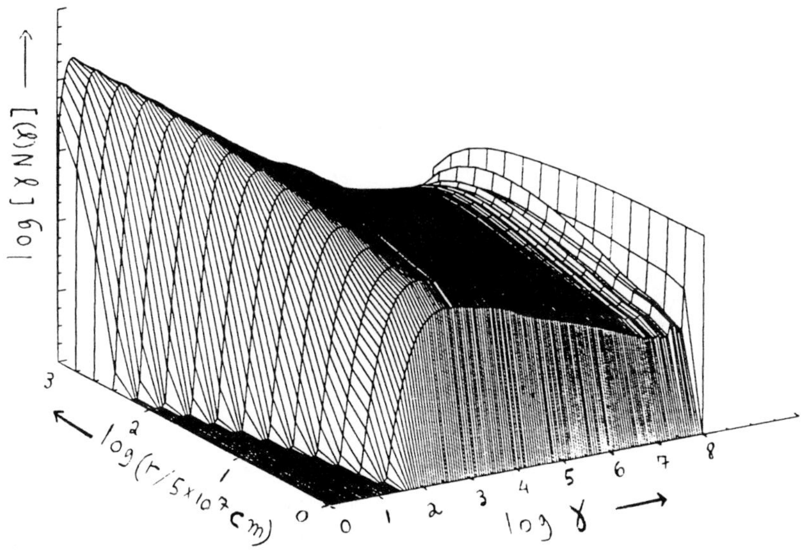

Fig. 1 – The differential pair number density per logarithmic energy interval resulting from the interaction of an ultrarelativistic electron beam (initial Lorentz factor $\gamma_{e,i} = 10^8$) with an isotropic blackbody radiation field (temperature $T = 10^5$ K).

Figure 1 shows the particle distribution per logarithmic energy interval as a function of the particle Lorentz factor and the distance from the injection point. (The scale of the vertical axis in Figs. 1 – 5 is arbitrary.) The Compton deceleration of the injected beam and the creation of new pairs by $\gamma\gamma$ collisions and (to a much lesser extent) by TPP are clearly visible in this plot. The increase in the pair number along the path of the beam is depicted by the solid curve in Figure 2, which demonstrates that, for a sufficiently large soft photon intensity, an ultrarelativistic beam can act as an efficient *pair multiplier*, leading to a many-fold increase in the number of pairs. In fact, if the cascade reaches saturation (as is the case in this example), the pair multiplication factor is $\sim \gamma_{e,i}/\gamma_{KN}$ ($\approx 5 \times 10^3$ in this case).

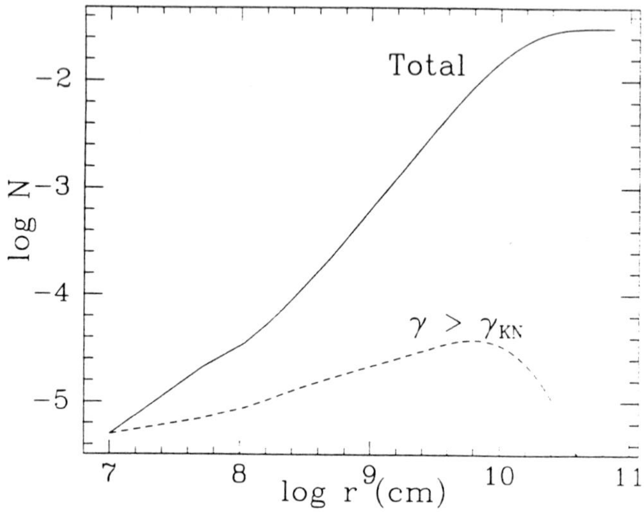

Fig. 2 – Variation of the pair number with distance from the injection point.

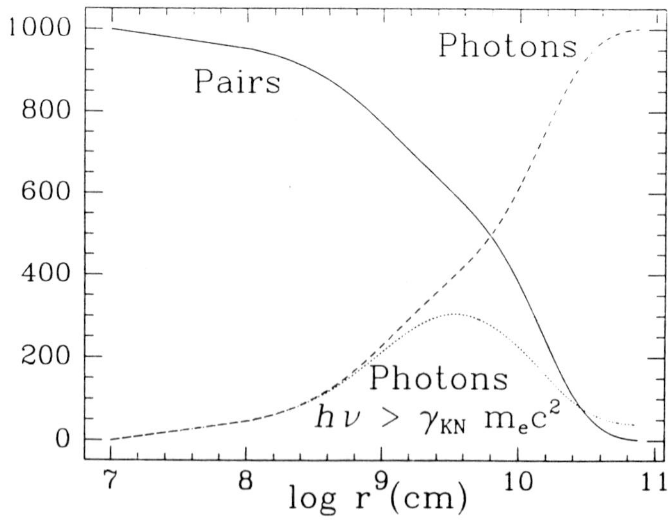

Fig. 3 – The spatial variation of the energies of pairs and photons.

The dashed curve in Figure 2 traces the evolution of pairs whose energy is above the Klein-Nishina threshold and which therefore can induce further pair creation. Initially most of the pairs are above this limit and, even after the low-energy pairs start to **dominate,** larger, additional pairs with $\gamma > \gamma_{KN}$ are produced by TPP and by the annihilation of upscattered photons. Eventually, however, the high-energy pairs disappear and the cascade ends. (This is seen to occur on a scale of a few times 10^{10} cm, which is consistent with the estimates given in Table 1.) The transfer of energy between the high-energy and the low-energy pairs is mediated by high-energy ($h\nu > \gamma_{KN} m_e c^2$) photons, as illustrated in Figure 3. The latter figure also shows that the pairs, in turn, mediate the transfer of energy from high-energy photons to low-energy photons, which essentially occurs in the course of the final deceleration of the particles in the Thomson scattering regime ($\gamma_e < \gamma_{KN}$). In this sense the interaction region acts as a *γ-ray filter*, preventing the escape of γ-rays for sufficiently large photon-annihilation optical depths.

The evolution of the upscattered photon distribution is shown in Figure 4, which clearly demonstrates the progressive increase in the number of low-energy photons as the pair-photon cascade develops. The spectra have been calculated on the assumption that virtually all the upscattered photons are beamed in the direction of the injected ultrarelativistic particles. For this reason, the emergent spectrum approaches the shape previously derived for a saturated pair-photon cascade under the assumption that both the photon and the (stationary) particle distributions are isotropic (see Zdziarski 1988 and references therein). In particular, at energies below the Klein-Nishina limit and well above $2.7kT$, where the injected blackbody spectrum can be approximated by a delta function. the flux density is a power law in frequency, $F_\nu \propto \nu^{-0.5}$. This spectrum, however, will in general *not* be relevant to observations of relativistic jet sources. This is because such sources are highly anisotropic and the emitted radiation is strongly Doppler boosted, so the observed spectrum is very sensitive to the inclination angle θ of the jet axis with respect to the line of sight and to multiple-scattering effects in the jet (see MK89).

Figure 5 illustrates this last point with a model fit to the UV through X-ray spectrum of 3C 273. a superluminal radio source that has a "miniblazar" optical emission component and that is also the only quasar detected by COS B in $\gtrsim$ 100 MeV γ-rays. The displayed spectrum has been calculated with a Monte Carlo code developed by J. Kartje and corresponds to a fully neutralized proton beam that is decelerated within the funnel of a thick accretion disk. It is evident from a comparison of Figures 4 and 5 that the model high-energy spectrum. which represents initially highly beamed photons that were scattered toward the observer by the $\gamma_e \approx \gamma_\infty$ electrons in the jet, is distinctly different from the forward-beamed, singly scattered spectrum that would be seen if the observer looked exactly down the jet. This simple scheme does not account for the COS B data, but it is conceivable that when the model is generalized to include pairs produced in a nonuniform soft photon field (which should lead to a range of high-energy spectral cutoffs) one could attempt to extend the fit to higher energies.

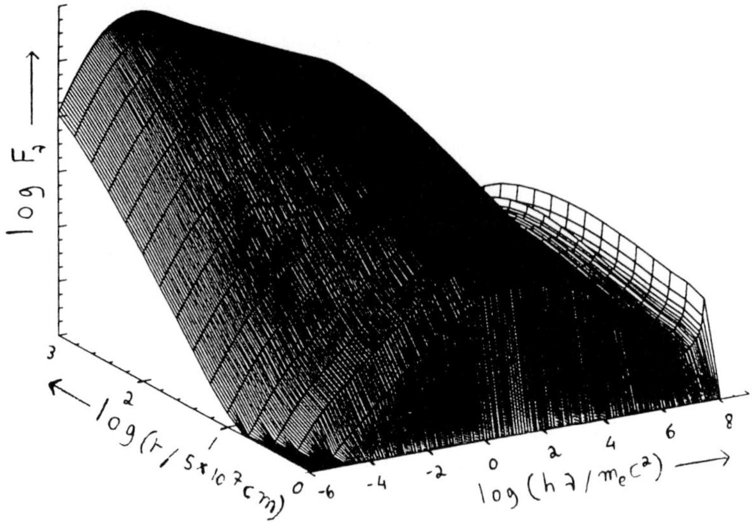

Fig. 4 – The photon number density per logarithmic energy interval (equivalently, the flux density) for the pair-photon cascade model of Fig. 1.

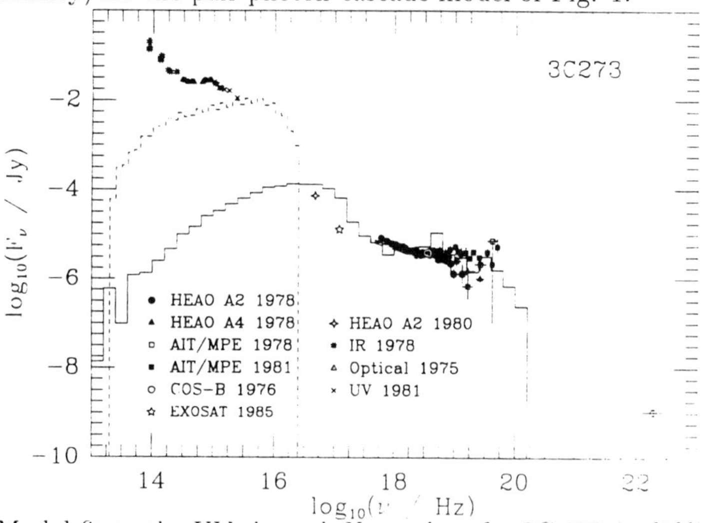

Fig. 5 – Model fit to the UV through X-ray data for 3C 273 (*solid* line), representing a thermal ($T_m = 2 \times 10^5$ K) disk spectrum (*dashed* line) and an upscattered disk component corresponding to the radiative deceleration of a pe^- jet from $\gamma_i = 10^4$ to $\gamma_\infty \approx 8$ within the funnel of a thick accretion disk (half-opening angle of 45°; see MK89 for details). The fit is for a jet inclination angle $\theta \approx 19°$.

4 IMPLICATIONS TO EXTRAGALACTIC RADIO SOURCES

One potential difficulty with the radiative deceleration scheme for superluminal sources is that, if the radio emission in these jets is dominated by the e^+e^- component, then the strong Compton drag in the immediate vicinity of the central black hole poses severe constraints on the maximum value of γ_∞ (Phinney 1987; MK89; Abramowicz, Ellis & Lanza 1990). Our model offers two alternative possibilities of alleviating this difficulty. In the first place, if the injected beam consists of protons and the soft photon intensity decreases sufficiently rapidly with distance from the center, then it is conceivable that the beam deceleration length will be of the order of the size of the emission region so that a **significant** fraction of pairs will be produced in regions where the radiative deceleration is less catastrophic (see §2.2). The second possibility is that the deceleration of the pairs is reduced on account of a large ($\gtrsim 1$) Thomson optical depth $\tau_\perp$ across the base of the jet that limits the penetration of ambient soft photons (see eq. [30] and §III*b* in MK89). Both of these suggestions need to be explored in detail before a firm conclusion can be reached. In particular, a nonnegligible $\tau_\perp$ may involve a feedback effect on the pair creation process (the more pairs produced, the larger the optical depth, which in turn implies less pair production), complicating the particle-photon cascade by rendering it *nonlinear.*

In discussing the formation of e^+e^- jets, one needs to address the question of *pair annihilation* (e.g., Rees 1984). Assuming a cold, pair-dominated jet, the annihilation length l_a is $\sim (32/3)(\gamma_\infty^2/\tau_\perp)R \approx 5\times10^{16}\,(\gamma_\infty/8)^2\,(R/7\times10^{13}\,\mathrm{cm})\,\tau_\perp^{-1}$ cm (where R is the jet's radius). In order for the e^+e^- component not to annihilate below the observed parsec scale, the outflow has to diverge over distances $r \lesssim l_a$ in such a way that the pair density decreases with r more rapidly than r^{-1}: one possibility is that the annihilation process itself regulates the jet expansion (see Becker & Begelman 1990). It is worth pointing out in this connection that, despite the presence of cold pairs in the jet, this scheme avoids the so-called "dead electron" problem (the accumulation of cold electrons in the source, which should lead to a detectable, but so far unobserved, steepening of the high-energy X-ray spectrum) that often afflicts stationary pair-plasma models, because in this case the pairs move rapidly out of the X-ray emission region. Another relevant comment is that the large-scale structure of the jet might be affected not only by pair annihilation but, at even larger distances ($\sim 1-100$ pc) from the nucleus, by the decay of the relativistic neutrons that are produced in photomeson reactions near the beam injection point (cf. Begelman et al. 1990).

In summary, although the work on the radiative deceleration model is still not complete, our preliminary results indicate that this scheme could provide viable clues to the transformation of ultrarelativistic beams into relativistic jets in extragalactic radio sources, incorporating kinematic, radiative, and structural constraints from observations of these objects. We are currently using our modified pair-plasma code to calculate the effects of proton-photon interactions and of anisotropic and nonuniform radiation fields, and we intend to further generalize the code to study nonlinear cascades. I expect that a more complete picture will emerge in the not-too-distant future.

I am grateful to my collaborators on this project, P. Coppi and J. Kartje, for valuable comments and assistance. This research was supported in part by NASA grant NAGW-1636.

REFERENCES

Abramowicz, M. A., Ellis, G. F. R., & Lanza, A. 1990, ApJ, 361, 47

Becker, P. A., & Begelman M. C. 1990, ApJ, 364, 203

Begelman, M. C., Rudak, B., & Sikora, M. 1990, ApJ, 362, 38

Blandford, R. D. 1976, MNRAS, 176, 465

Bridle, A. H., & Perley, R. A. 1984, ARAA, 22, 319

Burns, M. L., & Lovelace, R. V. E. 1982, ApJ, 262, 87

Coppi, P. S. 1991, MNRAS, in press

Laing, R. A. 1989, in Hot Spots in Extragalactic Radio Sources, ed. K. Meisenheimer & H.-J. Röser (Berlin: Springer), 27

Lovelace, R. V. E. 1976, Nature, 262, 649

Lovelace, R. V. E., MacAuslan, J., & Burns, M. L. 1979, in AIP Conf. Proc. 56, Particle Acceleration Mechanisms in Astrophysics, ed. J. Arons, C. McKee, & C. Max (New York: Amer. Inst. of Phys.), 399

Lovelace, R. V. E., Wang, J. C. L., & Sulkanen, M. E. 1987, ApJ, 315, 504

Mastichiadis, A. 1986, A&A, 169, 373

Mastichiadis, A. 1991, MNRAS, submitted

Mastichiadis, A., Marscher, A. P., & Brecher, K. 1986, ApJ, 300, 178

Melia, F., & Königl, A. 1989, ApJ, 340, 162 (MK89)

Phinney, E. S. 1983, Ph.D. Thesis, University of Cambridge

Phinney, E. S. 1987, in Superluminal Radio Sources, ed. J. A. Zensus & T. J. Pearson (Cambridge: Cambridge Univ. Press), 301

Porcas, R. W. 1987, in Superluminal Radio Sources, ed. J. A. Zensus & T. J. Pearson (Cambridge: Cambridge Univ. Press), 12

Rees, M. J. 1984, in IAU Symp. 110 on VLBI and Compact Radio Sources, ed. R. Fanti, K. Kellermann, & G. Setti (Dordrecht: Reidel), 207

Sikora, M., Kirk, J. G., Begelman, M. C., & Schneider, P. 1987, ApJ, 320, L81

Walker, R. C., Walker, M. A., & Benson J. M. 1988, ApJ, 335, 668

Wardle, J. F. C. 1977, Nature, 269, 563

Zdziarski, A. A. 1988, ApJ, 335, 786

Why do All the Extragalactic Jets Have Lorentz Factors Less Than Twenty ?

MAREK A. ABRAMOWICZ (1)

Space Telescope Science Institute, 3700 San Martin Drive
Baltimore, MD 21218, USA

(1) Permanent address: Nordita, Blegdamsvej 17, Copenhagen, Denmark. Also: ICTP, Strada Costiera 11, Trieste, Italy.

ABSTRACT

The observed Lorentz factors of extragalactic jets are never greater than 20. It was recently demonstrated (Abramowicz, Ellis and Lanza, 1989, 1990) that this may be explained as the effect of relativistic Compton drag, which gives an upper limit for the asymptotic Lorentz gamma factors of extragalactic jets, $\gamma_{\rm max} \approx (m_p/m_e)^{1/3} \approx 20$. The precise value of $\gamma_{\rm max}$ depends only very weakly on astrophysical details.

1. INTRODUCTION

It is known that an object moving toward us with a very high Lorentz factor γ_0, and very close to the line of sight, will be observed as having a superluminal transverse speed: $v_{\rm obs} \approx \gamma_0 c$. Superluminal motions have been detected in about 30 extragalactic jets emerging from active galactic nuclei by measuring proper motions of individual features seen in the VLBI radio maps which have been taken at several different times (see e.g. Porcas 1987). The data indicate that about one-half of the observed jets have speeds in the slow range ($1 < \gamma_0 < 5$) and the other half in the fast range ($5 < \gamma_0 < 20$). No jets with $\gamma_0 > 20$ are observed. The only possible exception is the jet in AO 0235+164, for which Baath, Zhang and Nicholson (1990) tentatively reported $\gamma_0 = 48$. However, these results were based on comparing two maps only, and the authors have doubts concerning identification of the individual features in these two maps. Not only the upper limit of $\gamma \approx 20$ emerges from the observational data, but also there are some impressive statistical differences between the jets having $\gamma < 5$ and those having $\gamma > 5$.

According to the present understanding, active galactic nuclei are most probably accretion disks around supermassive black holes. Extragalactic jets originate very close to the black hole and many theoretical models for the initial conditions for the jet motion, based on the black hole accretion processes, have been proposed (Rees, 1966, 1978, 1985). In particular, it was pointed out that the jets could be made of the electron-proton plasma, or of the electron-positron plasma, and that there are two kinds of jets, depending on the initial speeds.

The "slow" jets could be made of both electron-proton and electron-positron plasma. They start with very small, non-relativistic, speeds and are accelerated in the ordered magnetic field of the disk (Blandford 1976, Lovelace 1976, Lovelace, Berk and Contopoulos, 1991), or by radiation pressure gradient in the thick accretion disk funnels (Abramowicz and Piran, 1980, Sikora and Wilson 1981, Melia and Königl 1989), or by hydrodynamical effects similar to these occuring in the de Laval nozzle (Nobili, Calvani and Turolla, 1985). In all the calculated models of the slow jets the final speed (corresponding to superluminal motion observed in the VLBI region) was never greater than about $\gamma \approx 5$.

The "fast" jets are made of the electron-positron pairs. They start with extremely high speeds, up to $\gamma \approx 10^{10}$, due to pre-acceleration of electron and positrons by the black hole unipolar induction (Blandford and Znajek 1977, Macdonald and Thorne 1982, Phinney 1983).

Thus, depending on a particular jet model, and on the assumed details of a particular astrophysical situation (mass of the central black hole and its angular momentum, magnetic field of the accretion disk), the initial speeds postulated by theory cover 10 orders of magnitude, $1 < \gamma < 10^{10}$. The problem is why the *observed* speeds of jets fall in the rather narrow range $1 < \gamma_0 < 20$. Abramowicz, Ellis and Lanza (1989, 1990) suggested that this dramatic insensivity to the initial conditions is due to relativistic Compton drag. Elementary but very general discussion of the Compton drag shows that there is an upper limit for the asymptotic speed of jet,

$$\gamma_{\max} \approx \left(\frac{m_p}{m_e}\right)^{1/3}. \qquad (1)$$

The limit (1) is given just by the cubic root of the mass of the proton to mass of the electron ratio. This reflects the universal character of the physical processes involved. In this lecture I describe arguments used by Abramowicz, Ellis and Lanza to derive the above formula.

2. EQUATION OF MOTION FOR A SUPERLUMINAL JET

It is generally assumed that active galactic nuclei consist of accretion disks orbiting supermassive black holes located at the nuclei of their host galaxies. VLBI radio maps of jets correspond to regions very far from the central black hole and accretion disk. At these large distances the jet feels mostly gravity and radiation pressure of the center; each particle inside the jet is pulled in by the gravitational field of the black hole and pushed out by the radiation field of the disk. As the jet motion is mainly governed by competition between gravitational force and radiation pressure force, one important parameter which enters the equation of motion is the Eddington parameter Γ, defined as the ratio of the total luminosity of the disk to the Eddington luminosity of the jet, which is determined by the mass of the central black hole M, the inertia of the particles in the jet (m_p or m_e), and the Thomson cross-section σ_T. When $\Gamma = 1$ the gravitational and radiative forces would be exactly in balance for a particle with zero speed.

The only other relevant information about the radiation field (very far from the center) is the viewing angle α, which tells what fraction of the sky is occupied by the luminous disk. At the large VLBI distances the brightness distribution within the visible disk is irrelevant, and $\sin\alpha \ll 1$.

It is convenient to express the distance r from the center by the dimensionless gravitational redshift ξ, which is a monotonically increasing function of r, and goes to 1 when r goes to infinity,

$$\xi = \left(1 - \frac{r_G}{r}\right)^{1/2}, \tag{2}$$

Here $r_G = 2GM/c^2$ is the gravitational radius of the central black hole with the mass M. It is also convenient to express the speed v of the jet by the dimensionless Lorentz gamma factor,

$$\gamma = \frac{1}{\sqrt{1-\beta^2}}, \qquad \beta = \frac{v}{c}. \tag{3}$$

Abramowicz, Ellis and Lanza (1990) wrote down an explicit form of the equation of motion for the most general case of axially symmetric gravitational and radiation field. As an example they gave analytic expressions for all the structural functions appearing in the equation of motion (gravitational force in the comoving frame, components of the radiation tensor) in the special case of an isotropically radiating spherical star. Their formula is based on the earlier calculations of Jaroszyński, Abramowicz and Paczyński (1980). Vokrouhlický and Karas (1991) calculated the structural functions numerically for a thin accretion disk around a Kerr black hole. These disks have very sub-Eddington luminosities and for this reason, as we shall see later, are relevant only for slow jets. It would be very important to apply the formalism developed by Vokrouhlický and Karas to the case of thick accretion disks around Kerr black holes, which are directly relevant for fast, superluminal jets.

At a finite distance from the center the general equation of motion for the jet is very complicated, but far away, in the VLBI region relevant for the interpretation of the observational data, the equation for a superluminal jet ($\gamma \gg 1$, $\Gamma \gg 1$) is amazingly simple and it reads,

$$\frac{d\ln\gamma}{d\ln\xi} = \left(\frac{\Gamma}{\xi\gamma^3}\right)\left[(2-b)\gamma^4 + \frac{b}{4} - \left(\frac{\xi\gamma^3}{\Gamma}\right)\right],$$
$$b = 2\left(1 - \frac{1}{48}\sin^4\alpha\right) + O^6(\sin\alpha), \tag{4}$$
$$\sin\alpha = \frac{a\xi}{x}, \qquad x = \frac{r}{r_G}.$$

Here a is a constant of the order of unity connected to the radius of the luminous disk.

The quantity in the square bracket equals zero for

$$\gamma = \tilde{\gamma}(\xi, \Gamma, a). \tag{5}$$

The function $\tilde{\gamma}(\xi, \Gamma, a)$ is called the saturation speed. When the speed is smaller than the saturation speed, the radiation field accelerates the jet.

However, it is immediately clear from the equation of motion (4) that when the jet speed is greater than the saturation speed, $\gamma > \tilde{\gamma}$, the radiation field slows the jet down (the derivative $d\ln\gamma/d\ln\xi$ is negative). The slow down is due to the Compton drag; in the reference frame comoving with the jet the particles see an advective photon flux in the direction opposite to the motion. At a given location and fixed background radiation field the advective flux increases with the speed of the jet.

For $\Gamma \gg 1$ the saturation speed increases monotonically with the radius. Because $\tilde{\gamma} \neq$ const, the solution with $\gamma = \tilde{\gamma}$ everywhere is not possible.

3. THE UPPER LIMIT FOR THE SPEED OF JETS

The observational estimates based on fitting observed electromagnetic spectra of the accretion disks in active galactic nuclei to those calculated from theoretical disk models (Laor and Netzer 1990, Padovani 1989, Sun and Malkan 1989, Wandel 1989) point to disk luminosities not much different from the Eddington luminosity, $0.1 < \Gamma_{\rm disk} < 10$. The disk is made of ordinary plasma—a mixture of perfect gas and radiation. It contains electrons, which exchange momentum with photons through Thomson scattering, and protons which are strongly coupled to the electrons by the

Coulomb force. Momentum change is therefore determined by electrons, through the Thomson cross section σ_T, and inertia by protons through the proton mass, m_p. The Eddington flux ($\Gamma_{\rm disk} = 1$) corresponds to equilibrium between the radiative acceleration, which is proportional to σ_T/m_p, and the gravitational acceleration g. In the case of an electron-proton jet the situation is similar, and therefore $\Gamma_{e-p} = \Gamma_{\rm disk}$. However, in the case of an electron-positron jet the inertia is smaller by the factor m_p/m_e and therefore $\Gamma_{e-e+} = (m_p/m_e)\Gamma_{\rm disk}$. Thus, for the fast jets one has $\Gamma \gg 1$ because

$$0.1\left(\frac{m_p}{m_e}\right) < \Gamma < 10\left(\frac{m_p}{m_e}\right). \tag{6}$$

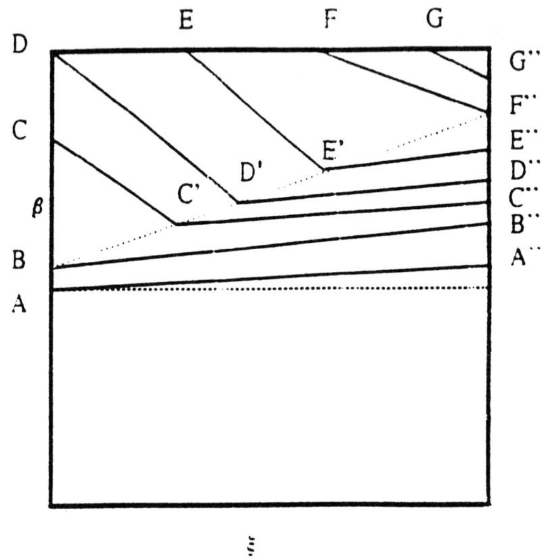

Figure 1 — Motion of jets for the $\Gamma \gg 1$ case.

Figure 1 shows schematically the seven topologically different solutions for the jet motion, $\beta = \beta(\xi)$ corresponding to a situation with $\Gamma \gg 1$. The saturation speed $\tilde{\gamma}(\xi)$ is shown by the dotted line BC′D′E′F″. The horizontal dotted line which starts at the point A, corresponds to $\beta = 0$.

The solutions AA″, BB″, CC″ start close to the central part of the accretion disk with the initial speeds smaller than the speed of light, the solution DD′D″ starts close to the center with the light speed, and all the other solutions start also with the light speed but further away from the center.

Let us consider first the solution BB″. It starts with the initial speed corresponding to the saturation speed at the center, $\beta_{\rm B} = \tilde{\beta}(\xi_0)$, and arrives at infinity with the speed corresponding to the point B″. Note, that all the solutions which start close to the center with initial speeds in the range

$$0 < \beta < \beta_{\rm B} \tag{7}$$

arrive at infinity with the speeds in the range

$$\beta_{\rm A''} < \beta < \beta_{\rm B''}. \tag{8}$$

These solutions correspond to the "slow" jets. It is interesting to note, that the point B″, which gives an upper limit for the observed speeds of the slow jets (in the case $\Gamma \approx 10 m_p/m_e$, which is the upper limit for Γ suggested by observations), corresponds to $\gamma_{\rm B''} \approx 5$.

Let us now consider any of the solutions between CC′C″ and FF″. Jets start with very high initial speeds,

$$\beta_{\rm B} < \beta \leq 1, \tag{9}$$

greater than the corresponding saturation speed. For this reason they are initially deccelerated by the Compton drag until they reach the saturation speed, and later they are accelerated again, but because the saturation speed is monotonically increasing function of radius (and ξ), they cannot reach the saturation speed again. Thus, they all arrive at infinity with the speeds corresponding to the range,

$$\beta_{\rm B''} < \beta < \beta_{\rm F''}, \tag{10}$$

which is considerably smaller than the initial range of speeds. The upper limit for the speed of jets in this class corresponds to the point F″. Note that the solutions of the GG″ type have final velocities greater than the one at the point F″. However, these solutions correspond to the jets which start very far from the center. There are no physical processes far away from the black hole which could pre-accelerate jets to these initial high speeds which are needed for this type of solution, and for this reason these solutions are astrophysically impossible.

It is thus clear that the upper limit for the possible speeds of jets is given by the point F″, corresponding to the saturation velocity at infinity. From equation of motion (4) one gets

$$\gamma_{\rm F''} = \left(\frac{\Gamma}{2}\right)^{1/3}. \tag{11}$$

This equation together with equation (5), which is the astrophysical condition for Γ immediately gives the upper limit for the speeds of jets (1), derived by Abramowicz, Ellis and Lanza.

The limit (1) does not depend on astrophysical details if the astrophysical range for Γ is roughly given by (5). Speeds of jets much higher than permitted by (1) are consistent with much higher disk luminosities. For example, if the jet in AO 0235+164 has indeed $\gamma_0 = 48$ this implies an intrinsic disk luminosity about 1000 times in excess of the Eddington one. Although such luminosities are theoretically possible (from thick accretion disks, see Abramowicz, Calvani and Nobili 1980 for an explanation), there is no clear observational evidence for them.

Therefore, one concludes that if the luminosities of accretion disks in active galactic nuclei do not substantially (more than factor of about 10) exceed the Eddington limit, the observed (apparent) speeds of superluminal jets are limited by the Compton drag and cannot exceed the light speed by more than about 20 times.

REFERENCES

Abramowicz M. A., Calvani M. and Nobili L. 1980, *Ap. J.*, **242**, 772.

Abramowicz M. A. and Piran T. 1980, *Ap. J. (Letters)*, **241**, L7.

Abramowicz M. A., Ellis G. F. R. and Lanza A. 1989, in *BL Lac Objects*, ed. L. Maraschi, T. Maccacaro and M.-H. Ulrich (Lecture Notes in Physics, No. 334: Berlin, Springer-Verlag), p. 358.

Abramowicz M. A., Ellis G. F. R. and Lanza A. 1990, *Ap. J.*, **361**, 470.

Baath L. B., Zhang F. J. and Nicholson G. 1990, *Astr. Ap*, Onsala Radio Observatory preprint.

Blandford R. D. 1976, *M.N.R.A.S.*, **176**, 465.

Blandford R. D. and Znajek F. J. 1977, *M.N.R.A.S.*, **179**, 433.

Jaroszyński M., Abramowicz M. A. and Paczyński B. 1980, *Acta. Astr.*, **30**, 1.

Laor A. and Netzer H. 1990, *M.N.R.A.S.*, **338**, 897.

Lovelace R. V. E. 1976, *Nature*, **262**, 649.

Lovelace R. V. E., Berk H. L. and Contopoulos J. 1991, *Ap. J.*, **379**, 696.

Macdonald D. and Thorne K. S. 1982, *M.N.R.A.S.*, **198**, 345.

Melia F. and Königl A. 1989, *Ap. J.*, **340**, 162.

Nobili L., Calvani M. and Turolla R. 1985, *M.N.R.A.S.*, **303**, 573.

Padovani P. 1989, *Astr. Ap.*, **209**, 27.

Phinney E. S. 1983, in *Astrophysical Jets*, ed. A. Ferrari and A. Pacholczyk (Dordrecht: Reidel), p. 201.

Porcas R. W. 1987, in *Superluminal Radio Sources*, ed. J. A. Zensus and T. J. Pearson (Cambridge: Cambridge University Press), p. 12.

Rees M. J. 1966, *Nature*, **211**, 468.
Rees M. J. 1978, *Observatory*, **98**, 210.
Rees M. J. 1985, in *Extragalactic Energetic Sources*, ed. V. K. Kapahi (Bangalore: Indian Academy of Science), p. 53.
Sikora M. and Wilson D. B. 1981, *M.N.R.A.S.*, **197**, 529.
Vokrouhlický D. and Karas V. 1991, *Astr. Ap.*, in press.
Wandel A. 1989, in *Proc. IAU Symp. 134, Active Galactic Nuclei*, (Dordrecht: Kulver).

Electron-positron beam production in compact radio sources

R. SCHLICKEISER, U. ACHATZ

Max-Planck-Institut für Radioastronomie

Auf dem Hügel 69, D-5300 Bonn 1, FRG

ABSTRACT

The study of the acceleration of energetic electrons and positrons in active galactic nuclei has shown that the concerted action of accretion, accretion shock, magnetic fields and MHD turbulence produces a nearly mono-energetic distribution of relativistic electrons and positrons (the 'pile-up'-distribution). Via excitation of oblique Langmuir waves the pile-up heats the thermal plasma, thereby causing an explosion so that the resulting blast wave carries the relativistic particles to the outer parts of the galaxy. The time evolution of the initially isotropic pile-up is studied for the two extreme cases of presence or absence of Alfvén waves in the galactic medium. In the latter case due to the outward decrease of the magnetic field adiabatic focussing and synchrotron radiation losses transform the isotropic pile-up into beams of relativistic electrons and positrons propagating along the magnetic field lines giving ready explanations both for the formation of superluminal VLBI jets and for the radio gap phenomenon.

1. INTRODUCTION

The nonthermal origin of the radiation from the core, jet and hot spot of extragalactic radio sources has been established beyond any doubt by polarization measurements from radio to optical frequencies. The synchrotron radiation process traces the combination of the relativistic electron density and some power of the magnetic field strength, so that the knowledge of these two constituents in these objects is the key to understand the physical processes in these objects, in particular the link between the activity of the central engine and the formation of beams, jets and hot spots. We have developed a "2-phase" kinetic plasma model that describes a time sequence of events related to the generation and evolution of energetic particle beams and jets in these sources (Achatz et al. 1990). According to our model the central object undergoes two repeating phases: in a "quiescent phase" relativistic electrons and positrons are efficiently accelerated at the central accretion shock, whereas in a short and violent "flaring phase" the accelerated particles are ejected into the surrounding medium.

2. PARTICLE ACCELERATION IN THE QUIESCENT PHASE BY ALFVÉN WAVES AND SHOCK WAVES FROM THE ACCRETION FLOW ($0 \leq t \leq t_b$)

We consider an accreting object surrounded by a magnetized low $\beta = 8\pi nkT/B_0^2 < 1$ plasma in the nucleus of a galaxy. The central object accretes the surrounding matter through a disk and also provides an ordered magnetic field perpendicular to the disk (see sketch in Fig. 1). Standing shocks develop in the neighbourhood of the central object's magnetosphere and below and above the accretion disk which act as blunt obstacles in the path of the accretion flow. At the accretion shocks and in the layer of shocked gas behind the shock (the *accretion shock magnetosheath*) energetic charged particles are efficiently accelerated via Fermi-type processes. The injection source of electrons and positrons may well be the $\gamma - \gamma$ pair production in the hot accretion disk (Zdiarski 1991). Energetic particles gain energy by resonant interactions with Alfvén waves which are associated with the accretion flow. There are various sources of Alfvén waves upstream from the accretion shock:

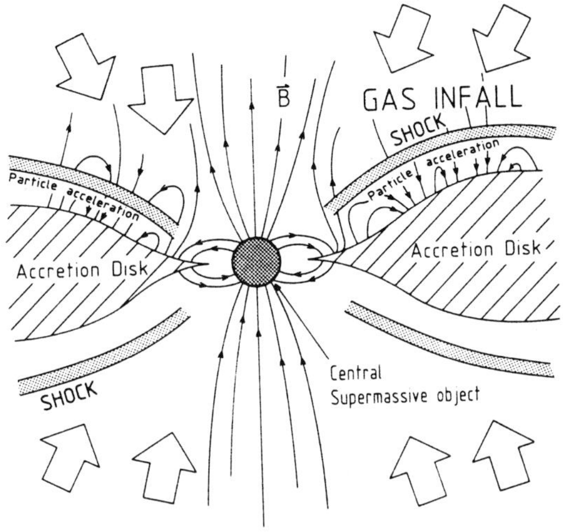

Fig. 1.
Sketch of the global structure of the central object of the active galaxy

a) stellar winds from solar-type stars in the central star cluster deliver Alfvén waves to the accretion flow;
b) infalling neutral accretion matter becomes ionized by the ultraviolet and soft X-ray radiation of the central object. These pick-up ions in the accretion flow generate Alfvén waves by virtue of their streaming (Lee and Ip 1987);
c) the streaming instability of ambient cosmic rays and of the upstream precursor distribution of accretion shock accelerated particles (Bell 1978);
d) cascading from the input at large eddies;
e) the electromagnetic ion beam instability from reflected particles at the accretion shock.

This upstream generated Alfvén wave turbulence becomes amplified in the interaction of the Alfvén waves with the accretion shock (McKenzie and Westphal 1969). Using the Rankine-Hugoniot relations the downstream Alfvén wave state can be calculated from the prescribed upstream Alfvén wave state. Starting from upstream backward moving (with velocity $U_1 - V_{A_1}$) Alfvén waves of magnetic field energy density $(\delta B_{b,1})^2$ only, Campeanu and Schlickeiser (1991) have demonstrated that in the interaction of these waves with a plane, super-Alfvénic accretion shock the downstream Alfvén waves are amplified by the factor

$$W \equiv \frac{(\delta B_{f,2})^2 + (\delta B_{b,2})^2}{(\delta B_{b,1})^2} = \frac{r}{4}(r^{1/2}+1)^2 \left(\frac{M+1}{M+r^{1/2}}\right)^2 (1+\psi^2) \tag{1}$$

where $r = U_1/U_2$ denotes the gas compression ratio, $M = U_1/V_{A_1}$ is the Alfvénic Mach number of the shock, and

$$\psi = \left(\frac{r^{1/2}-1}{r^{1/2}+1}\right) \frac{M+r^{1/2}}{M-r^{1/2}}. \tag{2}$$

The interaction leads to the presence of forward moving (velocity $U_2 + V_{A_2}$) and backward moving (velocity $U_2 - V_{A_2}$) Alfvén waves in the downstream region with respective energy densities $(\delta B_{f,2})^2$ and $(\delta B_{b,2})^2$, respectively, so that the cross helicity state of the Alfvénic turbulence changes from its upstream value $H_{c,1} = -1$ to

$$H_{c,2} = \frac{(\delta B_{f,2})^2 - (\delta B_{b,2})^2}{(\delta B_{f,2})^2 + (\delta B_{b,2})^2} = \frac{\psi^2 - 1}{\psi^2 + 1}. \tag{3}$$

A strong adiabatic shock ($M \to \infty$) in a nonrelativistic medium leads to $r \to 4$, $\psi \to \frac{1}{3}$, $H_{c,2} \to -0.8$, $W \to 10$. The incoming Alfvén waves are amplified by about a factor 10, and the cross helicity state $|H_{c,2}| \neq 1$ makes stochastic acceleration by the 2nd order Fermi process in the downstream magnetosheath region unavoidable (Dung and Schlickeiser 1990). If radiative cooling of the magnetosheath material occurs the gas compression ratio can be much larger than $r \gg 4$ (Drury 1983) leading to a downstream cross helicity state $H_{c,2}(r \gg 1) \to 0$ and much stronger amplification $W(r \gg 1) \to \frac{r^2}{2} \simeq 0(10^3)$ of the upstream Alfvén turbulence. The magnetosheath region behind the accretion shock clearly is a region of intense Alfvén wave turbulence where energetic particles are confined for time scales much longer than their acceleration time by resonant particle-wave interactions. Schlickeiser (1984, 1985) has investigated the acceleration of electrons and positrons under circumstances where the confinement time of the particles is much longer than their acceleration time and their synchrotron and inverse Compton radiation loss time scales: an almost monoenergetic distribution at Lorentz factor

$$\gamma_M = 68.5 \left(\frac{V_{A_2}}{100\,\mathrm{km\,s^{-1}}}\right)^2 \left(\frac{\kappa_2}{10^{24}\,\mathrm{cm^2\,s^{-1}}}\right)^{-1} \left(\frac{W_H + W_{ph}}{10^{-5}\,\mathrm{erg\,cm^{-3}}}\right)^{-1} \tag{4}$$

with a low-energy power law tail $N(\gamma \ll \gamma_M) \propto \gamma^{-1}$ results for the differential number density of electrons and positrons. γ_M is that Lorentz factor value where the Fermi acceleration time scale ($\tau_F \propto \kappa_2 V_{A_2}^{-2}$) exactly balances the radiation loss time scale ($\tau_R \propto \gamma^{-1}(W_H + W_{ph})^{-1}$), where κ_2 denotes the downstream spatial diffusion coefficient of the particles and W_H and W_{ph} are the energy densities of the ambient magnetic field and the target photons, respectively. The optically thin synchrotron spectrum from this "pile-up" electron and positron distribution has been detected in our galactic center, both from the central source Sgr A* and the extended ($\simeq$ 50 pc) central bridge and arc region, as well as from the centers of the galaxies M81 and M104 (Lesch et al. 1988). In optically thick environment the pile-up distribution leads to efficient synchrotron masering (Crusius and Schlickeiser 1988a, Crusius-Wätzel 1990) explaining H_2O and OH-megamasers in the centers of galaxies (Crusius and Schlickeiser 1988b).

Besides the synchrotron maser instability the pile-up is unstable against the excitation of oblique Langmuir waves (Lesch et al. 1989) which are first damped by nonlinear Landau damping but ultimately, at time t_b, heat the background electron/proton plasma via the modulation instability once a critical energy density $W_c = n_e(k_B T_e)^2/(m_e c^2)$ in Langmuir waves is reached. This is the end of the quiescent phase of the acceleration process. The time t_b sensitively depends on the local background plasma parameters.

3. FLARING PHASE ($t_b \le t \le t_b + \Delta t$) AND SUDDEN EXPANSION OF THE CENTRAL PLASMA

The almost instantaneous increase of the background gas entropy due to the rapid modulation instability heating leads to an explosive outward motion of the plasma, carrying the relativistic particles into the outer regions of the galaxy. This scenario is supported by the measurements of Babadzanhanyants and Belokon (1985) that in 3C 345 optical bursts are in close time correlation with the generation of compact radio jets. After the blast wave has left the central system in time $\Delta t \simeq (L/c) \simeq 3$ yrs (L/1 pc), the central source returns to its quiescent phase, and the same processes as described in Chapter 2 start again.

The outer regions of the galaxy are much more quiescent than the central magnetosheath region with a significantly lower level of Alfvén wave intensity. The further evolution of the injected isotropic **relativistic** electron and positron pile-up energy spectrum

$$f(\gamma, z, t = t_b) = \frac{N(\gamma, z, t = t_b)}{4\pi(m_e c)^2 \gamma_M^2} \simeq \frac{N_r}{4\pi(m_e c)^3 \gamma_M^2}\, \delta(\gamma - \gamma_M)\, \delta(z) \tag{5}$$

after its release from the central source depends on intensity level of resonant plasma waves $(\delta B_{\rm res})^2$ in the galaxy: if the ratio $(\delta B_{\rm res})^2/B_0^2$ is smaller than the critical value ($r_{\rm L}$: electron/positron Larmor radius, D: galaxy size)

$$I_{\rm c} = \frac{1}{3}(r_{\rm L}/D)^{1/3} = 8 \cdot 10^{-6} \left[\frac{\beta\gamma}{B\,({\rm G})\;D\,({\rm pc})}\right]^{1/3} \tag{6}$$

the electron/positron pitch-angle scattering time scale is longer than their propagation time, and the dynamics of the particles can be treated as scatter-free. In the opposite case $(\delta B_{\rm res})^2/B_0^2 \geq I_{\rm c}$ rapid pitch-angle scattering preserves the isotropy of the electron/positron distribution function, and it suffices to study the evolution of the isotropic part of the electron/positron phase space density. We consider both cases in turn.

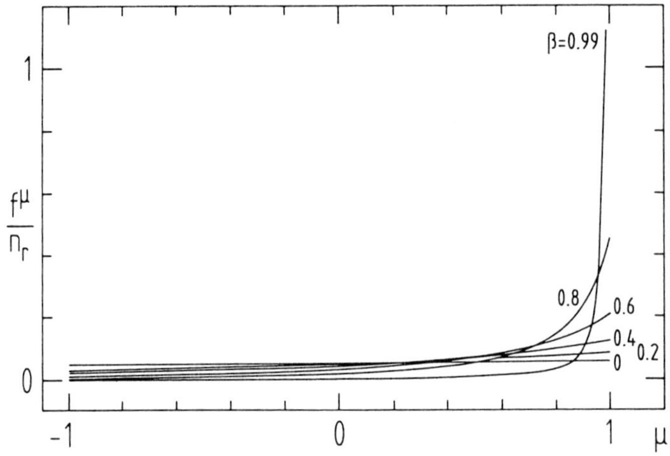

Fig. 2. The space- and Lorentz factor-averaged relativistic particle distribution as produced by synchrotron losses and adiabatic focussing at different values of $\beta = \tanh{(ct/2\ell)}$

3.1 Electron/positron beam formation in scatter-free transport

Synchrotron radiation losses and adiabatic focussing in the spatially decreasing central dipole magnetic field ($B \propto z^{-3}$) change the pile-up distribution function (5) with time into (Achatz et al. 1990)

$$f(t > t_{\rm b}, z, \mu, \gamma) = \frac{N_{\rm r}}{4\pi\, B(z=0)(m_{\rm e}c)^3} \frac{\gamma_0^2}{\gamma^4} \frac{(1-\mu_0)^2}{(1-\mu^2)}\, \delta(z_0)\, \delta(\gamma_0 - \gamma_{\rm M}) \tag{7}$$

where

$$\mu_0 = \frac{\mu - \tanh\left(\frac{ct}{2\ell}\right)}{1 - \mu \tanh\left(\frac{ct}{2\ell}\right)} \tag{8}$$

$$z_0 = z - 2\ell \ln \left[\cosh\left(\frac{ct}{2\ell}\right) + \frac{\mu - \tanh\left(\frac{ct}{2\ell}\right)}{1 - \mu \tanh\left(\frac{ct}{2\ell}\right)} \sinh\left(\frac{ct}{2\ell}\right) \right] \tag{9}$$

$$\gamma_0 = \frac{\gamma}{1 - \frac{2\ell\gamma}{c\tau_s} \tan\left(\frac{ct}{2\ell}\right) \frac{1-\mu^2}{1-\mu \tanh\left(\frac{ct}{2\ell}\right)}} \tag{10}$$

where $\ell = |-B/(dB/dz)| = z/3$ is the focussing length and $\tau_s = 5.2 \cdot 10^8\,(B_0/\mathrm{G})^2$ s the synchrotron loss time. In Fig. 2 we show the resulting space- and Lorentz factor-averaged distribution with respect to pitch angle μ for different times. Obviously all particles tend to $\mu \to 1$ at $t \to \infty$, i.e. at late times the distribution becomes a beam propagating along the magnetic field. At late times the Lorentz factor of the beam particles becomes $\bar{\gamma}(t \to \infty) = \frac{\gamma_M}{2\alpha} \ln(1 + 2\alpha)$ where $\alpha = \frac{2\ell\gamma_M}{c\tau_s} = 13 \left(\frac{B_0}{0.1\,\mathrm{G}}\right)^2 \left(\frac{\ell}{\mathrm{pc}}\right) (\gamma_M/10^4)$, indicating that a significant fraction of the initial energy remains in the beam, e.g. for $\alpha = 13$ one finds $\bar{\gamma}/\gamma_M = 0.13$, and for $\alpha = 30$ $\bar{\gamma}/\gamma_M = 0.07$. Since particles moving along the ordered magnetic field do not synchrotron radiate this mechanism readily explains the formations of the "VLBI-jets" and the occurrence of radio gaps. It was noted before by Sanders (1974) that beams generated by adiabatic focussing are able to explain the phenomenon of "superluminal motion". The further fate of this electron/positron beam, i.e. its stability, is discussed in Achatz and Schlickeiser (1991) and Sol et al. (1989). They have demonstrated that the electron/positron beam can propagate large ($\sim$ kpc) distances if the value of $\bar{\gamma}$ is below the critical value $\gamma_n = (m_p/m_e)(V_A/c) = 1.33 \cdot 10^4\, B(\mathrm{G})\, n_e^{-1/2}\,(\mathrm{cm}^{-3})$, which would account for the energy transport over large scales from the nucleus to the hot spot.

3.2 Low speed jets

At the presence of a significant level of low-frequency plasma waves the electron/positron distribution function stays isotropic due to rapid pitch-angle scattering. The evolution of the isotropic part of the particle's distribution function is determined by the "convection-diffusion" transport equation (Schlickeiser 1989). In a galactic medium of cross helicity state $H_c = +1$ the electrons/positrons convect with bulk speed $V = U_{\mathrm{Gas}} + V_A$ along the magnetic field and are subject to spatial diffusion, adiabatic deceleration, bremsstrahlung, synchrotron and inverse Compton losses. Assuming the injection of many flare events, i.e. $t_b \ll D/(U + V_A) \simeq 0(10^4\,\mathrm{yrs})$, Pohl et al. (1991) calculated the resulting quasi-steady equilibrium distribution function $N(E, z)$ for mono-energetic injection at γ_M (Eq. (5)) using the analytical solutions of Lerche and Schlickeiser (1981). A brief account of this work is given in Pohl (1991). In case of injection at large energies the resulting electron/positron distribution function approaches the power law $N(E) \propto E^{-3/2}$ which can nicely account for Valtaoja's (1990) "universal" optical thin synchrotron spectrum $I(\nu) \propto \nu^{-0.2}$ for flaring sources and for the spatial and spectral evolution of the galactic center jet (Sofue et al. 1989).

4. CONCLUSIONS

The 2-phase kinetic plasma model for the acceleration and propagation of electrons and positrons in centers of galaxies successfully explains the basic nonthermal features of compact extragalactic radio sources as flaring, beam formation, VLBI-jets, superluminal motion, radio gap, large-scale energy transport and hot spot formation. Its key ingredient is that the central accelerator generates the quasi-monoenergetic pile-up distribution for electrons and positrons.

5. REFERENCES

Achatz, U., Lesch, H., Schlickeiser, R.: 1990, A&A **233**, 391
Achatz, U., Schlickeiser, R.: 1991, these proceedings
Babadzhanyants, M.K., Belokon, E.T.: 1985, Astrofizika **23**, 459
Bell, A.R.: 1978, MNRAS **182**, 147
Campeanu, A., Schlickeiser, R.: 1991, A&A, submitted
Crusius-Wätzel, A.R.: 1990, ApJ **368**, 390
Crusius, A., Schlickeiser, R.: 1988a, A&A **196**, 327
Crusius, A., Schlickeiser, R.: 1988b, A&A **195**, L9
Drury, L.O.C.: 1983, Space Sci. Rev. **36**, 57
Dung, R., Schlickeiser, R.: 1990, A&A **240**, 537
Lee, M.A., Ip, W.H.: 1987, JGR **92**, 11041
Lerche, I., Schlickeiser, R.: 1981, ApJ Suppl. **47**, 33
Lesch, H., Schlickeiser, R., Crusius, A.: 1988, A&A **200**, L9
Lesch, H., Crusius, A., Schlickeiser, R.: 1989, A&A **209**, 427
McKenzie, J.K., Westphal, K.O.: 1969, Planet. Space Sci. **17**, 1029
Pohl, M.: 1991, these proceedings
Pohl, M., Reich, W., Schlickeiser, R.: 1991, in preparation
Sanders, R.H.: 1974, Nature **248**, 390
Schlickeiser, R.: 1984, A&A **136**, 227
Schlickeiser, R.: 1984, A&A **143**, 431
Schlickeiser, R.: 1989, ApJ **336**, 264
Sofue, Y., Reich, W., Reich, P.: 1989, ApJ **341**, L47
Sol, H., Pelletier, G., Asseo, E.: 1989, MNRAS **237**, 411
Valtaoja, E.: 1990, in: *Variability of Active Galaxies*, eds. W.J. Duschl, S.J. Wagner, M. Camenzind, Springer, Berlin, p. 117
Zdiarski, A.: 1991, these proceedings

Production of Relativistic e+ - e- Beams

G. HENRI AND G.PELLETIER

Laboratoire d'Astrophysique de l'Observatoire de Grenoble
CERMO B.P 53 X
38041 Grenoble Cedex
FRANCE

1 INTRODUCTION

The two main pieces of evidence for the presence of relativistic particles in AGN are first the ubiquitous detection of hard X-ray and probably γ - ray spectra, and second the observation of superluminal motions on VLBI scales, which seem to imply the presence of a plasma with relativistic bulk motion. Although a firm connection between these two phenomena has not been established so far, there seems to be some correlation, at least for some objects, between the apparition of a new VLBI knot and a structural change in the shape and luminosity of the hard X-ray component. Furthermore, the most fashionable models of X-ray emission in AGNs imply the copious production of e^+- e^- pairs via γ - γ interactions, reprocessing by Inverse Compton process the soft photons emitted for example by an accretion disk (Zdziarski *et al.* 1990). On the other hand, the superluminal components have been explained by the expansion of a relativistic e^+- e^- beam pervading a subrelativistic medium, either a wind or a jet probably emitted by an accretion disk (Sol *et al.* 1989, Pelletier *et al.* 1991). Such models will be referred as "two-flow models". The study of the stability of this kind of configuration has shown that a stable system can be obtained with respect to excitation of Langmuir and Alfvén waves, if the magnetic field is strong enough ($\omega_B > \omega_{pe}$) and the bulk Lorentz factor not too high ($\gamma < (m_e / m_p)^{1/2} \sim 43$). It is thus an attractive idea to search for a "unification scheme" between these two phenomena, relating the pair production in the compact region and the formation of a relativistic beam.

From a theoretical point of view, the strongest constraint for the acceleration of relativistic particles comes from the very high photon density around a compact object accreting near the Eddington limit, implying a very efficient Compton cooling. Defining as usual the compactness parameter of an object of size R and luminosity L as:

$$l_s = \frac{\sigma_T L}{4\pi m_e c^3 R} \tag{1}$$

the characteristic cooling time for a light particle is then

$$\tau_c \sim \frac{1}{l_s \gamma}\left(\frac{R}{c}\right) \tag{2}$$

that is much shorter than the typical escape time. A continuous, very efficient acceleration mechanism distributed all over the compact region is thus needed to maintain a relativistic particle distribution.

The same kind of constraint arises for the formation of a directed beam with relativistic bulk Lorentz factor. Phinney (1987) has shown that, in the photon field emitted by a near-Eddington accretion disk, any relativistic light particle emitted along the axis (if its energy lies below the quantum Klein-Nishina limit) would be very rapidly decelerated until it reaches a saturation velocity, for which the flux of the aberrated photon field vanishes. This velocity increases very slowly with the distance z (as $(\frac{z}{R_i})^{1/4}$, where R_i is the inner radius of the disk), until the acceleration becomes inefficient. Note that this value increases much less rapidly with distance than that associated with a central object of finite size R, which would increase as $\frac{z}{R}$. This is due to the fact that the main contribution of the radiation force comes from the external part of the disk, at a distance $r \sim z$ from the central object. The asymptotic Lorentz factor is then (Phinney 1987) $\gamma_{b\infty} \sim (l_s)^{1/7}$, that is at most 2-3. The only crucial assumption here is that the particle acceleration takes place only near the central object, and that the particle is then travelling in the photon field of an accretion disk, regardless of the acceleration mechanism. Obviously a long range acceleration mechanism is thus needed to obtain an e^+- e^- beam with high bulk Lorentz factor ($\gamma \sim 10$). Quantitative estimate shows that the minimum distance at which the acceleration must take place is of the order of $10^2 R_i$. The issue I will address now is how the two-flow model can provide such a mechanism.

2.PAIR CREATION WITH STOCHASTIC ACCELERATION

2.1 The model

We consider the following picture: an accretion disk around the central engine drives a powerful subrelativistic flow through an MHD process (Pelletier *et al.* 1991). Differential rotation in the disk and the jet can induce a moderate Alfvén turbulence, with an instability growth rate of the order of $r\frac{d\Omega}{dr} \sim \chi\Omega$ (Tagger *et al.* 1991). Through their coupling with magnetosonic modes, Alfvén waves can propagate toward the core and heat a pair plasma up to relativistic energies by stochastic 2nd order Fermi acceleration. Note that, since there is no stable orbit in the vicinity of the black hole ($r < 3\, r_g$ for a Schwarzschild metric), the MHD jet is likely to be hollow over a large distance due to the very large transverse diffusion time of protons. Because of the very low abundance of protons, no significant absorption of Alfvén waves will occur at frequencies smaller than the electron cyclotron frequency. Thus, contrary to the usual case, no energy threshold is expected for the acceleration. The pairs cool down by the inverse Compton (IC) process on soft photons, e.g, from the accretion disk, producing X and γ-rays that can at turn produce new pairs by photon-photon annihilation. Since the pairs are continuously accelerated, no particle injection is required and the particle distribution can be calculated self-consistently.

2.2 The Particle Distribution Function

The stochastic acceleration can be described by a diffusion coefficient in phase space. Assuming that the distribution function is nearly isotropic, and that the Alfvén waves energy spectrum is a power-law spectrum of index ν_A, with a maximum wavelength λ_{max}, the diffusion coefficient can be expressed in term of the Lorentz factor γ as:

$$D = D_0 \gamma^{\nu_A} \qquad (3)$$

with

$$D_0 = \left(\frac{V_*}{c}\right)^2 \eta_T \frac{c}{R} \left(\frac{\lambda_{max}}{r_L}\right)^{2-\nu_A} \qquad (4)$$

where V_* is the Alfvén velocity, $\eta_T = \frac{<\delta B^2>}{B^2}$ the turbulence parameter , and $r_L = \frac{m_e c^2}{eB}$. Numerical factors of order unity have been omitted. One can assume $\lambda_{max} \sim R$, the typical size of the core region, typically a few gravitational radii (lower estimates would give a higher value for D_0). The turbulence level can be estimated by balancing the instability growth rate ($\approx \chi\Omega$) and the turbulence transfer rate ($\approx \eta_T V_* / R$), which gives

$$\eta_T \approx \chi V_{Kepl} / V_*.$$

If $V_* \simeq c$ (see below) and near a black hole, values of 10^{-2} - 10^{-1} can easily be reached. On the other side, the IC cooling process is such that:

$$\dot{\gamma} = -A\, \gamma^2 \qquad (5)$$

where

$$A = \frac{4}{3} \sigma_T\, c\, n_S\, \varepsilon_S$$

for an isotropic photon field of density n_S and mean energy $\varepsilon_S\, m_e c^2$.
The distribution function $\rho(\gamma)$ obeys the Fokker-Planck (FP) equation:

$$\frac{\partial}{\partial t}\rho(\gamma) = \frac{\partial}{\partial \gamma}\gamma^{2+\nu_A} D_0 \frac{\partial}{\partial \gamma}\frac{\rho(\gamma)}{\gamma^2} + \frac{\partial}{\partial \gamma} A\gamma^2 \rho(\gamma) \qquad (6).$$

The stationary solution of this equation is the so-called "pile-up" distribution (Schlickeiser 1984):

$$\rho(\gamma) = \rho_0\, \gamma^2 exp\left[-\left(\frac{\gamma}{\bar{\gamma}}\right)^{3-\nu_A}\right] \qquad (7)$$

peaking around a characteristic energy:

$$\bar{\gamma} = \left[\frac{D_0}{(3-v_A)\,A}\right]^{\frac{1}{(3-v_A)}} \tag{8}$$

Numerical estimate of equation (8) with $v_A = 3/2$ (Kraishnan 1965) $B = 10^4\ G$, $R = 10^{13}$ cm and $\eta_T = 10^{-2}$ gives $\bar{\gamma} \sim 10^3$. Now pair production will occur if the energy of γ photons produced by the IC process lies above $m_e c^2$. By taking a δ - approximation of the IC differential cross section (Svensson 1987):

$$\frac{d\sigma}{d\varepsilon}(\,\varepsilon;\gamma,\varepsilon_s) = \sigma_T\,\delta\,(\varepsilon - \frac{4}{3}\gamma^2\varepsilon_s\,)\;\Theta\,(1 - \frac{4}{3}\,\gamma\varepsilon_s) \tag{9}$$

where ε_s, ε, and γ are respectively the initial and final photon energies and the initial particle Lorentz factor in units $m_e c^2$, and $\Theta\,(x\,)$ the step function accounting for the quantum Klein-Nishina decline of the cross-section, one gets the γ threshold for pair creation:

$$\gamma_{pc} = (\,3/4\varepsilon_s\,)^{1/2} \tag{10}$$

Assuming that the soft photons are emitted in the UV bump, $\varepsilon_s \sim 10^{-4}$ and $\gamma_{pc} \sim 10^2$. Under these conditions, copious pair production can take place if the plasma is optically thick to γ-photons, that is if the X-compactness $l_X \equiv \frac{\sigma_T L_X}{4\pi m_e c^3 R}$ is large, which is the case in compact objects emitting a fair fraction of the bolometric luminosity in X-rays.

The FP equation can be modified by adding a creation term in the right hand side; assuming saturated pair cascades ($\tau_{\gamma\gamma} \gg 1$), and that pairs of Lorentz factor γ are produced by photons of energy 2γ with X photons of energy $1/\gamma$ (twice the threshold in the c.m. frame), this term takes the following form:

$$P(\gamma) = \frac{3}{2}\,\frac{n_s\sigma_T}{\varepsilon_s}\,\frac{\rho(\gamma')}{\gamma'}. \tag{11}$$

with $\gamma' = (\,3\,\gamma/\,4\varepsilon_s\,)^{1/2}$.

It turns out (Henri and Pelletier, 1991) that the full FP equation possesses no stationary physical solution (i.e., normalizable and positive everywhere) if $\bar{\gamma}$ exceeds some threshold of the order of γ_{pc}, equation (10). Under these conditions, the pair creation becomes catastrophic with an exponential growth of pair density. Such a runaway behavior is characteristic of pair cascades with pair reacceleration. Here the very fast acceleration prevents pairs from cooling and annihilating. The excess pressure thus must drive a flow along the flux tube, since no other mechanism can balance the pair creation.

3 RELATIVISTIC BEAM FORMATION

We now address the question of relativistic expansion of the pair plasma. We assume that pitch angle scattering on Alfvén waves is efficient enough to keep the distribution function isotropic in the plasma comoving frame. We thus make use of relativistic hydrodynamics to describe the quasi-1D expansion along the flux tube of section area $S(z)$ (corresponding to the jet's interior). Neglecting pair annihilation, the mass transfer equation reads in fix coordinates:

$$\frac{\partial}{\partial t}\gamma_b n_* + \frac{1}{S}\frac{\partial}{\partial z}(S\beta_b\gamma_b n_*) = \nu^+\gamma_b n_* \qquad (12)$$

where β_b is the bulk velocity in c -units, γ_b is the corresponding Lorentz factor, n_* the comoving density and ν^+ the creation rate, defined as $\nu^+ \equiv \frac{1}{n_*}\int P(\gamma)d\gamma$.

To write down the energy-momentum transfer equation, we note that the flow cannot be considered as adiabatic since the acceleration and cooling times are both smaller than the characteristic escape time. Taking into account the interaction with radiation and turbulence, one gets the covariant equation

$$T^{\mu\nu}{}_{;\mu} = \mathfrak{F}^\nu \qquad (13).$$

where $T^{\mu\nu}$ is the matter energy-momentum tensor and $\mathfrak{F}^\nu$ the 4-force density vector describing the interaction between matter and radiation. In the comoving frame, one can assume that the plasma exchanges no net momentum with Alfvén waves, i.e., the corresponding $\mathfrak{F}_A^\nu$ is purely temporal, $\mathfrak{F}_A^\mu = (\mathfrak{F}_A^0, 0, 0, 0)$. On the other hand, it can exchange energy and momentum with photons, giving a contribution $\mathfrak{F}_r^\mu = (\mathfrak{F}_r^0, 0, 0, \mathfrak{F}_r^z)$. Performing a Lorentz transformation to come back in the observer frame, one can compact the 4 equations (13) into a single one for a stationary flow, namely:

$$\frac{p+e}{\gamma_b}\frac{\partial \gamma_b}{\partial z} + \frac{\partial p}{\partial z} = \frac{\mathfrak{F}_r^z}{\gamma_b}. \qquad (14)$$

where e and p are respectively the total energy density and internal pressure in the comoving frame. They can be calculated assuming that the distribution function, equations (7)-(8), remains approximately valid in the comoving frame.

Now as long as pair creation takes place, a stationary solution of equation (13) and (14) can exist only for a small creation rate ν^+, that is a value of $\bar{\gamma}$ slightly above, but close to the instability threshold without flow. For a quasi stationary flow, there must be some feedback on the acceleration mechanism that limits the value of $\bar{\gamma}$ below 10^2, a feature shared by all pair models with reacceleration (Done *et al.* 1990). One possibility is that the Alfvén velocity is decreased well below c by the increase of the rest frame plasma density $n_* \bar{\gamma} m_e c^2$, which occurs for $n_* \bar{\gamma} > 10^{10}\ cm^{-3}$ for $B = 10^4\ G$. As the jet opens, the magnetic field and probably the turbulence will decrease and $\bar{\gamma}$ will fall naturally below the pair creation threshold. When no creation longer occurs, equation (8) becomes:

$$S\beta_b\gamma_b n_* = C^{nst}. \qquad (15)$$

If the plasma is completely relativistic, that is $\beta_b \simeq 1$ and $e \simeq 3p \propto n_* \bar{\gamma}$ this can be combined with equation (14) to give:

$$\frac{\partial}{\partial z} \ln\left(\frac{\gamma_b^3 \bar{\gamma}}{S}\right) = \frac{3\mathfrak{F}_r^z}{e\gamma_b}. \tag{16}$$

The right hand term of this equation describes the effect of the radiation force felt by the plasma in its comoving frame, due to external photons from the accretion disk. This term becomes equal to zero for some value γ_r of γ_b, for which the flux associated to the aberrated radiation field vanishes. As mentioned above, this saturation velocity γ_r is nearly equal to $(\frac{z}{R_i})^{1/4}$ along the symmetry axis of a standard accretion disk.

Near the black hole, where the photon density is very high, $\mathfrak{F}_r^z$ will be very large unless $\gamma_b \simeq \gamma_r$ and the bulk Lorentz factor will thus be always close to γ_r. Radiation pressure will then become inefficient at a critical distance z_1, for which $\gamma_{b1} = \gamma_{r\,1} \sim (l_s\ \bar{\gamma}_1)^{1/7}$. This is essentially the same result as Phinney's, with an extra $\bar{\gamma}_1$ factor because we consider now a hot plasma instead of a cold one (O'Dell 1981, Phinney 1982). As mentioned above, $\bar{\gamma}_1$ can reach at best 10^2, and so the maximum value of γ_{b1} is around 4-5, still too low to account for VLBI observations. But now internal pressure is still available for the expansion if $\bar{\gamma}_1 \gg 1$. After the plasma has escaped the radiation dictatorship, one can neglect the right hand side of equation (12) to get the simple relation: $\gamma_b^3\ \bar{\gamma} \propto S$.

To further evaluate γ_b, one can assume a power law variation of $\bar{\gamma}$ and S with distance z, namely:

$$\bar{\gamma}(z) = \bar{\gamma}_1 \left(\frac{z}{z_1}\right)^{-\alpha} \qquad S(z) = S_1 \left(\frac{z}{z_1}\right)^{\varepsilon}. \tag{17}$$

One gets the final bulk Lorentz factor $\gamma_{b\infty}$ by setting $\bar{\gamma} = 1$ in equation (18),

$$\gamma_{b\,\infty} = \gamma_{b1} \bar{\gamma}_1^{\frac{1}{3}\left(1+\frac{\varepsilon}{\alpha}\right)}. \tag{18}$$

With a constant turbulence level along the jet and assuming that $B \propto S^{-1}$, which holds for a longitudinal field, one can show that $\alpha = \frac{\varepsilon}{2}$; taking a value of 10^2 for $\bar{\gamma}$, $\gamma_{b\infty}$ would be as high as 400. However, this is an absolute upper limit, and such high velocities would probably not be stable; on the other hand, in the extreme case where the turbulence damps much more rapidly than the jet opens, $\frac{\varepsilon}{\alpha} \simeq 0$ and $\gamma_{b\infty}$ ranges from 2 (if the plasma cools before escaping the radiation pressure) to about 20 (if it is so hot that pair creation still occurs at this point). This covers approximately the range of observed superluminal expansion gamma factors. Another possibility is that the maximal value of $\bar{\gamma}$ set up by the saturation process is actually lower than 10^2, of the order of 10. One would then get the same range of asymptotic $\gamma_{b\infty}$ values, for any geometry (in fact they could even be obtained through an adiabatic expansion).

In this picture, the final bulk Lorentz factor would depend essentially on the relative position of the compact accelerating region and the critical distance at which Compton drag becomes inefficient. A compact source close to the central engine would produce a less energetic beam than a source located farther away. Theoretical models predict that MHD jets can be refocused after some distance by the magnetic hoop stress. It is tempting to explain the compact hard source by this focussing point, which could be situated at distances comparable (~ $10^2 R_i$) to the critical one. This would explain the location of this hard source above an accretion disk, which is thought to produce the hard X-ray reflection component (Zdziarski *et al.* 1990). Also this model predicts a moderate beaming of the compact source, with a Lorentz factor γ of about 2 increasing along the jet up to values about 10. This seems to be supported by unifications schemes of BL Lacs objects (Padovani and Urry 1990,1991) as well as direct observations of directed ionizing radiation (Morganti *et al.* 1991). A possible difficulty is to account for the "universal" hard X-ray spectral index $\alpha_X \simeq 0.7$ since the distribution function (7) cannot provide such a power-law spectrum. However, it could be obtained by a spatial variation of $\overline{\gamma}$, equation (17) with an appropriate exponent that may be related to the physics of expansion.

In conclusion, the two-flow model provides an attractive framework to explain the production of a relativistic e^+ - e^- beam. The use of a subrelativistic flow offers a clue to the Compton drag problem, since it can carry away a fair fraction of the total accretion energy without suffering strong losses, up to distances where the photon density is low enough to allow relativistic bulk velocities. Thus superluminal motion could paradoxically give indirect evidence for a powerful subrelativistic jet. This model could also nicely explain the location of the hard source as well as the beaming of the X-rays.

References

Done, C., Ghisellini, C., and Fabian, A.C. 1990, *M.N.R.A.S.*, **245,** 1.

Henri, G. and Pelletier G. 1991, *Ap. J. (Letters),* in press.

Kraishnan, R.H., 1965, *Phys. Fluids,* **8,** 1385.

Morganti, R., Robinson, A., Fosbury, R.A.E., di Serego Alighieri, S., Tadhunter, C.N. and Malin, D.F. 1991 *M.N.R.A.S.*, **237,** 411; Fosbury, R.A.E., these proceedings.

O'Dell, S.L. 1981, *Ap. J. (Letters)*, **294,** L147.

Padovani, P.and Urry, C.M. 1990, *Ap. J.* , **356,** 75; 1991, *Ap. J.* , **368,** 373.

Pelletier, G., Pudritz, R., Ferreira, J. and Rosso, F. 1991, these proceedings.

Phinney, E.S. 1982, *M.N.R.A.S.*, **198,** 1109.

Phinney, E.S. 1987 in Superluminal Radio Sources, ed. J.A. Zensus and T.J. Pearson (Cambridge University Press) 301.

Schlickeiser, R. 1984, *Astr. Ap.*, **136,** 227.

Sol, H., Pelletier, G., and Asséo, E. 1989, *M.N.R.A.S.*, **237,** 411.

Svensson, R. 1987, *M.N.R.A.S.*, **227,** 403.

Tagger, M., Pellat R., Corotini, F.C. 1991, *Ap. J.*, submitted.

Zdziarski, A.A., Ghisellini, G., George, I.M., Svensson, R., Fabian, A.C., and Done, C. 1990, *Ap. J. (Letters)*, **363,** L1.

MHD ACCRETION-EJECTION FLOW FOR AGN

G. PELLETIER, R. PUDRITZ*, J. FERREIRA, F. ROSSO

Laboratoire d'Astrophysique de l'Observatoire de Grenoble. France.

*also at McMaster University, Ontario, Canada.

1. WHY A TWO-FLOW MODEL

The most popular viewpoint about extragalactic jets is the assumption that they are made of a relativistic plasma with a relativistic bulk motion; the jet velocity u_j is greater than $c/\sqrt{3}$. So the Doppler beaming explains the observed superluminal motions, Doppler amplification explains asymmetries, the "Compton catastrophe" is avoided. A bulk Lorentz factor γ_b of order 10 is required to account for the VLBI observations. The radiation power emitted by the jets is produced by converting at most 10 percent of their kinetic energy flux through turbulence and shocks. So the kinetic energy flux is of order 10^{46} erg/s. The radio lobes need to be supplied by some $10^{-1} M_{\odot}/year$.

The jet production must overcome the Compton drag due to soft photons emitted by the accretion disk (Phinney 1987). This is the major difficulty for the production of relativistic jets.

It turns out that it is much easier to comply with observations if we assume two kinds of flows (Sol et al. 1989), namely a large scale subrelativistic flow and a parsec scale relativistic beam. In a jet of a strong double radiosource like Cygnus A, the luminosity is much smaller than the kinetic energy flux before the terminal shocks at the hot spots: $L \ll P_K = \frac{1}{2}\dot{M}_j u_j^2$. Thus the kinetic energy flux is almost constant along the jet, therefore the jet velocity is almost constant since the mass flux $\dot{M}_j$ is constant. The energy budget in hot spot shocks with synchrotron spectrum diagnosis indicates that the flow velocity is likely nonrelativistic: $u_j = 0.2 - 0.4c$ (Pelletier and Roland 1986, Roland et al. 1988, Meisenheimer et al. 1989).

Moreover internal Faraday rotation measurements in optical jets, together with synchrotron spectrum analysis (Fraix-Burnet et al 1989) lead to similar estimates for the jets velocities and indicate that the main component of the jet is a non relativistic thermal plasma, the relativistic plasma being more tenuous. Beyond kpc scale no relativistic motion has been detected and morphology tends to rule out relativistic dynamics.

The relativistic knots observed at VLBI scale can be explained with a low energy

budget (Sol et al. 1989, Pelletier and Roland 1989). Jet asymmetries could have an intrinsic physical origin (see Roland's talk in this meeting).
We propose that the subrelativistic flow can be a self-collimated MHD wind driven by the accretion disk, whereas the relativistic knots can result from the escape of the electron positron responsible for the hard X-ray emission (Henri and Pelletier 1991).

2. THE ANGULAR MOMENTUM TRANSPORT BY THE MHD JET

The most promising model to explain the origin of jets is probably the MHD one which makes use of the unavoidable magnetic field carried by the accretion flow. It is assumed that a bipolar poloidal magnetic field is anchored on the accretion disk and its opened field lines are frozen in a plasma wind. Because of differential rotation, a toroidal component grows:

$$B_\phi \overrightarrow{u_p} = (\Omega - \Omega_*) r \overrightarrow{B_p}, \tag{1}$$

where Ω is the angular velocity of the plasma and Ω_* is the angular velocity of the field lines. That toroidal component contributes to three important phenomena:

i. it is responsible for a magnetic braking that transfers angular momentum.

ii. it allows a poloidal propulsion because it participates to a Poynting flux:

$$\overrightarrow{P} = \frac{\overrightarrow{E} \times \overrightarrow{B_\phi}}{4\pi} = -\Omega_* r \frac{B_\phi \overrightarrow{B_p}}{4\pi} \tag{2}$$

iii. because of its tension effect, it allows a self-collimation of the jet.
The self-consistent problem can be reduced to two nonlinear partial differential equations for the two unknown functions, the reduced flux function $\psi(r,z)$ and the local Alfvén Mach number $m(r,z)$. Dynamics is described by a Bernoulli equation:

$$\frac{1}{2}\frac{(\nabla\psi)^2}{r^2} + \frac{\partial U}{\partial m^2} = 0. \tag{3}$$

The confinement is described by a Grad-Shafranov equation:

$$div\left(\frac{m^2-1}{r^2}\nabla\psi\right) + \frac{\partial U}{\partial \psi} = 0 \tag{4}$$

These two equations can be derived from a Lagrangian (Rosso 1991) involving the function $U(m^2, \psi; r, z)$ that depends on ψ through Ω_*, the angular momentum invariant $l(\psi)$ and the invariant $\mu(\psi)$ that gives the mass flux per unit magnetic flux:

$$\mu(\psi) \propto \frac{\partial \dot{M}_j}{\partial \psi}.$$

Whereas Ω_* is determined on the accretion disk, the two other functions are a priori undetermined. The three critical surfaces do not provide enough constraints to determine completely these functions. We have developed a theory (Pelletier and Pudritz 1991) based on the assumption that the opened field lines extract all the angular momentum from the accretion disk (Pudritz 1985). From the local angular momentum transport equation, namely,

$$div(\rho\Omega r^2\overrightarrow{u_p} - r\frac{B_\phi\overrightarrow{B_p}}{4\pi}) = r div\overrightarrow{\tau}_{visc}\,|_\phi, \qquad (5)$$

we obtain by integrating (5) over z the following equation:

$$\{\dot{M}_a(r)\frac{d}{dr}\Omega r^2 + 2\frac{d\dot{M}_j}{dr}(2\Omega r^2 - \Omega_* r_A^2)\}\,|_{disk\,surf.} = \frac{d\Gamma_{visc}}{dr}. \qquad (6)$$

The Alfvén radius r_A is such that $l(\psi) = \Omega_* r_A^2$; it is the "lever arm" of the angular momentum extraction. To get (6), we approximate B_ϕ by its ideal MHD value outside the disk.

The magnetic braking is more efficient than the viscous torque if the following inequality is fulfilled:

$$\frac{B^2}{4\pi p} \gg \frac{\alpha h}{r_A};$$

the condition has been obtained by using the Shakura-Sunyaev prescription, namely: $\tau_{r\phi} = \alpha p$; h is the half width of the disk. This condition is easily achieved in AGNs; it requires a magnetic field of 1-10 kG at few gravitational radii. It turns out that if $r_A \gg r_0$ (r_0 being radial coordinate of the foot of the field line on the disk), then

$$\frac{\dot{M}_j}{\dot{M}_a} \sim \frac{r_0^2}{r_A^2} \ll 1;$$

this implies that the accretion rate is approximately constant. When the former strong inequality is not achieved, the ejection rate is of the order of the accretion rate which, therefore, varies with the radius. This is likely the case in the inner region about a Schwarzschild black hole where the field surfaces widen very little.

The asymptotic velocity is obtained from Bernoulli equation (3):

$$u_{p\infty} \simeq \sqrt{2}\Omega_* r_A. \qquad (7)$$

Thus in the inner region about a Schartzschild black hole where the magnetic surfaces are almost cylindrical, we do not expect a terminal velocity much greater than the Keplerian velocity. Since the inner radius of the accretion disk is 3 times the gravitational radius, the motions remain subrelativistic. A detailed analysis leads to the conclusion that a relativistic jet cannot be produced by an accretion disk about a Schwarzschild black hole; we found typically $u_{p\infty} \simeq 0.4c$. However the analysis is deeply different about a fastly rotating black hole (Camenzind 1986). But the Compton drag becomes a serious difficulty to reach the expected Lorentz factor.

The energetic budget can be obtained directly from the angular momentum equation (6) by multiplying it by Ω and integrating from the inner magnetic surface to anyone labeled by ψ, and we get the kinetic power:

$$P_K \equiv \frac{1}{2}\int_{\psi_i}^{\psi} d\psi' \frac{d\dot{M}_j}{d\psi'} u_{p\infty}^2 \simeq \frac{1}{4}\frac{G\dot{M}_a M_*}{r_i}(1-\frac{r_i}{r_0}) \tag{8}$$

for $r_A \gg r_0$.
For this latter extreme, one half of the gravitational energy is converted into jet kinetic energy, independently of the magnetic field that plays only the role of a lever, provided that ideal MHD applies above the disk. Dissipation above the disk would reduce the efficiency, of course.
We analysed the flow with two control parameters; the one controls the jet geometry:

$$\lambda \equiv \frac{\Omega_*^2 r_A^2}{V_{A|A}^2}$$

(λ must be greater than unity to get a superAlfvénic flow)($V_{A|A}$ is the Alvén velocity at the critical Alfvén surface); the other one caracterises the ability of the accretion disk to support an opened force free field:

$$\varepsilon \equiv \frac{\dot{M}_a}{3}\frac{\Omega_e r_e^3}{(\Phi_{max}/2\pi)^2} \sim \frac{u_r \Omega h}{V_A^2}|_{r_e},$$

(the label e refers to the external radius r_e of the outflow on the disk). The opened force free field is maintained above the disk for ε smaller than some value of order unity, and rB_ϕ and I_p scale like $\varepsilon\Omega r$. The constraint on ε limitates the region of the disk where an outflow can exist.
A wide class of constant λ solutions can be derived, which is weakly restrictive because the Alfvén radius is weakly sensitive to λ:

$$r_A \simeq \lambda^{\frac{1}{6}}[\frac{(\Phi/2\pi)^2}{\dot{M}_j\Omega_*}]^{\frac{1}{3}}. \tag{9}$$

There exist two kinds of fast flows. i) On one hand, there are FM-flows whose velocity reaches the fast magnetosonic (FM) velocity and cannot overtake it; these flows achieve cylindrical collimation. For these flows, the ratio $B_\phi/B_p \simeq M_A/M_{FM}$ is large. ii) On the other hand, there are super FM-flows that undergo recollimation down to a radius greater but of the order of the Alfvén radius. The Alfvén Mach number is lowered in favour of the magnetosonic Mach number. This recollimation could trigger oblique shocks responsible for hard X-rays emission. The ratio $B_\phi/B_p \simeq M_A/M_{FM}$ is greater but of order unity.
Approximate analytical solutions have been derived far from the disk (Rosso 1991) by inserting trial functions into the Lagrangian; which allows to solve for the critical

surfaces with a system of differential equations instead of the system of partial differential equations. Sausage type solutions have been found also in the super FM-regime with a periodicity of order r_e/ε.

3. THE STRUCTURE OF AN ACCRETION DISK DRIVING A JET

An accretion disk is deeply remodeled when it drives a jet, especially when the magnetic field is strong enough to extract all the angular momentum. This latter case is interesting because no viscosity is necessary, and the outflow controls the accretion rate (see eq.(6)). However a quasi stationary structure cannot exist without some kind of diffusion of the accreted plasma across the magnetic tubes; this could result from a turbulent resistivity or a kind of ambipolar diffusion (Konigl 1989). For the structure we are dealing with, turbulent diffusion by Alfvén waves is appropriate; we know from the recent work by M. Tagger et al. (1991) that these waves are naturally excited in an accretion disk at a rate $\propto r\frac{\partial\Omega}{\partial r}$.

Moreover in this model, the relevent heating is the absorption of magnetic energy, although the ultimate source of energy is the Keplerian rotation whose shear is the cause of the turbulence.

There are two kinds of possible accretion disks of that sort. There are either ion supported tori such that:

$$\rho\Omega^2 r^2 \sim p \sim \frac{B_\phi^2}{8\pi}. \tag{10}$$

These nonKeplerian tori are probably subject to instability. There are also possible geometrically thin or slender disks quasi Keplerian, which we investigated in detail (Ferreira et al. 1991). These structures avoid the development of a prohibitive toroidal field, nevertheless sufficient to extract the angular momentum. However the poloidal field is not very bended and thus when the field lines have an angle smaller than 30° with respect to the axis (Blandford and Payne 1982) a corona is necessary to start the outflow. Compton reprocessing, reconnections etc. are currently invoked to explain X-ray emission of the disk; so a corona seems unavoidable in any case.

A closed system of equations for the disk structure has been solved under the assumption of self-similar form for various quantities such as the density, the poloidal velocities, the temperature, the toroidal magnetic field. For instance we seeked for a density of the form:

$$\rho(r,z) = \rho_0(r) f(z/h(r)), \tag{11}$$

h(r) is the typical half width of the disk. It turns out that the magnetic diffusivity ν_m must scale like:

$$\nu_m = \alpha_m V_A h, \tag{12}$$

where the dimensionless coefficient α_m is a function of $z/h(r)$ only and is less or equal to unity. Relation (12) plays a role analogous to the well-known Shakura-Sunyaev prescription in the standard theory of accretion disks.

The closed system is based on 6 basic equations: 1) continuity equation, 2) the equation for the transport of angular momentum (equation (5) with no viscous contribution), 3) the equation for vertical equilibrium, in which the Laplace force can be relevent for some class of solutions, 4) the power balance equation in which the luminosity is maintained by the turbulent Joule effect, 5) the equation for the magnetic surfaces in which the diffusion balances the convection:

$$\nu_m r^2 div(\frac{\nabla\psi}{r^2}) = \vec{u_p}.\nabla\psi \tag{13}$$

and 6) the induction equation for $b \equiv rB_\phi$:

$$r^2 div(\nu_m \frac{\nabla b}{r^2}) = r^2 div(b\vec{u_p} - \Omega r^2 \vec{B_p})/r^2. \tag{14}$$

The vertical structure can be mapped as function of the parameter ε_m defined by

$$\varepsilon_m \equiv \frac{u_r h^2}{\nu_m r}|_{r_e};$$

ε_m is a small parameter, because, from eq(13), the magnetic Reynolds number, $R_m \equiv ru_r/\nu_m$, is of order unity. Very well behaved solutions are obtained for $\varepsilon_m = 0.1$ for instance: a wide bunch of stream lines actually turns upwards and forms the jet, while the density and temperature profiles decrease smoothly to provide a comfortable basis to a hot optically thin corona.

4. CONCLUSION

The assumption of an MHD flow driven by a Keplerian disk about a slowly rotating black hole comply with kpc observations of jets like those of Cygnus A. The flow is subrelativistic.

This MHD flow constitutes a "magnetic cauldron" for high energy physics: hard X-rays, $e^+ - e^-$ pair creation, relativistic clouds (Henri and Pelletier 1991). Alfvén turbulence is the convenient fire to heat the cauldron. That kind of turbulence is naturally produced in the accretion disk (Tagger et al. 1991) where it is responsible for the necessary anomalous transport and can propagate vertically outside the disk. This MHD flow constitues a channel for $e^+ - e^-$ beams. In case of a shot of a relativistic cloud at some angle with respect to the axis, the cloud can flow along helical field lines. New observations by Schalinsky et al. (1991), like those of S5-1928+738 which displays superluminal motions at VLBI scale in a jet that is symmetric at kpc scales, should allow to discriminate the two viewpoints: either a single intrinsic relativistic jet or a relativistic beam channeled by a more massive subrelativistic flow. Indeed the observations show a relativistic flow that has a strong bending (which seems uncompatible with an intrinsicly relativistic dynamics), the knots having the

same apparent motion in a given segment. The superluminal motions are compatible with a jet axis at an angle smaller than 17^0 with respect to the line of sight and the velocity of the jet at kpc scale would be $0.43 - 0.45c$.
So the MHD accretion-ejection structure with its moderate Alfvén turbulence is the powerhouse for the hard X-rays and the relativistic beams (Henri and Pelletier 1991).
There are three important questions about the MHD accretion-ejection structure that are still opened, namely:
i. How a fastly rotating black hole changes the scheme, when one takes into account radiation drags?
ii. How is the dynamo effect quenched in the accretion disk?
iii. How the magnetic field is shared between opened field lines and closed field lines? What are the relevent boundary conditions?
Max Camenzind and coworkers are deeply involved in this investigation, and we hope that they will address solutions to these issues soon. Many astrophysicists are currently working to master and test the MHD model in analytical approaches, numerical simulations and observations (especially on young stellar objects). Crucial clues are expected in the near future to support the theory.

5. REFERENCES

Blandford R.D. and Payne D.G., 1982, MNRAS, **199**, 883.
Camenzind M., 1986, Astron. Astrophys., **156**, 137.
Ferreira J., Pelletier G., 1992, Astr. Astrophys., in preparation.
Fraix-Burnet D., Le Borgne J.F., Nieto J.L., 1989, Astron. Astrophys., **224**, 17.
Henri G., Pelletier G., 1991, Ap.J. Letters, *in press*; and proceeding of this meeting.
Konigl A., 1989, Ap. J., **342**, 208.
Lovelace R.V.E., Wang J.C.L., Sulkanen L.E., 1987, Ap.J., **315**, 504.
Meisenheimer K., Roser H.J., Hiltner P.R., Yates M.G., Longair M.S., Chini R., Perley R.A., 1989, Astron. Astrophys., **219**, 63.
Pelletier G., Pudritz R., 1991, Ap. J. (submitted)
Pelletier G., Roland J., 1989, Astron. Astrophys., **224**, 24.
Phinney S., 1987, in Superluminal Radiosources, ed. J.A. Zensus and T.J. Pearson (Cambridge University Press) 301.
Pudritz R.E., 1985, Ap. J., **293**, 216.
Rosso F., 1991, rapport de magistère ENS Lyon.
Roland J., Pelletier G., Muxlow T.W.B., 1988, Astron. Astrophys., **207**, 16.
Schalinsky C., Witzel A., Hummel C.A., Krichbaum T.P., Quirrenbach A., Johnston K.J., 1991, in Variability of Blazars, ed. Valrauja.
Sol H., Pelletier G., Asseo E., 1989, MNRAS, **237**, 411.
Tagger M., Pellat R., Coroniti F., 1991, Ap. J. Letters (submitted).

Magnetically Driven Jets and Winds

R.V.E. LOVELACE[1] and J. CONTOPOULOS[2]

[1]Department of Applied Physics, Cornell University, Ithaca, New York, 14853

[2]Department of Astronomy, Cornell University, Ithaca, New York, 14853

ABSTRACT

A simple but general theory for the formation and propagation of jets and winds from magnetized accretion disks is derived from the equations of ideal magnetohydrodynamics. The basic equations of the theory follow from the conservation of mass, momentum, angular-momentum, and energy for the jet flow. The theory allows the self-consistent, determination of the jet's axial velocity, radius, angular rotation rate, and temperature from the disk surface to infinity. The magnetic field has an essential role in that it acts, in opposition to gravity, to "drive" matter through the slow magnetosonic critical point. At large distances, the external medium acts, in opposition to gravity, to "drive" the flow through the fast magnetosonic critical point. Further, the external medium acts to collimate the flow leading under some conditions to relativistic terminal velocities.

1 INTRODUCTION

High-velocity outflows or jets of matter are observed to emanate from the nuclei of active galaxies and quasars (see, for example, Hughes 1990). On a smaller scale, the apparently singular galactic object SS433 exhibits oppositely directed high-velocity jets (Margon 1984). Both low-mass T-Tauri and high-mass pre-main sequence stars exhibit bipolar mass outflows (Bally and Lada 1983). In active galaxies and quasars, the activity, including the jets, is thought to be powered by accretion onto a massive black hole with the accretion occurring through a disk of ionized gas. Highly ordered magnetic fields are deduced to exist in the nuclear regions of some of these sources due to the rapid time variations of the intensity and polarization of the synchrotron emission (see, for example, Angel and Stockman 1980; Witzel et al. 1988).

The origin of the jets and outflows remains a fundamental open question. One idea for the origin of the jets is based on the occurrence of narrow, essentially empty vortex funnels in accretion flows of matter with angular momentum (Lynden-Bell 1978; Paczynski and Witta 1980; Abramowicz, Calvani, and Nobili 1980). However, non-axisymmetric instabilities of "fat" accretion disks are predicted to destroy the narrow funnels (Papaloizou and Pringle 1986; Zurek and Benz 1986).

An alternative idea for the origin of jets, which is discussed further here, is that the jets are magnetically driven (Uchida and Shibata 1985; Sakurai 1985; Lovelace, Berk, and Contopoulos 1991 - hereafter LBC). We follow our earlier approach (Lovelace, Wang, and Sulkanen 1987) where the interior of the disk is treated as viscous and resistive, with a stationary **B** field resulting from a balance of field diffusion against radial advection by the accretion flow. We assume even field symmetry so that magnetic flux function $\Psi(r,z) = \Psi(r,-z)$, where the poloidal field is given by $\mathbf{B}_p = \nabla \times (\Psi\hat{\phi}/r)$. The space exterior to the disk, where the jets accelerate and propagate, is treated using ideal magnetohydrodynamics. The approach we describe here involves averaging the different physical variables over the cross-section of the jet at a given axial distance z from the equatorial plane (Lovelace, Mobarry, and Contopoulos 1989; Koupelis and Van Horn 1989; and LBC). The basic nature of a non-relativistic solution close to the accretion disk is shown in Figure 1 (from LBC).

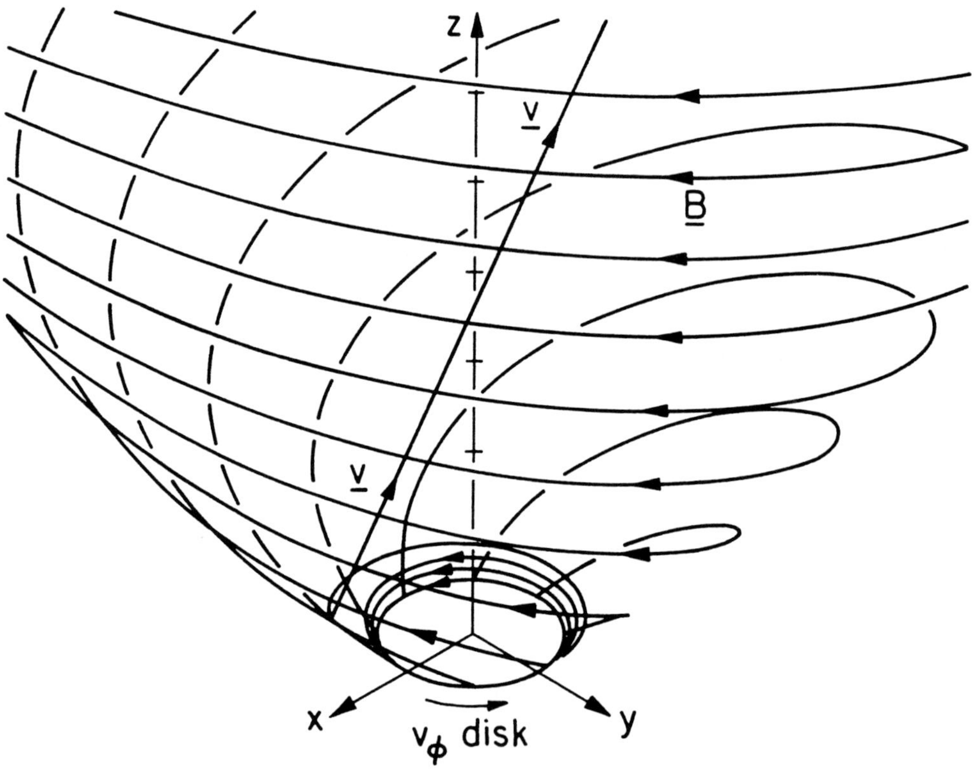

Figure 1 *Perspective view of eight field lines and one streamline for a non-relativistic jet solution. The z-axis is tipped towards the viewer by 30°. Close to the disk the flow velocity of the jet is almost azimuthal with a magnitude close to that of the disk. The twist of the magnetic field corresponds to a magnetic torque acting to slow down the disk's rotation.*

2 THEORY

We use a cylindrical, inertial coordinate system (r, ϕ, z) with the origin at the central object and the z-axis perpendicular to the disk. The outflows or jets are considered to be axisymmetric ($\partial/\partial\phi = 0$) and stationary ($\partial/\partial t = 0$). The basic equations are those of ideal magnetohydrodynamics (MHD). Close to the central object, we assume that the flow is non-relativistic and we treat the gravitational field as being Newtonian. We are mainly interested in the axial (z) variations of the jet parameters such as the jet velocity and radius, and the magnetic field components. Therefore, we consider a simple factorization of the radial and axial dependences,

$$v_r(r,z) = \left[\frac{r}{\mathrm{a}(z)}\right]\left[\frac{d\mathrm{a}(z)}{dz}\right]v_z(z) \ ; \qquad v_\phi(r,z) = r\omega(z) \ ;$$
$$v_z(r,z) = v_z(z) \ ; \qquad \rho(r,z) = \bar{\rho}(z)f(\alpha) \ . \tag{1}$$

Here, a(z) is the jet radius; ω(z) is the angular rotation rate of the jet at the axial distance z; $\bar{\rho}(z)$ is the average density of the jet; $\alpha \equiv [r/\mathrm{a}(z)]^2$ is a dimensionless measure of radial distance in the jet; and $f(\alpha) \geq 0$ is a dimensionless mass-density "profile function" normalized so that its integral over α is unity. The mass flow in the jet across any $z = const$ surface is a constant so that

$$\dot{M}_j = \int_{z=const} d^2x \ \rho v_z = \pi[\mathrm{a}(z)]^2\bar{\rho}(z)v_z(z) = const \tag{2}$$

The magnetic field obeys the "frozen-in" field equation so that we may take

$$B_r(r,z) = \alpha^{\frac{1}{2}}\left(\frac{dR(Z)}{dZ}\right)B_z(r,z) \quad ,$$
$$B_\phi(r,z) = \alpha^{\frac{1}{2}}\beta_i\left(\frac{\Omega(Z)-1}{V_z(Z)R(Z)}\right)B_i(\alpha) \quad ,$$
$$B_z(r,z) = \left(\frac{1}{R(Z)}\right)^2 B_i(\alpha) \quad , \tag{3}$$

for $\alpha < \alpha_o$, within the jet, and $\mathbf{B} \approx 0$ for $\alpha > \alpha_o$, outside the jet, where $\alpha_o = O(1)$. Here, $Z \equiv z/a_i$ is a dimensionless axial distance; $R \equiv a(z)/a_i$ is the dimensionless jet radius; $a_i \equiv a(z=0)$ is the "initial" jet radius; and $B_i(\alpha)$ is the axial field profile in the disk. Further, $V_z \equiv v_z(z)/c$ is the dimensionless jet velocity with c the speed of light; $\Omega \equiv \omega(z)/\omega(z=0)$ is the dimensionless rotation rate of the jet; and $\beta_i \equiv \omega(z=0)a_i/c$.

An equation for the conservation of the axial momentum flow of the jet can readily be obtained (LBC) by integrating the general momentum conservation equation over a volume of the jet between Z and $Z + dZ$. The result is:

$$\frac{d}{dZ}\mathcal{F}_P = \frac{F_z}{V_z} + \pi P_{ex}\frac{dR^2}{dZ} \quad , \tag{4}$$

where F_z is the axial gravitational force on the jet matter within Z to $Z + dZ$; P_{ex} is the pressure of the matter and weak magnetic field outside of the jet; and $\mathcal{F}_P$ is the dimensionless axial momentum flow of the jet. That is,

$$(a_i B_o)^2 \mathcal{F}_P = \int_{z=const} d^2x (T_p^{zz} + T_f^{zz})$$
$$= \dot{M}_j c\left(\gamma V_z + \frac{\tau}{V_z}\right) + \frac{1}{4}\int_0^{a(z)} r dr (B_r^2 + E_r^2 + B_\phi^2 - B_z^2 - E_z^2) .$$

Here, T_p^{ij} is the stress tensor for the matter and T_f^{ij} is that for the electromagnetic field; $\tau \equiv kT(z)/(\mu_o c^2)$ is the dimensionless jet temperature (with μ_o the mean particle mass); and $B_o \equiv \max(\mid B_i(\alpha) \mid)$ is a reference value for the poloidal magnetic field in the disk; and

$$\gamma = \left\{1 - V_z^2\left[1 + \left(\frac{dR}{dZ}\right)^2\right] - \beta_i^2(\Omega R)^2\right\}^{-\frac{1}{2}}$$

is the Lorentz factor for the jet. From equation (3) we obtain

$$\mathcal{F}_P = \mathcal{F}_M(\gamma V_z + \frac{\tau}{V_z})$$
$$+ \frac{1}{2}\mathcal{F}_o\left[\left(\frac{R'}{R}\right)^2 + \beta_i^2\left(\frac{\Omega - 1}{V_z}\right)^2 - \frac{1}{R^2} + \beta_i^2 - \eta_E \beta_i^2 (R')^2\right] \quad , \tag{5}$$

where $\mathcal{F}_M$ is the dimensionless mass flow rate, $R' \equiv dR/dZ$, and $\mathcal{F}_o$ and $\eta_E \geq 1$ are dimensionless constants of order unity.

The gravitational force F_z in equation (4) depends of course on the radial distribution of matter within the jet (LBC). We assume that this distribution is peaked near the jet's edge ($\alpha \sim 1$) so that $F_z = -Z\delta(R^2 + Z^2)^{-\frac{3}{2}}$, where $\delta = GM\mathcal{F}_M\eta_R/(a_i c^2)$ is a dimensionless measure of the effective gravitational force, and $\eta_R \leq 1$ allows for the influence of radiation pressure. In addition to equations (2) and (4), we have conservation equations for the flow of angular-momentum and energy in the jet. These four conservation equations allow one to solve for the jet velocity, radius, rotation-rate, and temperature from $Z = 0$ to infinity. The jet solutions are however constrained by the fact that the mass, angular-momentum, and energy carried by the $\pm z$ jets must be equal to the loss of these quantities by the disk. Figure 2 shows a representative self-consistent non-relativistic jet solution (LBC).

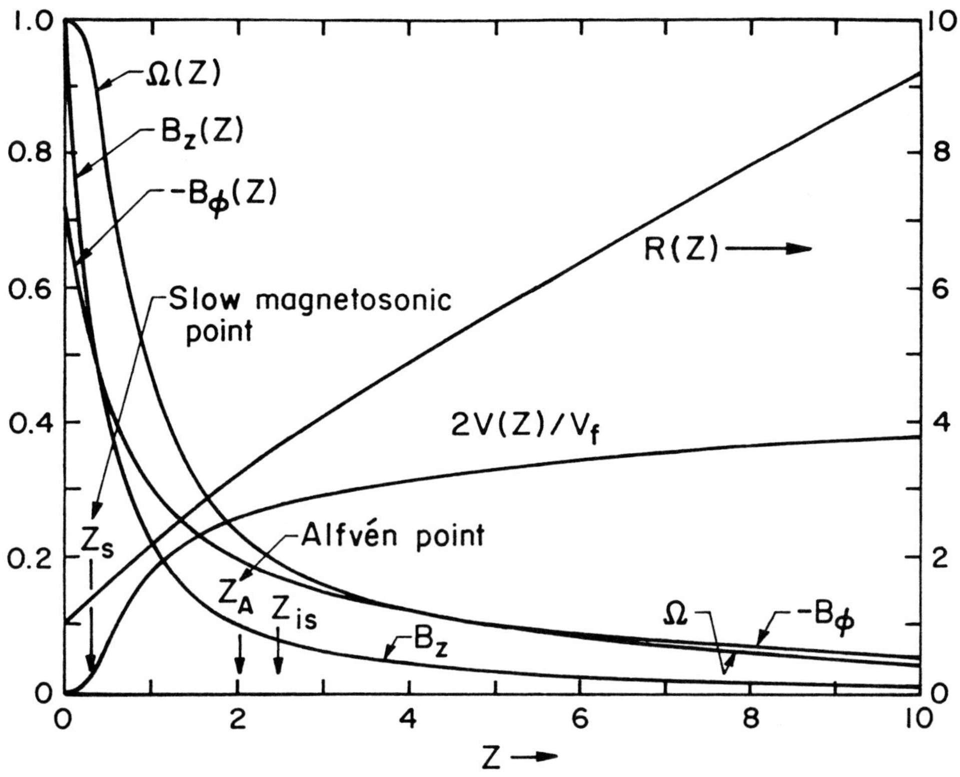

Figure 2 *Axial dependences of the basic jet variables close to the central object for a non-relativistic solution. In the figure, the fields are normalized so that the axial field is* $B_z = 1/R^2$ *and the toroidal field is* $B_\phi = -(1-\Omega)/(V_z R)$. V_z *is normalized to the fast magnetosonic speed,* V_f. *For* $Z < Z_{is}$ *the jet is treated as isothermal, while for* $Z > Z_{is}$ *it is treated as isentropic. That* B_ϕ *is negative corresponds to the magnetic torque of the jet removing angular momentum from the disk. Also, that* B_ϕ *is negative corresponds to an outward Poynting energy flux. Notice, that the jet carries no net current so that* $rB_\phi(r,z) \to 0$ *for* $r \to \infty$.

The key parameter of the theory is $\mu \equiv \mathcal{F}_M/\mathcal{F}_o = c\dot{M}_j/(a_i^2 B_o^2 \mathcal{F}_o)$, which is a measure of the "mass-loading" of the jet. For $\mu \ll 1$, the initial jet propagation is field dominated - the energy and angular momentum fluxes are initially carried by the electromagnetic field. In the opposite limit of μ not small compared with unity, the matter and EM fields are important at all distances from the disk. In general, the maximum asymptotic ($Z \to \infty$)jet speed or Lorentz factor increases as μ decreases. Figure 3 shows an example of a relativistic jet solution.

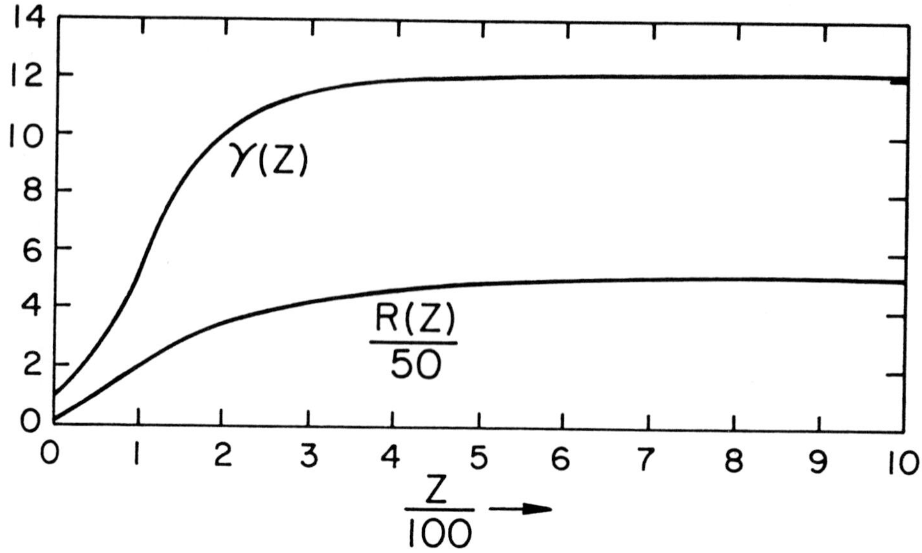

Figure 3 *Axial dependences of jet radius and Lorentz factor for* $\mu = 10^{-2}$, $P_{ex} \propto (1 + Z^2/H^2)^{-\frac{3}{2}}$, *and* $H = 200$.

REFERENCES

Angel, J.R.P., and Stockman, H.S. 1980, *Ann. Rev. Astron.* and *Astrophys.*, **8**, 321
Abramowicz, M.A., Calvani, M., and Nobili, L. 1980, *Ap. J.*, **242**, 772
Bally, J., and Lada, C.J. 1983, *Ap. J.*, 265, 824
Hughes, P.E. 1990, *Astrophysical Jets* (Cambridge: Cambridge University Press)
Koupelis, T., and Van Horn, H.M. 1989, *Ap. J.*, **342**, 146
Lovelace, R.V.E., Mobarry, C.M., and Contopoulos, J. 1989, in *Accretion Disks and Magnetic Fields* in *Astrophysics*, G. Belvedere, Ed. (Kluwer Academic Publishers), p. 71
Lovelace, R.V.E., Berk, H.L. and Contopoulos, J. 1991, *Ap. J.*, October (to appear)
Lynden-Bell, D. 1978, *Phys. Scripta*, **17**, 185
Margon, B. 1984, *Ann. Rev. Astron.* and *Astrophys.*, **22**, 507
Paczynski, B. and Witta, P.S. 1980, *Astron.* and *Astrophys.*, **88**, 23
Papaloizou, J.C.B., and Pringle, J.E. 1985, *M.N.R.A.S.*, **213**, 899
Sakurai, T. 1985, *Astron.* and *Astrophys.*, 152, 121
Uchida, Y., and Shibata, K. 1985, *Pub. Astron. Soc. Japan*, **37**, 515.
Witzel, A., et al. 1988, *Astron. and Astrophys.*, **206**, 245
Zurek, W.H., and Benz, W. 1986, *Ap. J.*, **308**, 123

Magnetic Fields And Reconnection In Extragalactic Jets

M.M. ROMANOVA[1] and R.V.E. LOVELACE[2]

[1]Institute for Space Research, U.S.S.R. Academy of Sciences, Moscow, U.S.S.R.

[2]Department of Applied Physics, Cornell University, Ithaca, New York, 14853

ABSTRACT

Extra-galactic radio jets are investigated theoretically taking into account that the magnetic field is dragged out from the central source by the jet flow. Thus, magnetohydrodynamic models of jets are considered with *zero net poloidal current* and *zero net poloidal flux.* At large distances the magnetic field is predominantly toroidal and the field naturally has a cylindrical neutral layer. Shear-stretching of reconnected toroidal magnetic field can in one limit give a strong axial field in the vicinity of the neutral layer. In another limit the field remains toroidal throughout the volume of the jet. Driven collisionless reconnection at neutral layers may lead to acceleration of electrons to relativistic energies owing to a weak electrostatic field. A simple model for particle acceleration at neutral layers in electron/positron and electron/proton plasmas is discussed.

1 INTRODUCTION

Magnetic fields have an important role in the appearance and also probably the dynamics of jets in radio galaxies and quasars (see, for example, Begelman, Blandford, and Rees 1984; Asseo and Sol 1987). Recent observations have revealed many details in the observed brightness and projected magnetic field patterns (for example, Bridle and Perley 1984; Perley 1985; Clarke, Burns, and Feigelson 1986; Owen, Hardee, and Cornwell 1989; Reid et al. 1989). Most jets are observed as center-brightened; however, recent VLA observations of the nearby jets in Cen A (Clarke, Burns, and Feigelson 1986) and M 87 (Owen, Hardee, and Cornwell 1989), which have been done with unprecedented angular resolution and dynamical range, revealed prominent *limb-brightening* all along the jet, including the regions of bright knots and the parsec-scale region of the jet in M 87 (Reid, et al. 1989). However, the actual three-dimensional structure of the magnetic fields is still unknown. The question of this structure and its variation along and across the jet, remains a fundamental open problem.

Another fundamental problem is the acceleration and/or reacceleration of electrons in jets. It is widely believed that the observed radiation is incoherent synchroton radiation of relativistic electrons in a weak magnetic field. The correlation between the regions of high brightness and polarization (for example, Perley 1985; Bridle 1986) suggests that particle reacceleration is connected with the ordering or amplification of magnetic fields. Different mechanisms of particle acceleration have been studied, for example, first-order shock acceleration and second-order Fermi acceleration in magnetohydrodynamical turbulence, magnetic reconnection, etc., but it is not known what mechanism accelerates electrons in jets.

It is often supposed that acceleration of electrons in jets is due to the Fermi mechanism of particle acceleration (Christiansen, Pacholczyk and Scott 1976; Axford, Leer and Scadron 1977; Krimsky 1977; Bell 1977, 1978; Blandford and Ostriker 1978; Eilek and Hughes 1990). In both the cases of shocks and turbulent plasma the acceleration is effective if the particle gyroradius is larger than the size of typical magnetic field inhomogeneities or than the scale of the transition layer of the shock. Both values have a scale of the order of the thermal ion gyroradii, so that the most efficient acceleration is of ions. As for electrons, an electron can be accelerated effectively if its gyroradius is comparable to or larger than the thermal ion gyroradius, that is, electron Lorentz factors $\gamma_e \gtrsim m_p/m_e$. So, electrons must be highly relativistic prior to the acceleration. Sometimes, limb-brightening of jets is interpreted as particle acceleration at the oblique shocks (for example, Clarke, Burns and Feigelson 1986). However, acceleration of electrons at oblique shocks is also inefficient and pre-acceleration of electrons to relativistic energies is also necessary.

In this paper we consider models where the jet magnetic field is dragged out from the central object by the jet flow. The models imply that the jet magnetic field is predominantly toroidal and that this field naturally has a cylindrical neutral layer where $\mathbf{B} = 0$. In such configurations, boundary phenomena connected with spontaneous and driven reconnection of the magnetic field together with shearing by the flow may have important consequences in determining the apparent magnetic field and in accelerating electrons to relativistic energies. Previously, Königl and Choudhuri (1985) have discussed the possibility of electron acceleration in radio jets due to resistive tearing of neutral layers.

2 MAGNETIC FIELD MODEL

We assume that the jet originates from the inner region of a magnetized accretion disk around a massive black hole. We assume that the magnetic field of jet is dragged out from the disk by the jet flow. Rotation of disk leads to twisting of poloidal magnetic field lines and to the appearance of a toroidal component of magnetic field.

To study the bulk motion of the jet, we first consider axisymmetric ($\partial/\partial\phi = 0$), and stationary ($\partial/\partial t = 0$) jet flows in a cylindrical, inertial coordinate system (r, ϕ, z) with the z-axis coincident with the jet axis, and with z=0 the location of the central object. As a zeroth approximation we neglect dissipative effects and turbulence, and assume that the flow is non-relativistic. The basic equations are those of ideal magnetohydrodynamics (MHD). Their solution is discussed by Lovelace, Berk, and Contopoulos (1991, hereafter - LBC) and by Romanova and Lovelace (1991, herafter - RL). At large distances, $z > 1pc$, the jet radius a(z) is much larger than its initial radius, $R \equiv a(z)/a_i \sim 3 \times 10^6(\theta/0.1)(z/kpc)(10^8 M_\odot/M)$, where M is the black hole mass, θ is the half angular width of the jet at the axial distance z and $a_i \sim 6GM/c^2 \approx 10^{14}cm(M/10^8 M_\odot)$.

Near the accretion disk the magnetic field has comparable poloidal and toroidal components (LBC, RL). Prior to the outward propagation of the jet, the poloidal field may be assumed to have a dipole-like configuration which is the lowest order multipole allowed with reflection symmetry about the equatorial plane (z=0). As, say, the +z jet propagates away from the disk it drags this dipolar field with it. Thus, the jet carries *zero net poloidal magnetic flux.* The current flow is assumed to be confined to the jet channel (RL) so that the jet carries *zero net poloidal current.* As a result of the difference between the angular velocity of the disk and the matter of the jet, the poloidal field is twisted and a significant toroidal field is generated. For example, if the poloidal field threading the disk has $B_z > 0$ at small r and $B_z < 0$ at large r, then at large axial distances the jet will have $B_\phi < 0$ at small r and $B_\phi > 0$ at large r due to the larger (counter-clockwise) rotation of the disk relative to the rotation of the jet. From this discussion, it is clear that the jet magnetic field will have neutral layer at $r = r_o(z)$ where $\mathbf{B}(r_o, z) = 0$.

At large distances, the axisymmetric poloidal field of the jet is negligible compared with the toroidal field, that is, $\mid B_p/B_\phi \mid \sim 1/R \ll 1$. With the estimate $B_i \sim 10^3 G[\dot{M}/(M_\odot/yr)]^{\frac{1}{2}}[10^8 M_\odot/M]$ for the poloidal field in the disk, we obtain $B_\phi \sim 0.3mG(0.1/\theta)(1kpc/z)[\dot{M}/(M_\odot/yr)]^{\frac{1}{2}}$, where $\dot{M}$ is the mass accretion rate onto the black hole. Figure 1 shows an example of the radial dependences of the field components and the jet velocity.

3 RECONNECTION AND SHEARING

We consider that the dominant axisymmetric magnetic field of the jet is toroidal and has a cylindrical neutral layer at $r = r_o(z)$, where $B_\phi(r_o, z) = 0$. Neutral layers are known to be unstable to spontaneous or driven reconnection under different conditions in a variety of laboratory and space plasma configurations. By reconnection we refer here to a change in topology of the magnetic field which leads to field lines connecting spatially separated plasma elements which were not initially connected. Because of the low densities of astrophysical jets the reconnection is collisionless.

In the vicinity of the neutral layer the equilibrium field can be represented by the Harris (1962) solution $A_z^o = -B_o\Delta \ln[\cosh((r - r_o)/\Delta)]$, where A_z^o is the vector potential, B_o is the equilibrium toroidal field far from the neutral layer, and $\Delta \ll r_o$ is the neutral layer thickness. Omitting a possible z dependence, the perturbation

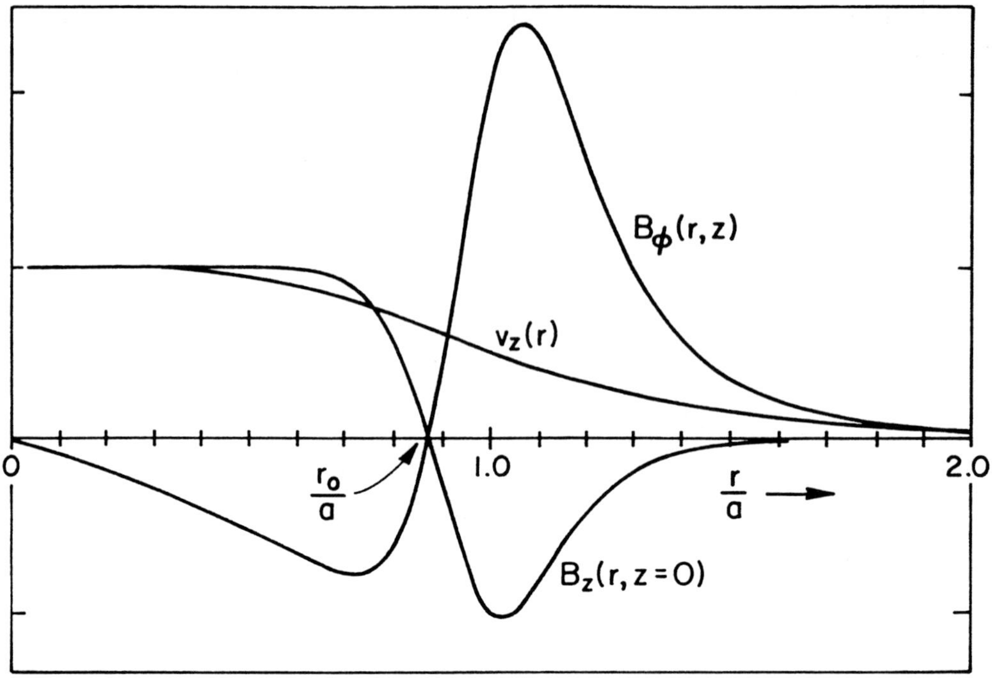

Figure 1 *Radial profiles of the magnetic field components and the axial flow velocity of the axisymmetric model jet. For the distances of interest the axial field* $B_z(r,z)$ *is negligigle compared with the toroidal field* $B_\phi(r,z)$. *The magnetic field is zero on a cylindrical neutral layer,* $r = r_o$.

is taken to be $\delta A_z = B_1 \Delta \cos(m\phi)/\cosh((r - r_o)/\Delta)$, where $B_1 \ll B_o$ is the perturbation amplitude, and m = 1,2, $\cdots$ is the toroidal mode number. The total vector potential, $A_z^o + \delta A_z$, evidently corresponds to a magnetic field with "0" points at $r = r_o$ and $\phi = 0, 2\pi/m, 4\pi/m, \cdots$ and "X" points at r_o and $\phi = \pi/m, 4\pi/m, \cdots$ There are m magnetic islands surrounding each "0" point with the radial extent of an island being $r \approx r_o \pm \Delta(2B_1/B_o)^{\frac{1}{2}}$.

The jet will in general have shear in the axial velocity across the neutral layer, $\partial v_z/\partial r\Big|_{r_n} \neq 0$. Magnetic inhomogeneities (for example, magnetic islands), formed as a result of reconnection or resonantly driven Kelvin-Helmholtz instabilities (Ray and Erskovich 1983; Choudhury and Lovelace 1986), are stretched by the jet flow and form elongated loops in the axial direction. This process leads to the formation of an axial component of the magnetic field in the boundary layer of the jet. The growth of the axial field will of course be limited once it becomes dynamically significant for the flow. This limit is reached when $B_z^2 \approx 4\pi\rho(\partial v_z/\partial r|_n)^2(\Delta r)^2$, where Δr is the radial half-width of the magnetic islands. The relative significance of the axial field can be measured by comparing it with the axisymmetric toroidal field within a distance Δr of the neutral layer,

$$\frac{(B_z^2)_{max}}{(B_\phi^2)_{max}} = \frac{4\pi\rho_n\left(\frac{\partial v_z}{\partial r}\Big|_n\right)^2}{\left(\frac{\partial B_\phi}{\partial r}\Big|_n\right)^2} \equiv \beta_n,$$

If $\beta_n \gtrsim 1$, the shear-stretching of the reconnected field can give rise to an appreciable axial magnetic field in the vicinity of the neutral layer. We suggest that this limit may correspond to the jets in powerful sources, the FR type II jets (Fanaroff and Riley 1974). In the opposite limit, $\beta_n \ll 1$, the shear stretching gives only a small axial field so that the toroidal magnetic field dominates throughout the volume of the jet. This limit may correspond to the weak sources which show mainly a transverse magnetic field, the FR type I jets.

Independent of the value of β_n, reconnection of the toroidal field across the neutral layer can give an important off-diagonal contribution to the momentum flux-density, $T_{rz} = - < \delta B_r \delta B_z > /4\pi$, which acts as a "magnetic friction" or viscosity. An analogous magnetic friction has been discussed in connection with the anomalous viscosity in accretion disks (see, for example, Eardley and Lightman 1975; and Coroniti 1981).

4. DRIVEN COLLISIONLESS RECONNECTION AND ELECTRON ACCELERATION

For simplicity, we consider a "planar piece" of the cylindrical neutral layer. We use a Cartesian coordinate system with the z-axis parallel to the jet axis, the y-axis along the jet radius, and the x-axis along the toroidal direction. There are two regions of oppositely directed magnetic field, $\mathbf{B} = B_x(y)\hat{\mathbf{x}}$, with $B_x\left(y \gtrless 0\right) \gtrless 0$. The current-density is $\mathbf{J} = J_z(y)\hat{\mathbf{z}}$, with $J_z \leq 0$. In driven reconnection, described by Alfvén (1968), plasma inflows towards the neutral layer, $v_y\left(y \gtrless 0\right) \lessgtr 0$. Thus, there is an associated electrostatic field, $\mathbf{E} = -\frac{1}{c}(\mathbf{v} \times \mathbf{B}) = +(v_y/c)B_x\hat{\mathbf{z}}$, and particles reaching the neutral layer are accelerated in the $+z$ (ions) or $-z$ (electrons) directions (along the jet). However, the one-dimensional Alfvén (1968) model of driven collisionless reconnection is over simplified in the respect that the current-density is independent of x. Thus, we assume that the current density varies with x so that there are 0 and X points along the x-axis. This plasma problem is two dimensional and an analogy can be made with the classic MHD reconnection problem treated by Sweet (1958), Parker (1963), Syrovatskii (1971), and others. As in the MHD reconnection problem, most of the volume of the plasma can be treated using ideal MHD. We replace the small volume where resistivity is important in the classic problem with the small region in which the electron and ion orbits are non-adiabatic.

For a more detailed treatment, we follow the considerations of Syrovatskii (1971) which is supported by the numerical simulations of Biskamp (1986). Thus, we consider a stagnation point of the magnetic field and of the flow, with a current sheet in the $x-z$ plane with $J_z \sim J_z(x)\delta(y)$ for $|x| < d$. To lowest order in x, $B_y(x, y=0) \propto x^3$. Thus, the two field lines which pass through the neutral point $(x, y=0)$ osculate at this point in contrast with the usual case of an X point where they cross at a finite angle (Biskamp 1986). The particle motion in the y and z directions in the current sheet is similar to that in the Alfvén (1968) model. The y-motion is magnetically confined to the current sheet by the B_x field. The motion in the $\pm z$ directions undergoes a steady acceleration due to the E_z field. In the x direction the particles are expelled from the sheet by the $v_z B_y$ force. For an electron/proton plasma an electron is expelled much more rapidly than an ion and

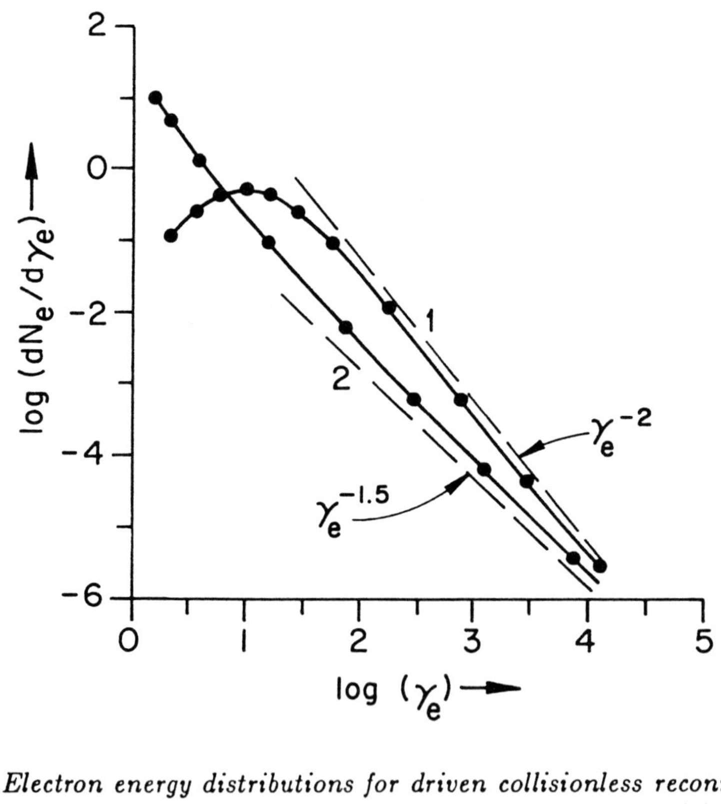

Figure 2 *Electron energy distributions for driven collisionless reconnection in an electron/proton plasma (1) and in an electron/positron plasma (2). The electron energy is $E_e = m_e c^2 \gamma_e$, where γ_e is the Lorentz factor.*

this gives rise to an ambipolar electrostatic field (E_x^a, E_y^a) as discussed by RL. For an electron/positron plasma there is no ambipolar field.

An electron which enters the current sheet at, say, x_o is found, by numerical integration, to exit from the sheet with a final energy $E_e = m_e c^2 \gamma_e(x_o)$. The inflowing plasma is uniform so that the number of electrons dN_e accelerated to Lorentz factors between γ_e and $\gamma_e + d\gamma_e$ is simply $dN_e \propto dx_o$. Thus, the electron energy distribution is $dN_e/d\gamma_e \propto | d\gamma_e/dx_o |^{-1}$ (RL). Figure 2 shows the electron energy distributions for an electron/proton plasma and for an electron/positron plasma. Recall that $dN_e/d\gamma_e \propto \gamma_e^{-p}$ corresponds to a synchrotron spectrum $S_\nu \propto \nu^{-\frac{(p-1)}{2}}$. Of course, these model results need to confirmed and studied further by self-consistent computer simulations.

REFERENCES

Alfvén, H. 1968, *J. Geophys. Res.*, **73**, 4379

Asseo, E., and Sol, H. 1987, *Phys. Rep.*, **148**, 301

Axford, W.I., Leer, E., and Scadron, G. 1977, *in Proceedings of the 15th Internat. Cosmic Ray Conference* (Bulgarian Academy of Sciences, Plovdiv, Bulgaria), Vol. 2, p. 273

Begelman, M.C., Blandford, R.D., and Rees, M.J. 1984, *Rev. Mod. Phys.*, **56**, 255

Bell, A.R. 1977, *M.N.R.A.S.*, **182**, 147

Bell, A.R. 1978, *M.N.R.A.S.*, **179**, 573

Biskamp, D. 1986, *Phys. Fluids*, **29**, 1520

Blandford, R.D., and Ostriker, J.P. 1978, *Ap. J. Lett.*, **221**, L29

Bridle, A.H. 1986, *Can. J. Phys.*, **64**, 353

Choudhury, S.R., and Lovelace, R.V.E. 1986, *Ap. J.*, **302**, 188

Christiansen, W.A., Pacholczyk, A.G., and Scott, J.S. 1976, *Ap. J.*, **210**, 311

Clarke, D.A., Burns, J.O., and Feigelson, E.D. 1986, *Ap. J. Lett.*, **300**, L41

Coroniti, F.V. 1981, *Ap. J.*, **244**, 587

Eardley, D.M., and Lightman, A.P. 1975, *Ap. J.*, **200**, 187

Eilek, J.A., and Hughes, P.E., 1990, *in Astrophysical Jets*, Ed. P.E. Hughes (Cambridge: Cambridge University Press) p. 428

Fanaroff, B.L., and Riley, J.M. 1974, *M.N.R.A.S.*, **167**, 31

Harris, E.G. 1962, Nuovo Cimento, **23**, 115

Königl, A., and Choudhuri, A.R. 1985, *Ap. J.*, **289**, 173

Krimsky, G.F. 1977, *Docl. Acad. Nauk SSSR*, **234**, 1306

Lovelace, R.V.E., Berk, H.L., and Contopoulos, J. 1991, *Ap. J.*, in press (LBC)

Owen, F.N., Hardee, P.E., and Cornwell, T.J. 1989, *Ap. J.*, **340**, 698

Parker, E.N. 1963, *Ap. J. Suppl.*, **77**, 8

Perley, R.A., 1985, *in Proceedings of IAU Colloquium No. 89, Radiation Hydrodynamics in Stars and Compact Objects*, ed. D. Mihalas and K.-H.A. Winkler (Springer-Verlag)

Ray, T.P., and Erskovich, A.I. 1983, *M.N.R.A.S.*, **204**, 821

Reid, et al. 1989, *Ap. J.*, **336**, 112

Romanova, M.N., and Lovelace, R.V.E. 1991, *submitted to Astron. and Astrophys.* (RL)

Sweet, P.A. 1958, *Nuovo Cimento*, Suppl; **8**, 188

Syrovatskii, S.I. 1971, *Sov. Phys. JETP*, **33**, 933

The Stability of Force-Free MHD Jets

STEFAN APPL AND MAX CAMENZIND

Landessternwarte Königstuhl, D-6900 Heidelberg, Germany

1 INTRODUCTION

Most promising for jet formation are models that collimate the outflows in rotating magnetospheres. This drives a poloidal electrical current of typically 10^{18} Ampères. This current is balanced by a return current of equal size. Such a circuit model was first proposed by Benford (1978). The accelerated plasma passes through the critical MHD-points and emerges at supermagnetosonic speed. It was shown by Heyvaerts and Norman (1989) that jets carrying electrical currents are asymptotically cylindrical. VLBI observations indicate intrinsic opening angles of $1°$ or even less. Similar static plasma configurations studied in fusion research generally suffer from violent MHD instabilities. Therefore, there has been a major concern that such jets might not persist but instead be disrupted by nonaxisymmetric MHD instabilities. We investigated the stability properties of a cylindrical force-free jet propagating at supermagnetosonic velocity. The details of the calculations will be given elsewhere (Appl and Camenzind, 1991a).

2 THE EQUILIBRIUM

Emerging from the collimation zone the jet possesses a nearly force-free configuration with a particular current distribution. Dissipative processes will alter the topological structure by breaking-up and reconnection of field lines. The jet will most likely settle into a state of minimal energy. For a low-β plasma relaxation theory provides such a configuration (Taylor, 1986). It is given by a linear force-free field $\nabla \wedge \mathbf{B} = \alpha \mathbf{B}$. α is constant which is determined by the initial and boundary conditions. Imposing cylindrical symmetry, solutions to the above force-free condition are given by the Bessel functions $B_\phi = B_0 J_1(\alpha r)$ and $B_z = B_0 J_0(\alpha r)$. This kind of equilibrium was studied in fusion research, however without plasma motion, as a model for a socalled *reversed field pinch* (RFP). It is characterized by high magnetic shear and by B_ϕ and B_z being of the same size, unlike the Tokamak. It owes its name to the fact that the longitudinal field reverses its sign for $\alpha r > 2.40$, where R is the jet radius. Such equilibria were applied to jets first by Königl and Choudhuri (1985).

Assuming that current and return current flow within the jet automatically fixes the value of $\alpha = 3.83/R$. This corresponds to the first zero of J_1. This defines a force-free equilibrium with zero net current that does not depend on the exact magnetic configuration as it arises after the collimation, and as we are dealing with a minimum energy configuration we do not have to worry about the details of the dissipative processes involved. It therefore offers a good model for jets in AGNs. The process of relaxation occurs on the Alfvén time scale. Therefore, within several Alfvén crossing times, which are of the order of the light crossing time, the jet will settle into such a configuration.

For the plasma flow we assume a constant velocity within the jet and zero outside. The jet is confined by a thermal external medium without magnetic fields. As parameters we choose the fast magnetosonic Mach number $M = v/v_{\rm f}$ ($v_{\rm f}^2 = v_{\rm a}^2 + c_{\rm s}^2$, $v_{\rm a}$ and $c_{\rm s}$ are the Alfvén and sound velocity, respectively), and the ratio $v_{\rm A} = v_{\rm a}/v_{\rm f}$ of the Alfvén and fast magnetosonic velocity. The latter gives a measure of the plasma-β, $\beta = 8\pi p/B^2$, and simultaneously for the electric current. It was often argued that a magnetic field stabilizes referring to calculations with a homogeneous magnetic field comparing jets with the same *sonic* Mach number. However, for decreasing plasma-β, the sonic Mach number is no more a physical Mach number, it is only a parameter. It is the fast magnetosonic Mach number that bears a causal content and that determines if compressible shocks can evolve or not. It is this Mach number that has to be held fixed when comparing hydrodynamic and MHD jets.

3 STABILITY ANALYSIS

3.1 The Eigenvalue Problem

Linearizing the ideal compressible MHD equations for the above infinite cylindrical equilibrium and expanding the perturbed quantities into normal modes of the form $\exp\{i(m\phi + kz - \omega t)\}$, the resulting ordinary differential equation can be reduced to the following first order system

$$\begin{aligned} AS\frac{1}{r}\frac{d}{dr}(r\xi_r) &= C_1\xi_r - C_2P' \\ AS\frac{d}{dr}P' &= C_3\xi_r - C_1P' \end{aligned} \tag{1}$$

for the radial displacement ξ_r of the fluid elements and the total perturbed pressure $P' = (p + B^2/8\pi)'$. The equilibrium quantities and the Fourier variables are contained in the coefficients C_i and the factors A and S. Together with Sommerfeld radiative boundary conditions this forms an eigenvalue problem for the complex frequency ω, whose imaginary part is the growth rate. The wavenumber k was taken to be real. This can be converted into the spatial interpretation by means of the group velocity $v_{\rm g} = \partial\Re\omega/\partial k$ using the relation $k_{\rm g} = \Im\omega/v_{\rm g}$.

The quantities A and S have the form

$$\begin{aligned} A &= \rho(\omega - \mathbf{k}\cdot\mathbf{v})^2 - (\mathbf{k}\cdot\mathbf{B})^2/4\pi \quad , \\ S &= (B^2/4\pi + \gamma p)(\omega - \mathbf{k}\cdot\mathbf{v})^2 - \gamma p\,(\mathbf{k}\cdot\mathbf{B})^2/4\pi \quad , \end{aligned} \tag{2}$$

(ρ density, p pressure, γ adiabatic index, $\mathbf{k}$ wavenumber of perturbation). When A or S the latter vanishes, the differential equations become singular. A and S are functions of the cylindrical radius, and the zeros correspond to the radii where the phase velocity of the mode equals the local Alfvén or slow magnetosonic velocity, respectively. The singularities give rise to continuous spectra of highly localized modes (Appert et al., 1974). As each of the factors can vanish only when ω is real, the continua are stable.

3.2 The Mode Structure

We shall compare the results of the stability analysis to those of hydrodynamic jets and flows with a purely longitudinal magnetic field with the same velocity profile and the same parameters. To get an idea about possible modes we recall the mode structure of current-free jets. These modes are classified according to the azimuthal wavenumber into the axisymmetric $m = 0$ pinch and the nonaxisymmetric $|m| = 1$ kink modes. Hydrodynamic jets are subject to Kelvin-Helmholtz (KH) instabilities, which are driven by the kinetic energy of the flow. They exist as socalled fundamental modes that are present for any Mach number and wavelength (Ferrari et al., 1981). When grown to nonlinear amplitudes, they are able to disrupt the jet as a whole. The second class, the reflection modes, are waves inside the jet that grow resonantly when reflected at the jet boundaries and propagate like in a waveguide. They exist only for supersonic flows and for wavelengths sufficiently small to fit within the jet. They are labeled by the number of nodes. Simulations show that they tend to saturate in oblique shocks (Norman et al., 1982) and do not threaten the entire jet.

Current-carrying equilibria of the type described in Sect. 2, however without plasma flow and with fixed boundary conditions, were investigated in fusion research (Bodin and Newton, 1980). Most dangerous for these configurations are current-driven instabilities, in particular the nonaxisymmetric free-boundary mode.

Jet flows as considered here contain the combined effects of KH and current-driven instabilities and generally exhibit a very complicated mode structure. We do not expect these equilibria to be absolutely stable but ask instead: how do they compare to hydrodynamic jets and MHD jets with a homogeneous magnetic field? We are interested in the most unstable modes, i.e. those with the shortest growth rate. For supermagnetosonic velocities the kinetic energy is by definition dominant, and hence the relevant instability is of KH type, modified by the magnetic structure.

3.3 Results

The inverse growth length for fundamental and reflection modes were calculated for the force-free equilibrium and compared with the results for a jet with longitudinal magnetic field and a hydrodynamic jet. The densities inside and ouside the jet are taken to be equal, the adiabatic indices are $\gamma = \frac{4}{3}$ and $\gamma = \frac{5}{3}$, respectively. Normalization is w.r.t. the jet radius.

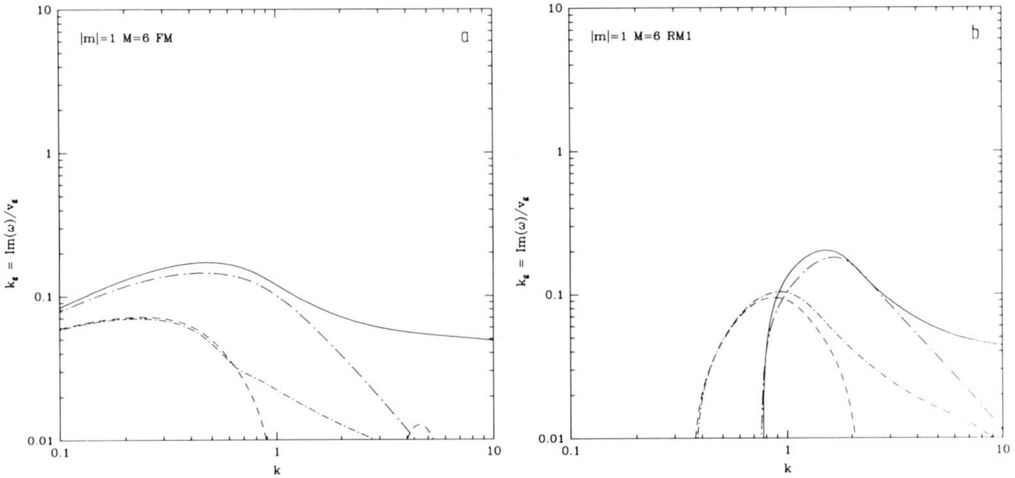

Fig. 1. Inverse growth length of nonaxisymmetric instabilities for a jet with $M = 6$ for different equilibria: hydrodynamic (solid), longitudinal magnetic field (long dash-dotted) force-free jet ($m = 1$ dashed; $m = -1$ short dash-dotted). a) fundamental mode, b) first reflection mode

The pinch fundamental mode is absolutely stabilized for magnetically dominated flows. Figure 1 displays the nonaxisymmetric fundamental and first reflection mode of a jet with Mach number $M = 6$. Both diagrams show that the hydrodynamic jet is the most unstable one followed by the configuration with a homogeneous magnetic field. There are nearly no differences at long wavelengths, but the magnetized jet is effectively stabilized against short wavelength instabilities. A stabilization by the longitudinal field at small k is only observed for subalfvénic flows. Far more stable than the two current-free equilibria is the force-free jet. For long wavelengths the $m = 1$ and $m = -1$ modes are nearly degenerate, unable to resolve the magnetic structure. For wavelengths of the order of the jet radius the difference shows up (see also Fig. 2). In addition, for the fundamental and reflection modes, the most unstable wavelength is larger by a factor of two for the helical configuration compared to their current-free counterparts. The magnetic shear apparently provides additional rigidity to the beam, even compared to a longitudinal field.

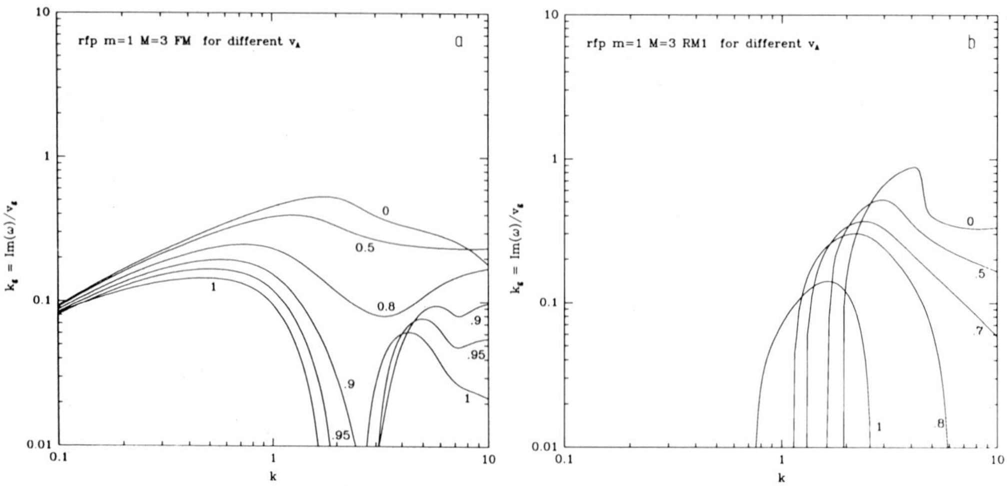

Fig. 2. Inverse growth length of nonaxisymmetric instabilities for a force-free jet with $M = 3$ for various values of v_A. a) fundamental mode, b) first reflection mode

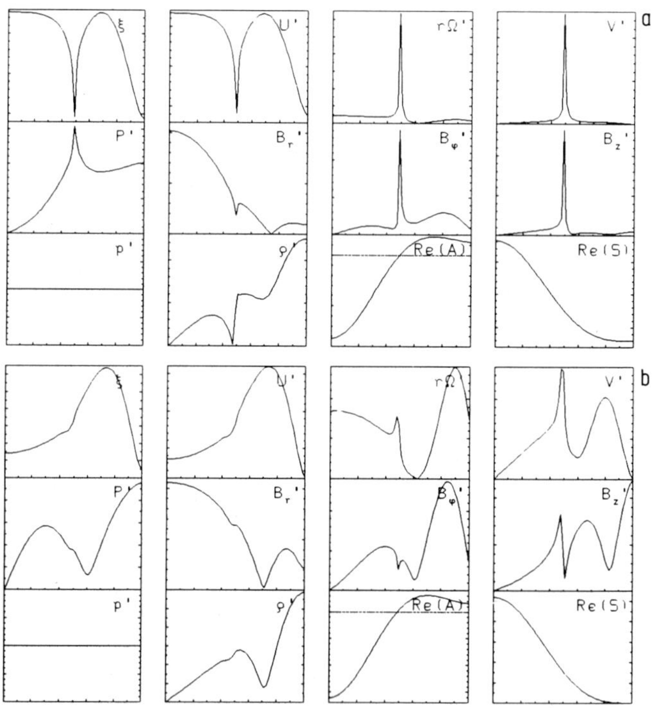

Fig. 3. The perturbed physical quantities for the localized $m = 1$ fundamental mode with $M = 3$, $v_A = 1$ as a function of r/R. a) Stable Alfvén mode $k = 2.56$, $\Im\omega = 0.001$. b) Unstable mode at maximum of growth rate, $k = 4.29$, $\Im\omega = 0.167$

The effect of the plasma-β is shown in Fig. 2 for a jet with $M = 3$. The parameter v_A that labels the curves in Fig. 2 is related to the plasma-β through $\beta = 2(1-v_A^2)/\gamma v_A^2$. Already a moderate magnetization leads to a considerable stabilization in comparison to the hydrodynamic jet. For a magnetically dominated equilibrium the fundamental modes stabilizes in some wavelength interval (see also Fig. 3). The stabilization of short wavelength instabilities, however, is only efficient at low β. For the reflection modes this stabilization sets in earlier. In this domain, however, a longitudinal field stabilizes at already higher plasma-β.

The stable gap in Fig. 2 is a consequence of the existence of the continuous part of the spectrum. This is demonstrated in Fig. 3, which shows the perturbed physical quantities of the fundamental $m = 1$ mode with $v_A = 1$. Figure 3a corresponds to $k = 2.56$ where the mode stabilizes, and Fig. 3b to $k = 4.29$, where it is maximally unstable. The stable mode (Fig. 3a) shows extremely localized perturbations at the cylindrical radius where the factor A vanishes. We are therefore dealing with the Alfvén continuum. The unstable mode in Fig. 3b consists in a superposition of a localized and a global instability. It has a high enough growth rate to be excited easily and will produce radially localized features. This mode is of the same type as indicated by the vanishing real part of A.

4 APPLICATIONS FOR EXTRAGALACTIC JETS

To discuss some applications for extragalactic jets we divide the instabilities into three groups. First, nonaxisymmetric long wavelength instabilities lead to a lateral displacement of the ridge line of the jet. We showed in the last section that a high current (not simply a longitudinal field) and high Mach number both produce oscillations of long wavelength. Many sources (e.g. Birkinshaw, 1990) show such nonaxisymmetric displacements: the radio galaxy Cyg A ($\lambda \simeq 20-80R$), NGC 6251 ($\lambda \simeq 15R$), M87 ($\lambda \simeq 10R$), the VLBI jet of the quasar 0836+71 ($\lambda \geq 50$) (Krichbaum et al., 1990). Being manifestations of the nonlinear developments of the fundamental kink mode, these would require very high Mach numbers ($M > 50$ in some cases) if the jets were hydrodynamic. Force-free jets need Mach numbers smaller by a factor of two. If FR 1 and FR 2 jets differ in electric current and/or Mach number, then the wavelength of lateral oscillations and knot spacings should be smaller in FR 1 sources than in FR 2, and vice versa.

Secondly, instabilities with wavelengths much shorter than the jet radius are commonly believed to produce small scale turbulence. The suppression of such instabilities by magnetically dominated jets implies that those can propagate effectively without much dissipation of kinetic energy.

Finally, the presence of localized modes offers some nice consequences. The stable continua can be understood as "passive modes" that do not grow by themselves but have to be excited at their resonant frequency externally, e.g. by coupling to other modes or radiation. More interesting are the localized instabilities described in the last section. They have considerable growth rates and possess no counterparts in hydrodynamic jets. Such a mode will evolve into a radially localized zone of dissipation. A nonaxisymmetric instability as shown in Fig. 3b will produce helical filaments as observed in M87 (Macchetto, 1991). At the same time these modes are very good candidates for triggering the relaxation process described in Sect. 3.1.

In general, we find that jets with a force-free magnetic field configuration are more stable than hydrodynamic jets and even than those with a purely longitudinal field. They thus provide an attractive model for radio jets. Furthermore, such configurations are able to explain features as the helical strands seen in M87, which are difficult to explain otherwise. The favourable stability properties support models that collimate jets by the combined action of rotation and magnetic fields (Camenzind, 1990).

The difference between FR 1 and FR 2 radio jets may, at least partially, be attributed to a different importance of the plasma-β, which is for a helical field equivalent to the relative size of the electric current. We conclude with the following suggestion: FR 1 jets probably carry a moderate current in contrast to high currents and low plasma-β in FR 2 sources, the latter being also more relativistic. This makes plausible why FR 1 jets appear to be turbulent and FR 2s propagate effectively, generally having low surface brightness. It would be interesting to test if there is a tendency for FR 2 jets to show larger wavelength oscillations than in FR 1, indicative of a difference in current and/or Mach number. The equipartition values for the magnetic fields as derived from observations are much bigger for FR 2 than for FR 1 sources supporting the above suggestion. This behaviour fits nicely into unified schemes according to which the nuclei of FR 2 sources harbour rapidly rotating black holes and FR 1 galaxies slow rotators.

Acknowledgement. This work was supported by the Deutsche Forschungsgemeinschaft (SFB 328).

REFERENCES

Appert, K., Gruber, R., Vaclavik, J. 1974. *Phys. Fluids* **17**, 1471.

Appl, S., Camenzind, M. 1991. *Astron. Astrophys.*, submitted

Benford, G. 1978. *Mon. Not. R. astr. Soc.* **183**, 29

Birkinshaw, M., 1990, in *Beams and Jets in Astrophysics*, Ed. P. A. Hughes, Cambridge University Press, Cambridge

Bodin, H.A.B., Newton, A.A. 1980. *Nucl. Fusion* **20**, 1255

Camenzind, M. 1990. *Rev. Mod. Astron.* Ed. G.Klare, Vol. **3**, 234, Springer Verlag, Heidelberg

Ferrari, A., Trussoni, E., Zaninetti, L. 1981. *Mon. Not. R. astr. Soc.* **196**, 1051

Heyvaerts, J., Norman, C. 1989. *Astrophys. J.* **347**, 1055

Königl, A., Choudhouri, A.R. 1985. *Astrophys. J.* **289**, 173

Krichbaum, T.P., Hummel, C.A., Quirrenbach, A., Schalinski, C.J., Witzel, A., Johnston, K.J., Muxlow, T.W.B., Qian, S.J. 1990. *Astron. Astrophys.* **230**, 271

Macchetto, D. 1991, these proceedings

Taylor, J.B. 1986. *Rev. Mod. Phys.* **58**, 741

Stability of electron-positron beams

U. ACHATZ, R. SCHLICKEISER

Max-Planck-Institut für Radioastronomie

Auf dem Hügel 69, D-5300 Bonn 1, FRG

ABSTRACT

The instability of relativistic electron-positron beams, recently discussed as a component of extragalactic jets, against the excitation of electromagnetic waves moving along a common axis of beam and magnetic field is examined. It is found that highly relativistic beams ($\gamma > \frac{m_p}{m_e} \frac{v_A}{c}$) are very unstable so that under usual conditions they cannot propagate. If one goes to lower energy, however, first the electron component becomes stable leading to the formation of currents. If the Lorentz factor is below a critical value γ_c depending on density and magnetic field strength no perturbation by parallel electromagnetic waves is possible any more.

1. INTRODUCTION

Recently electron-positron beams as a component of extragalactic jets have been discussed as a way to transport energy from the galactic nucleus to the kpc-scale radio jets with differing success (Sol et al., 1989; Achatz et al., 1990). The work of Sol et al. has shown that, provided the only possible instability disrupting such a beam is by electrostatic waves, it is possible to make predictions concerning the onset of this instability which are well in agreement with the observations. However, Achatz et al. found that highly energetic beams are very unstable against the excitation of electromagnetic waves which would decelerate them quickly. This does not apply to Sol et al. who examined beams with Lorentz factors ≈ 10. Thus it seems necessary to extend the examined energy range also for electromagnetic waves to the mildly relativistic range.

2. STABILITY ANALYSIS OF AN ELECTRON-POSITRON BEAM IN A HOMOGENEOUS MAGNETIC FIELD B_0

For an analysis of the stability of the situation we considered a cold e^+e^- beam of density n_r propagating with a velocity u_0 along a homogeneous magnetic field B_0 in a cold e^-p background plasma of density n_b. If the beam is unstable it excites plasma waves. Thus one has to find the dispersion relation $\omega = \omega(k)$ of these plasma waves which are perturbations of the form

$$\delta\vec{E} = \delta\vec{E}_0 \; e^{i(\vec{k}\cdot\vec{x}-\omega t)}. \tag{1}$$

It has been shown (Kennel and Wong, 1967; Tademaru, 1969) that waves propagating at an angle to the magnetic field are thermally damped by the background plasma, so that we concentrate on parallel waves. One finds essentially two implicit dispersion relations (Baldwin et al., 1969; Achatz et al., 1990), one for electrostatic waves ($\delta\vec{E} \parallel \vec{B}_0$)

$$0 = 1 - \frac{\omega_{pp}^2}{\omega^2} - \frac{\omega_{pe}^2}{\omega^2} - 2\frac{n_r}{n_b}\frac{\omega_{pe}^2}{\gamma^3}\frac{1}{(\omega - ku_o)^2} \tag{2}$$

and one for electromagnetic waves ($\delta\vec{E} \perp \vec{B}_0$)

$$\begin{aligned} 0 &= c^2k^2 - \omega^2 + \omega_{pe}^2\frac{\omega}{\omega - \Omega_e} + \omega_{pp}^2\frac{\omega}{\omega - \Omega_p} \\ &+ \frac{n_r}{n_b}\frac{\omega_{pe}^2}{\gamma}\left[\frac{\omega - ku_0}{\omega - ku_0 - \Omega_e/\gamma} + \frac{\omega - ku_0}{\omega - ku_0 + \Omega_e/\gamma}\right] \\ &= D_-(\omega) \end{aligned} \tag{3}$$

where $\omega_{pi} = \sqrt{4\pi n_b q_i^2/m_i}$ is the plasma frequency and $\Omega_i = q_i B_0/mc$ the gyrofrequency of particle species i. In (3) the first term in the brackets arises from the electrons of the beam and the second from the positrons.

Sol et al. (1989) have examined the instability against excitation of electrostatic waves and find that the beam stays stable as long as the magnetic field exceeds a critical value B_c. We will assume that this condition holds and look at the excitation of electromagnetic waves.

2.1 Strong beams

If the beam density is large, i.e. $\frac{n_r}{n_b} \gg \gamma/2$, the beam itself is stable. There are complex roots of (3) but these arise from instabilities of the protons and electrons of the background which are swept up and accelerated by the strong beam.

2.2 Weak beams

If

$$\frac{n_r}{n_b} \ll \gamma, \tag{4}$$

the contributions from the beam in (3) are perturbations to the implicit dispersion relation in an e^-p plasma. Its solutions are, depending on the wave number k, Alfvén, ion cyclotron, Whistler and electron cyclotron waves. All of these are stable, i.e. $\Gamma = Im(\omega) = 0$. The relativistic electrons and positrons which under the given weak-beam condition do not affect each other are however able to trigger each its own instability which occurs if the resonance condition

$$Re(\omega) = \omega_r = ku_0 + \epsilon\Omega_e/\gamma \tag{5}$$

($\epsilon = -1$ for positrons, $+1$ for electrons) can be **satisfied**, which means that a zero of the unperturbed dispersion relation is prevented by a coincidence with a singularity from the beam contributions (cf. Figure 1).

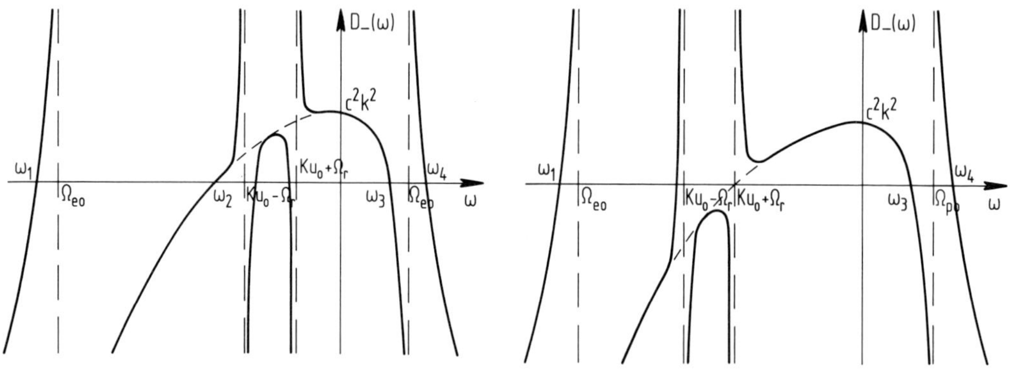

Fig. 1. Dispersion function $D_{-}(\omega)$ for transversal waves in an electron-positron beam, stable (left) and unstable case (right). $\Omega = |\Omega_e/\gamma|$.

2 .2.1 Highly relativistic beams

In the low-frequency regime $|\omega| \ll |\Omega_e|$ (3) transforms to

$$0 = c^2k^2 + \omega_{pp}^2 \frac{\omega^2}{\Omega_p(\omega - \Omega_p)} + \frac{\omega_{pe}^2}{\gamma}\frac{n_r}{n_b}\left[\frac{\omega - ku_0}{\omega - ku_0 - \Omega_e/\gamma} + \frac{\omega - ku_0}{\omega - ku_0 + \Omega_e/\gamma}\right] \tag{6}$$

The waves are assumed to be subluminal ($\omega^2 \ll c^2k^2$). The relativistic particles are a perturbation to the usual cold plasma dispersion relation if

$$\frac{n_r}{n_b} \ll \gamma\frac{m_e}{m_p} \tag{7}$$

Achatz et al. (1990) have shown that under the condition

$$\gamma \gg \frac{v_A}{c} \frac{m_p}{m_e} \tag{8}$$

low-frequency waves ($|\omega| \ll \Omega_p$) are excited which become Alfvén waves (phase-velocity $= v_A$) if $n_r \to 0$. The growth rate for this process is

$$\Gamma = \frac{n_r}{n_b} \frac{\omega_{pe}}{\sqrt{\gamma}} \left[\frac{v_A}{2c} \left(1 + \gamma \frac{n_r}{n_b} \frac{m_e}{m_p} \frac{c^2}{v_A^2} \right)^{-1/2} \right]^{1/2} \tag{9}$$

if

$$\gamma \frac{n_r}{n_b} \frac{m_e}{m_p} \ll 2 \left(\frac{v_A}{c} \right)^3 \qquad \text{or} \qquad \gamma \frac{n_r}{n_b} \frac{m_e}{m_p} \gg \frac{1}{4}$$

and

$$\Gamma \approx \frac{|\Omega_e|}{\gamma} \left[\frac{1}{2} \gamma \frac{n_r}{n_b} \frac{m_e}{m_p} \right]^{1/3} \tag{10}$$

if

$$2 \left(\frac{v_A}{c} \right)^3 \ll \gamma \frac{n_r}{n_b} \frac{m_e}{m_p} \ll \frac{1}{4}$$

If the Alfvén velocity is subluminal, i.e. $v_A \ll c$, the main effect of the excited waves is to isotropize the beam particles in pitch-angle space thereby causing them to suffer synchrotron losses and decelerate the beam itself to a bulk velocity of the order of v_A. This is described by a diffusion equation for the distribution function of particle species a in phase-space

$$\frac{\partial f_a}{\partial t} = \frac{\partial}{\partial \mu} \left(D_{\mu\mu} \frac{\partial f_a}{\partial \mu} \right) \tag{11}$$

where $\mu = \cos\varphi$ ($\varphi =$ pitch angle) and $D_{\mu\mu}$ depends on the wave intensities, and an equation for the time dependence of these waves (forward- and backward moving)

$$\frac{\partial I_\pm}{\partial t} = 2\,\Gamma_\pm(f_a)\, I_\pm \tag{12}$$

with a growth rate $\Gamma_\pm$ which depends on the shape of the distribution function f_a. Generally it is

$$f_a(\mu, t) = N(\gamma)\, \delta(\gamma - \gamma_0)\, f_a^\mu(\mu, t) \tag{13}$$

where

$$f_a^\mu(\mu, 0) = \delta(\mu - 1) \tag{14}$$

$$f_a^\mu(\mu, t \to \infty) = 1/2 \tag{15}$$

It is possible to derive the final wave intensity after the isotropization

$$I_\pm(|k|, t \to \infty) = \frac{1}{2}\left\{\left[\left(\frac{T}{|k|^2}\right)^2 + i_0^2|k|^{-2q}\right]^{1/2} \pm \frac{T}{|k|^2}\right\} \tag{16}$$

where

$$T = \frac{n_r}{n_b}\left(\frac{m_e}{m_p}\right)^{1/2}\frac{\omega_{pe}}{c}B_0^2 \tag{17}$$

and the initial spectrum was assumed to have the form

$$I_\pm(|k|, t = 0) = i_0|k|^{-q}, \qquad 1 < q \leq 2 \tag{18}$$

Reinserting this result into (11) in order to get a lower limit for the distance the beam covers before having become decelerated, one gets for typical values of the respective densities in extragalactic jets

$$\ell \approx \frac{B_0^2}{\pi T} = 2.4 \cdot 10^{13}\mathrm{cm}\left(\frac{n_r}{10^7\ \mathrm{cm}^{-3}}\right)^{-1}\left(\frac{n_b}{0.1\ \mathrm{cm}^{-3}}\right)^{1/2} \tag{19}$$

so that highly relativistic beams having a length larger than ℓ cannot propagate. This certainly assumes that nonlinear processes are not effective in damping the resonant waves.

The other interesting limit is the high-frequency regime ($|\omega| \gg \Omega_p$) of subluminal plasma waves. One derives from (3) the dispersion relation:

$$0 \approx c^2k^2 + \omega_{pp}^2 + \omega_{pe}^2\frac{\omega}{\omega - \Omega_e} + \frac{n_r}{n_b}\frac{\omega_{pe}^2}{\gamma}\left[\frac{\omega - ku_0}{\omega - ku_0 + \Omega_e/\gamma} + \frac{\omega - ku_0}{\omega - ku_0 - \Omega_e/\gamma}\right]. \tag{20}$$

Neglecting the beam contributions one finds the root

$$\omega = \Omega_e\frac{c^2k^2 + \omega_{pp}^2}{c^2k^2 + \omega_{pe}^2} \tag{21}$$

which in the limit $|k| \gg \omega_{pe}/c$ describes electron cyclotron waves ($\omega \to \Omega_c$) and in the long-wavelength limit ($|k| \ll \omega_{pe}/c$) gives the Whistler dispersion relation

$$\omega \approx \frac{\Omega_e}{\omega_{pe}^2}(c^2k^2 + \omega_{pp}^2). \tag{22}$$

It is easy to convince oneself from Figure 1 that only the positrons can cause an instability, i.e. remove a zero of the unperturbed dispersion relation. This is the case when resonance condition (5) is fulfilled. Combining (5) with (22) one finds furthermore that the resonant wave number is

$$k_{1,2} = -\frac{\Omega_p}{v_A}\left\{\frac{c}{2v_A} \pm \sqrt{\frac{c^2}{4v_a^2} - \frac{m_p}{\gamma m_e} - 1}\right\} \tag{23}$$

Evidently a wave-particle interaction can only happen if the radicand in (23) is positive, i.e.

$$\gamma > \gamma_c := \frac{m_p}{m_e}\frac{1}{\frac{c^2}{4v_A^2} - 1} \approx \frac{m_p}{m_e}\frac{4v_A^2}{c^2} = 3.9 \cdot 10^2 \left(\frac{B}{\mathrm{mG}}\right)^2 \left(\frac{n_b}{10^{-3}\,\mathrm{cm}^{-3}}\right)^{-1} \tag{24}$$

Beams with $\gamma < \gamma_c$ cannot excite Whistler waves. In order to be in the correct frequency regime one has to make sure that $\Omega_p/v_A < |k| < \sqrt{m_p/m_e}$. k_1 satisfies this if $\gamma < \frac{m_p}{m_e}\frac{v_A}{c}$ which is just complementary to (8). k_2 is within the required boundaries if $\frac{c}{2v_A} < \sqrt{\frac{m_p}{m_e}}$ and $\gamma < \left[\frac{c}{v_A}\sqrt{\frac{m_e}{m_p}} - 1\right]^{-1/2}$ which can become very large.

Under the quoted conditions the maximal growth rate can be found by using (5) and (20) giving

$$\Gamma = \frac{|\Omega_e|}{\gamma}\sqrt{\frac{n_r}{n_b}}\frac{|\frac{\omega_r}{\Omega_e} - 1|}{\sqrt{1 - \frac{n_r}{\gamma n_b}}} \approx \frac{|\Omega_e|}{\gamma}\sqrt{\frac{n_r}{n_b}} = 1.7 \cdot 10^{-3}\mathrm{s}^{-1}\left(\frac{\gamma}{10^4}\right)^{-1}\sqrt{\frac{n_r}{n_b}}\left(\frac{B}{\mu\mathrm{G}}\right) \tag{25}$$

As for an interaction with electron cyclotron waves, (25) shows that $\Gamma \to 0$ as $\omega_r \to \Omega_e$. Just as important is that electron cyclotron waves are heavily damped in a non-cold plasma (e.g. Stix, 1962), so that their contribution to destabilizing the beam can probably be neglected.

2 .2.2 Mildly relativistic beams

From the above said one finds some interesting physics concerning beams with a Lorentz factor below

$$\gamma_h := \frac{m_p}{m_e}\frac{v_A}{c} = 4.2 \cdot 10^2 \left(\frac{B}{\mathrm{mG}}\right)\left(\frac{n_b}{10^{-3}\,\mathrm{cm}^{-3}}\right)^{-1/2}. \tag{26}$$

They cannot excite Alfvén waves so that only Whistler waves remain. However, electrons are not even in interaction with these so that we meet with the phenomenon that the positron component of the beam becomes decelerated but the electrons are not, leading to the formation of a current. This would be a possibility to produce magnetic fields as they are observed in extragalactic jets as discussed by Rose (1987, 1989). This is changed even more if the Lorentz factor is below the critical value

γ_c. Then no excitation of parallel transversal waves whatever is possible. Given that electrostatic waves are damped by the modulational instability and that oblique transversal waves are thermally damped, this would imply that electron-positron beams with $\gamma < \gamma_c$ are stable and can propagate freely. In order to give this some physical substance we have applied it to the model of Sol et al. (1989) who predict a first radio knot in the source 3C120 at a distance of $z = 3''$ from the galactic nucleus. By taking the same physical parameters (from Walker et al., 1987), i.e. $n_b = 9.9 \cdot 10^{-4}\,\tau^{-0.03}\mathrm{cm}^{-3}$ and $B = 4.1 \cdot 10^{-5}\,\tau^{-0.97}\,\mathrm{G}$ where τ is the jet width in arcsec, we find

$$\gamma_c = 0.66\,\tau^{-1.97}. \tag{27}$$

This means that a $\gamma = 10$ beam becomes unstable at $\tau = 0.3''$ corresponding to a distance $z = 2.3''$ from the nucleus which is somewhat shorter than the value given by Sol et al. (1989). This does not mean that the model of Sol et al. is wrong since as long as $\gamma < \gamma_h$ the electron component is stable. We find

$$\gamma_h = 17\,\tau^{-0.96} \tag{28}$$

so that $\gamma_h = 10$ at $\tau = 1.7''$ corresponding to $z = 13''$ which is above the predicted knot distance.

3. CONCLUSIONS This work shows that electron-positron beams can be discussed as a component of extragalactic jets as long as their energy is below a critical value $\gamma_h = (m_p/m_e)(v_A/c)$, so that the electrons of the beam can propagate freely or even better if $\gamma < \gamma_c = \frac{m_p}{m_e}\frac{4v_A^2}{c^2}$ in which case the beam is not affected by parallel transversal waves. This certainly assumes that oblique waves are not important in destabilizing the beam which will have to be examined in future.

4. REFERENCES

Achatz, U., Lesch, H., Schlickeiser, R.: A&A **233**, 391 (1990)
Baldwin, D.E., Bernstein, I.P., Weenink, M.P.H.: in *Advances in Plasma Physics* Vol. III, eds. A. Simon, W.B. Thompson, Wiley, New York, p. 1 (1969)
Kennel, C.F., Wong, H.V.: J. Plasma Phys. **1** (1), 75 (1967)
Rose, W.K.: Ap.J. **313**, 146 (1987)
Rose, W.K.: Ap.J. **337**, 9 (1989)
Sol, H., Pelletier, G., Asseo, E.: MNRAS **237**, 311 (1989)
Stix, T.H. *The theory of plasma waves*, McGraw Hill, New York (1962)
Tademaru, E.: Ap.J. **258**, 131 (1969)
Walker, R.C., Benson, J.M., Unwin, S.C.: Ap.J. **316**, 546 (1987)

Streaming Particle Beams in Active Galactic Nuclei

H. SOL

UPR 176 - CNRS, DARC - Observatoire de Paris, 92195 Meudon Cedex, France

1. INTRODUCTION

As emphasized by Hughes and Miller in 'Beams and jets in astrophysics' [1], the term 'beam' is often used to refer to "figments of the theorist's imagination" and can have in fact several meanings usually related to the first parts of extragalactic jets, close to the central engine, at or even below the VLBI scale: ultrarelativistic particles, photon beams, streaming particles... In this contribution the term refers to beams of relativistic particles interpenetrating some background plasma and streaming through some ambient medium and magnetic field. Some possible observational signatures of the beams are proposed, especially for the case of BL Lac objects.

2. STABILITY PROPERTIES AND RADIATION FROM RELATIVISTIC BEAMS

It seems that close to the central engine the magnetic field is roughly parallel to the jet [2–5]. In collaboration with G. Pelletier and E. Asséo [6] we found that energetic beams can be stable relatively to the excitation of Langmuir waves as long as this parallel magnetic field is larger than a critical value $B_c = 3.2 \times 10^{-3} n_p^{1/2}$, where n_p is the density of the ambient plasma ($B = B_c$ corresponds to equality between the ambient plasma frequency ω_p and the cyclotron frequency ω_c). In collaboration with G. Pelletier [7] we strenghtened this conclusion by showing that ambient density fluctuations can broaden the sharp resonance of monokinetic beams and deture the beam instability (see [7] or [8] for more details). Studies of the beam stability relatively to the excitation of alfvenic waves [7,9,10] concluded that electron–positron beams can be stable as long as their bulk Lorentz factor γ is smaller than $\gamma_A = \left(\frac{\omega_c}{\omega_p}\right)\left(\frac{m_p}{m_e}\right)^{1/2}$, while electron–proton beams or more energetic electron–positron beams (with $\gamma > \gamma_A$) are rapidly destroyed. Excitation of whistler waves can be important for highly relativistic beams but does not affect electron–positron beams with bulk Lorentz factor smaller than $\gamma_W = \left(\frac{4v_a}{c}\right)\gamma_A$, where v_a is the Alfven velocity [10].

All these results lead to the idea that electron–positron beams with bulk Lorentz factors smaller than $(m_p/m_e)^{1/2} = 43$ can propagate rather far from the central engine if there is a strong enough magnetic field. We proposed in [7,8] that they are responsible for VLBI features and superluminal components since the Lorentz factors deduced from observed superluminal motion correspond to the range of beam stability. Radiation from such beams can be ascribed to different mechanisms. It can be due to synchrotron emission of a high energy tail of beam particles ejected from the central engine as for instance in the pile–up scenario proposed in [9,11]. Heating of ambient medium and subsequent synchrotron emission is also expected in regions where $B < B_c$ since the beams excite Langmuir wave packets which finally collapse to the Debye's sphere and accelerate particles [6,12]. Another possibility is the coherent or incoherent Compton radiation from beam particles scattered by the Langmuir wave packets [13–18]. Collective emission can be much more efficient than synchrotron one and could strongly reduce the energy budget in compact radio sources. The interaction of the beams with interstellar clouds also leads to radiative processes [14].

3. A TENTATIVE SKETCH FOR BL LAC OBJECTS

Although consistent with VLBI features observed in quasars, such a scenario can not be directly applied to VLBI jets observed in BL Lac objects since they show an average angle of about 6° between the electric vectors and the VLBI elongations, suggesting a magnetic field nearly perpendicular to the beam for optically thin synchrotron radiation [19–22]. This is usually explained by transverse shocks which locally compress an initially tangled magnetic field [23]. However it is not yet clear why such shocks would not occur in VLBI quasar jets and how they can be constrained in BL Lac objects. Another possibility is to assume that VLBI jets in BL Lac objects are emitted in a direction roughly perpendicular to the ambient magnetic field. The high energy tail of beam particles radiate in this transverse magnetic field for instance at the edge of the beam, while the beam itself either (i) clears its path through the ambient medium due to ram pressure or to some microscopic processes [13] or (ii) propagates through this external field due to an inner longitudinal self–magnetic field or to plasma polarization fields and associated $E \times B$ drift [24]. In the absence of a significant longitudinal magnetic field, the beam will be more affected by beam–plasma or Kelvin–Helmholtz instabilities and will more rapidly fade out. This can qualitatively account for intrinsic properties of VLBI jets in BL Lac objects, which appear to be smaller and to have lower superluminal speeds than VLBI jets in quasars [22,25,26].

REFERENCES

[1] Beams and jets in astrophysics, 1991, P.A. Hughes, ed., Cambridge University Press.
[2] Asséo, E., Sol, H., 1987, Physics Reports, 148, 307.
[3] Saikia, D.J., Salter, C.J., 1988, ARAA, 26, 93.
[4] Magnetic fields and extragalactic objects, 1987, E. Asséo, D. Grésillon, eds, Cargèse workshop june 1987, Editions de Physique.
[5] Galactic and intergalactic magnetic fields, 1990, R. Beck, P.P. Kronberg, R. Wielebinski, eds, Kluwer academic publishers.
[6] Pelletier, G., Sol, H., Asséo, E., 1988, Phys. Rev. A, 38, 2552.
[7] Pelletier, G., Sol, H., 1991, MNRAS, in press.
[8] Sol, H., Pelletier, G., 1991, in Physics of AGN, to be edited by S.J. Wagner and W.J. Duschl, Springer Verlag.
[9] Achatz, U., Lesch, H., Schlickeiser, R., 1990, AA, 233, 391.
[10] Achatz, U., Schlickeiser, R., these proceedings.
[11] Schlickeiser, R., 1984, AA, 136, 227.
[12] Sol, H., Pelletier, G., Asséo, E., 1989, MNRAS, 237, 411.
[13] Benford, G., 1984, in Physics of energy transport in extragalactic radio sources, A.H. Bridle, J.A. Eilek, eds, p. 185 and 1985, in Unstable current systems and plasma instabilities in astrophysics, M.R. Kundu, G.D. Holman, eds, Reidel publishing company, p. 131.
[14] Rose, W.K., Beall, J.H., Guillory, J., Kainer, S., 1987, ApJ, 314, 95.
[15] Baker, D.N., Borovsky, J.E., Benford, G., Eilek, J.A., 1988, ApJ, 326, 110.
[16] Weatherall, J.C., Benford, G., 1991, ApJ, 378, 543.
[17] Benford, G., these proceedings.
[18] Lesch, H., these proceedings.
[19] Gabuzda, D.C., Wardle, J.F.C., Roberts, D.H., 1989, ApJ Lett, 336, L59, and ApJ, 338, 743.
[20] Roberts, D.H., Wardle, J.F.C., Brown, L.F., Gabuzda, D.C., Cawthorne, T.V., 1990, in Parsec-scale radio jets, J.A. Zensus, T.J. Pearson, eds, p. 110.
[21] Gabuzda, D.C., these proceedings.
[22] Mutel, R.L., 1990, in Parsec-scale radio jets, J.A. Zensus, T.J. Pearson, eds, p. 98.
[23] Hughes, P.A., Aller, H.D., Aller, M.F., 1989, ApJ, 341, 68.
[24] Mostovych, A.N., Manka, C.K., Ripin, B.H., Stamper, J.A., 1988, in "Topical conference on Plasma Astrophysics", Santa Fe, September 1988.
[25] Cawthorne, T.V., these proceedings.
[26] Mutel, R.L., these proceedings.

Nonlinear Interaction of the Relativistic Wind with the Ambient Plasma

YA.N.ISTOMIN

Lebedev Physical Institute, USSR Academy of Sciences,
117924, Moscow

In the plasma–beam system there exist the electromagnetic modes in which the refractive index approaches to infinity in the region of plasma resonance when the wave frequency ω coincides with plasma frequency ω_p. It is easy to see from the example of the longitudinal wave frozen in the stream and propagated along the stream velocity $\vec{u}$:

$$1 - \frac{\omega_p^2}{\omega^2} - \frac{\omega_s{}^2}{(\omega - ku)^2} = 0, \tag{1}$$

or $(\omega - ku)^2 = \omega_s^2/\varepsilon$, where ω_s is the plasma frequency of streaming plasma ($\omega_s^2 = \sum_i 4\pi n_s{}^i e_i^2/m_i\gamma_i^3$; γ_i is the Lorenz–factor of moving particles), ε is the plasma permittivity. It is known [1] that such dispersion leads to full absorbtion of the wave in the region of particular point of the refractive index ($k^2 \to \infty$). In our case it is the region of the plasma resonance when $\omega = \omega_p$. Since the resonance is defined by the value of plasma density, then the wave pressure can change significantly the character of propagation and absorbtion of the wave. In some cases the charged streams can be powerful, so the nonlinear action of electromagnetic wave, frozen in the stream, can strongly change the plasma dynamics. It is very important because of weak interaction of plasma with particles due to the strong relativity and low plasma density. It deals to the interaction of winds from stars, pulsars, active galactic nuclei with the ambient plasma.

There are four eigen modes of electromagnetic wave in the media with permittivity ε in which propagates the charge particle stream of the velocity $\vec{u}$ [2]. The transverse mode with the electric field, being orthogonal to the stream velocity $\vec{u}$ ($\vec{E}\vec{u}=0$) has the usual dispersion relation

$$n^2 = \varepsilon - \gamma^2\omega_s^2/\omega^2. \tag{2}$$

The another transverse mode in which the electric vector has the component along the stream velocity in the limit of small ε ($\varepsilon << 1$) satisfies the relation

$$n^2 = \varepsilon - \gamma^2 \frac{\omega_s^2}{\omega^2} \frac{1-\beta^2}{1-\beta^2\cos^2\theta}, \quad \beta = u/c, \ \cos\theta = \vec{k}\vec{u}/ku. \tag{3}$$

And, finally, two modes propagating with stream in the same limit are described as follows

$$(1-\vec{n}\vec{\beta})^2 = \frac{\omega_s^2}{\varepsilon\omega^2}(1 + \beta^2\gamma^2\sin^2\theta). \tag{4}$$

In the limit $\beta \longrightarrow 0$ the relation (4) describes the usual Langmuir oscillations. In the general case the modes (4) are potential only when θ=0. For the not very high frequencies $\omega<\gamma\omega_s$ the transverse modes (2) and (3) do not propagate and excite. So the quasilongitudinal modes (4) are the single modes excited by central source and propagated together with plasma streaming. Their refractive indexes become infinity in the point of plasma resonance.

The wave equation which corresponds the dispersion equation (4) has the form

$$\Delta(1+i\beta\nabla_{||})^2E_{||} = [\Delta_{||} + (1+\beta^2\gamma^2)\Delta_{\perp}]\frac{\omega_s^2}{\varepsilon\omega^2}E_{||}, \tag{5}$$

where signs $||$ and $\perp$ denote correspondingly directions along and orthogonal to the velocity $\vec{u}$. We passed also to the dimensionless coordinates $\vec{r}\,' = (\omega/c)\vec{r}$. The plasma permittivity ε can be represented in the convenient view if we introduce two characteristic densities N_p and N_c by which correspondingly the plasma frequency coincides with the wave frequency and collision frequency ν is equal to ω:

$$N_p = \frac{m_e\omega^2}{4\pi e^2} = 3.14\cdot10^{-10}\omega^2 \text{ cm}^{-3};$$

$$N_c = 3.5\cdot10^4T_e^{3/2}\omega/(\Lambda/10) \text{ cm}^{-3}.$$

Here T_e is the electron temperature in eV units. Then

$$\varepsilon = 1 - \frac{n}{1+i\mu n},$$

where n is the dimensionless plasma density, $n = N/N_p$; μ is the collision parameter, $\mu = N_p/N_c$. For the not very high frequencies $\omega \lesssim 10^{14}T_e^{3/2}\text{sec}^{-1}$ $\mu<<1$.

The force of high frequency pressure $\vec{f}$ acts on the media with permittivity ε [3]

$$\vec{f} = \frac{\text{Re}\varepsilon-1}{16\pi}\nabla|E|^2 + \frac{\text{Im}\varepsilon}{8\pi}|E|^2\nabla\varphi, \tag{6}$$

where $|E|$ and φ are the module and phase of the electric field of wave. The hydrodynamic equations for the plasma on which the additional force (6) acts together with the nonlinear wave equation (5) describe the interaction of plasma with the wave field which is more strong in the region of plasma resonance ($\text{Re}\varepsilon$=0).

Let us consider the one–dimension case when all quantities depend on the coordinate z directed along the stream. For the rest plasma the hydrodynamic equations lead to the condition of the pressures balance

$$-\frac{1}{16\pi}\frac{n}{1+\mu^2n^2}\frac{\partial}{\partial z}|\mathcal{E}|^2 +$$

$$\frac{1}{8\pi}\frac{\mu n^2}{1+\mu^2n^2}|\mathcal{E}|^2(1/\beta+\partial\psi/\partial z) - \partial P/\partial z = 0, \qquad (7)$$

$$E = E_{\|} = |\mathcal{E}|e^{iz/\beta+i\psi}$$

In the limit of $\mu<<1$ $(n<\mu^{-1})$ and for the adiabatic low $P=P_pn^{\Gamma}$ (Γ is the adiabatic index) the equation (7) means that the full pressure is constant

$$\frac{\Gamma}{\Gamma-1}n^{\Gamma-1}+\frac{1}{2}r^2 = \text{const}, \qquad (8)$$

where $\mathcal{E} = [8\pi P_p]^{1/2}re^{i\psi}$, r is the dimensionless wave amplitude. By this the equation (5) gives two equalities

$$\frac{\partial^2 r}{\partial t^2} - r(\partial\psi/\partial t)^2 = -\frac{1-n+\mu^2n^2}{(1-n)^2+\mu^2n^2}r;$$

$$\frac{\partial}{\partial t}(r^2\,\partial\psi/\partial t) = \frac{\mu\;n^2}{(1-n)^2+\mu^2n^2}r^2. \qquad (9)$$

Here t is the coordinate which is proportional to z, $t = (\omega_s/\omega)z$. The equations (9) are equivalent to the motion of unit mass particle in central symmetric potential

$$V(r) = \int_0^r \frac{(1-n)+\mu^2n^2}{(1-n)^2+\mu^2n^2}r'dr' \qquad (10)$$

with orbital momentum $L = r^2(\partial\psi/\partial t)$ changing in time. Note that L varies only when n=1, i.e. at the resonance. Really, since at $\mu<<1$ the right hand side of equation (9) is proportional to δ–function

$$\frac{\mu n^2}{(1-n)^2+\mu^2n^2} = \pi\delta(1-n),$$

then the orbital momentum changes with jump on the quantity $\pi r_1^2/|\partial n/\partial t|_{n=1}$. The value r_1 is the dimensionless wave amplitude at the resonance

$$r_1^2 = \frac{2\Gamma}{\Gamma-1}(n_0^{\Gamma-1}-1). \qquad (11)$$

Thus, L becomes constant except the resonant moment when it changes by the quantity

$$L_{i+1} - L_i = \pi\Gamma r_1/|\partial r/\partial t|_{r=r_1}. \tag{12}$$

Also the total energy changes by jump ($W = 1/2\ \dot{r}^2 + 1/2\ L^2/r^2 + V(r)$)

$$W_{i+1} = W_i + (L_{i+1}^2 - L_i^2)/2r_1^2. \tag{13}$$

Note that the potential V(r) (10) has a shape of hole with the logarithmic minimum at n=1 (see figure 1)

$$V(r_1) = -\Gamma\ln(\mu^{-1}) + \int_1^{n_0} \ln|1-n|\Gamma(\Gamma-2)n^{\Gamma-3}dn -$$

$$-\Gamma n_0^{\Gamma-2}\ln|n_0-1|,$$

so the amplitude r begins to oscillate around the value $r=r_1$ which corresponds to the resonance (n=1).

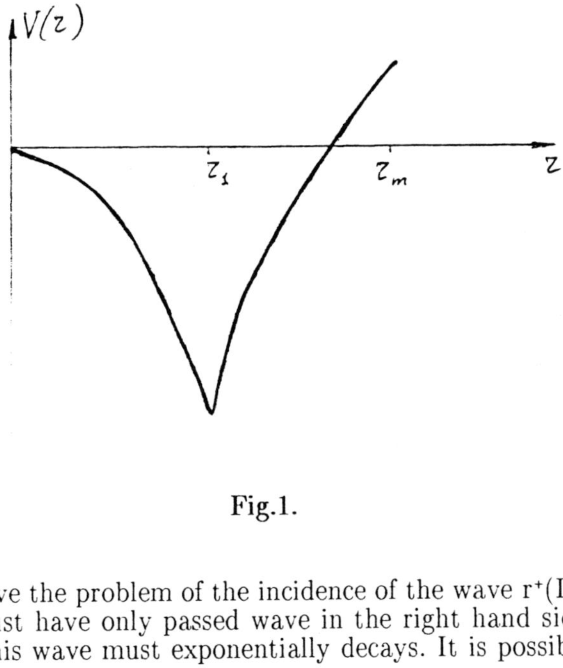

Fig.1.

When we solve the problem of the incidence of the wave $r^+(L^+=-(r^+)^2)$ on the plasma layer we must have only passed wave in the right hand side, where plasma density $n=n_0>1$. This wave must exponentially decays. It is possible only if $L^- \equiv 0$, i.e.

$$\frac{dr}{dt} = -(n_0-1)^{-1/2}r \qquad (W=0).$$

This problem is solving by the time reversing, when we start from the initial condition L=W=0 at point r=0. Then, the quantity L begins to deminish every time when the resonance takes place. L becomes negative while the energy W grows until the motion in the region $r \geq r_m$ becomes possible (fig.2).

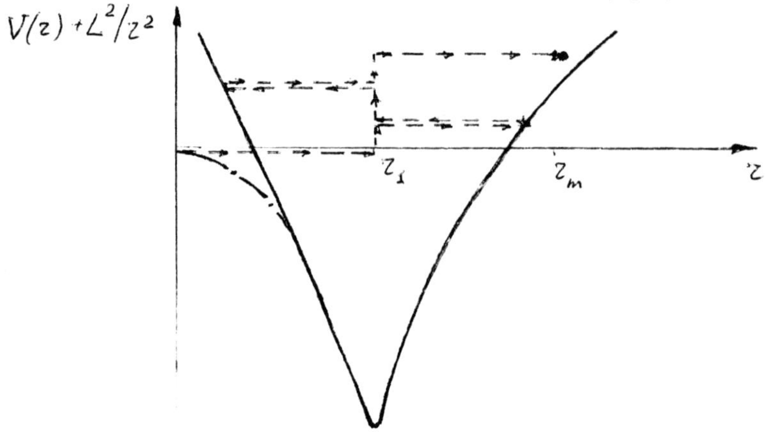

Fig.2.

Here r_m is the maximum value of the wave amplitude when the plasma density turns to zero

$$r_m^2 = \frac{2\Gamma}{\Gamma - 1} n_0^{\Gamma - 1}.$$

At this point there are fixed also two values $L = -[(r^+)^2 - (r^-)^2]$ and $(\partial r/\partial t)|_{r=r_m}$ which with the help of equalities

$$\left.\frac{\partial r}{\partial t}\right|_{r=r_m} = -2\,\frac{r^+ r^-}{r_m}\sin(\Delta\varphi);$$

$$r_m^2 = (r^+)^2 + (r^-)^2 + 2r^+r^-\cos(\Delta\varphi)$$

let us to determine the amplitude of refractive wave r^- and its phase shift $\Delta\varphi$. Since $(\partial r/\partial t)|_{r=r_m} \neq 0$, then $(\partial n/\partial t)|_{n=0}$ also does not equal to zero. It means that the density profile has the shape of shock with oscillations around the resonant value n=1. The number of oscillations is defined by the value of absorbtion μ. From (12) it is seen that the momentum changing $\Delta L \simeq \pi\Gamma r_1/(2\Gamma\ln(\mu^{-1}))^{1/2}$ decreases when μ decreases too. Also the energy's growth deminishes $\Delta W \propto 1/\ln(\mu^{-1})$. Thus we need to lift from the level W=0 to $W \simeq V(r_m)$, then the number of resonance crossing, i.e. number of oscillations will be proportional to $\ln(\mu^{-1})$. The described picture means that at the resonance region occurs the finite absorbtion of wave energy. In order to absorb all the wave it is necessary to pass through the resonance several times. However, at the big values of incident wave r^+ (or r_m) the total absorbtion of the passed wave takes place at one resonance crossing. It is due to the large reflection coefficient ($R = r^-/r^+ > 0.9$). On the language of the potential (10) it means that $V(r_m)<0$. Because of

$$V(n_0) = \int_0^{n_0} \frac{\Gamma n^{\Gamma-2}}{1-n} \, dn,$$

then $V(n_0)<0$, if $n_0 \gtrsim |(\Gamma-1)/(2-\Gamma)|^{1/(2-\Gamma)}$, i.e.

$$r_m^2 \gtrsim \frac{2\Gamma}{\Gamma-1} \left| \frac{\Gamma-1}{2-\Gamma} \right|^{(\Gamma-1)/(2-\Gamma)} .$$

The results of calculations of plasma density profiles, wave amplitude, reflection coefficient R and phase shift $\Delta\varphi$ for the case $\Gamma = 5/3$, $\mu=10^{-8}$ are presented on the figure 3.

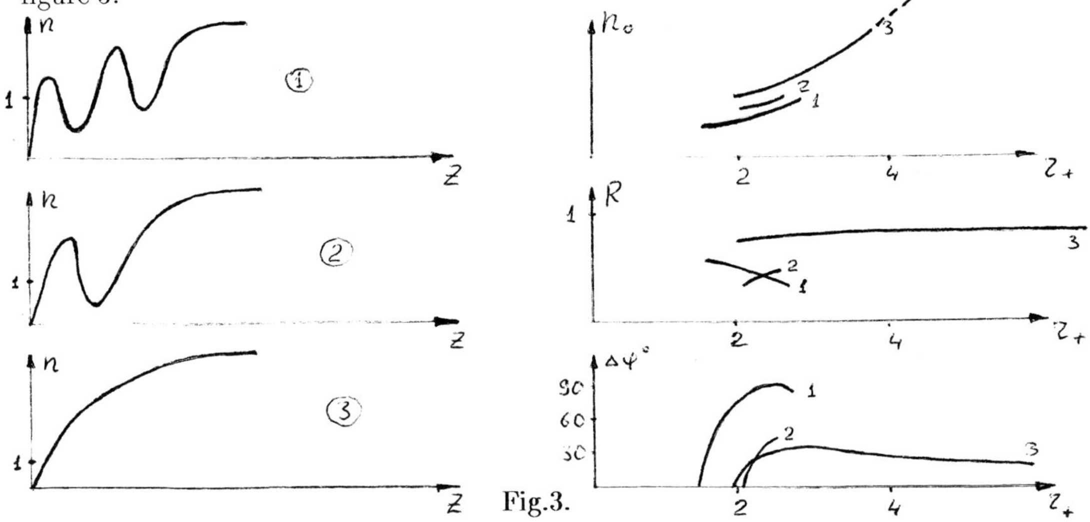

Fig.3.

Thus we see that the plasma can't exist in the region of strong wave which frozen in the relativistic wind.

The author thanks Pariev V.I. for the helpful discussions.

References

1. Ginzburg V.L. The Propagation of Electromagnetic Waves in Plasma. Pergamon Press, Oxford (1970).
2. Istomin Ya.N., Luk'yanov A.V. Sov. Phys. JETP **70**, 891 (1990).
3. Johnston T.W. RCA Review **21**, 4 (1960).

Physics of Bright Jets and their Velocities

G.V. BICKNELL

Mt. Stromlo and Siding Spring Observatories
Australian National University

1. FUNDAMENTAL DIFFERENCES BETWEEN CLASS I AND CLASS II JETS

Fanaroff and Riley (1974) showed that extragalactic radio sources divided into two distinct morphological types (edge-darkened vs edge-brightened) at a monochromatic power $P^{\mathrm{tot}}_{1.4\mathrm{GHz}} \sim 10^{24.5}\mathrm{W\ Hz^{-1}}$ and suggested that the transport of energy in the two types of sources (now known as Fanaroff-Riley class I and class II) is different. It is now realized that the type of jets envisaged in the Scheuer (1974) and Blandford and Rees (1974) theories are most relevant to the powerful, class II radio sources. As the resolution and signal to noise of the observations increased it was realized that the jets in class I radio sources are more prominent. Bridle (1979) also highlighted other marked differences between jets in class I and class II radio sources: Class I jets spread more rapidly and the brightness ratios of class I jets in the same galaxy are generally closer to unity. Jets in class II radio galaxies are frequently one-sided. The magnetic field in class I jets tends to be perpendicular for a large fraction of its length. In class II jets it is usually parallel for the whole length.

There is a consensus that jets in the more powerful of the class II sources have velocities $\gtrsim 0.1c$. The larger spreading rates of class I jets suggest that they are turbulent implying that they decelerate. Thus it is not surprising that it is generally accepted that they have lower velocities. The purpose of this paper is to outline the physical theory which supports this idea and to briefly discuss the implications of the theory for other aspects of radio galaxies.

2. DYNAMICAL MODEL OF TURBULENT CLASS I JETS

2.1 General considerations

It was Fanti (see Fanti et al, 1982) who originally had the insight that the deceleration due to entrainment could be responsible for the usually slow surface brightness decrease of class I jets. The following is an outline of the theory (Bicknell 1986b) required to actually calculate the run of jet velocity.

Turbulence has two major effects on a jet: it produces spreading and deceleration. The former is due to the turbulent diffusion of momentum from the centre to the edge of the jet and the latter is a result of the conservation of momentum. The correct formal approach to turbulent flow is to consider statistical averages of the flow variables ρ, P, $\mathbf{v}$ etc (denoted by bars and tildes in the following) and to average the fluid dynamic equations over an ensemble. The following equation for the jet momentum density represents the end result of this process (plus some manipulation of the dynamical equations):

$$\frac{\partial(\bar{\rho}\,\tilde{v}_z^2)}{\partial z} + \frac{1}{r}\frac{\partial(r\,\bar{\rho}\,\tilde{v}_r\,\tilde{v}_z)}{\partial r} = -\frac{1}{r}\frac{\partial\,(r\langle\,\rho\,v_r'\,v_z'\rangle)}{\partial r} - \frac{dP_{\rm ism}}{dz}\left(1 - \frac{\bar{\rho}}{\rho_{\rm ism}}\right)$$

The first term on the right represents the turbulent diffusion of momentum transverse to the jet due to the turbulent stress $\langle\,\rho\,v_r'\,v_z'\rangle$.

The absence of depolarization in class I jets indicates that they are lighter than the surrounding medium and this has another important consequence which can be seen in the above equation. The last term, the buoyancy term, is positive for light jets in galactic atmospheres implying an outward buoyant driving force. The collimation properties appear to be influenced quite strongly by the competition between turbulent stresses and buoyancy.

2.2 Semi-empirical turbulent jet model.

Unfortunately it is practically impossible to close the above momentum equation and other dynamical equations with physically realistic turbulent closure terms. However, progress is possible if we abandon the objective of determining both the spreading rate and the deceleration in one theory but determine the latter from the former. This is facilitated by the use of integral equations for the mean flow. For example, the momentum equation above can be integrated across the jet to yield:

$$\frac{d}{dz}\int \bar{\rho}\tilde{v}_z^2\,r\,dr = -\frac{dP_{\rm ism}}{dz}\int_0^\infty \left(1 - \frac{\bar{\rho}}{\rho_{\rm ism}}\right)\,r\,dr$$

In order to make use of this equation one needs to make some assumption for the profiles of various quantities, e.g. $\tilde{v}(r,z) = v_c(z)\,V(r/a(z))$ where $v_c(z)$ is the central jet velocity, $V(r/a(z))$ is the profile function and $a(z)$ is the half-radius based on the velocity. In Bicknell (1986b) I used top-hat profiles, motivated by simplicity and the general observation that the observed polarization profiles do not indicate a large amount of shearing. This assumption is not critical. What is more important is that the integrals of various combinations of profile functions vary slowly with respect to the central values of the dynamical variables. For a top-hat jet the integrated momentum equation becomes:

$$\frac{d}{dz}\left(\bar{\rho}_c(z)\tilde{v}_c^2(z)a^2(z)\right) = -a^2(z)\frac{dP_{ism}(z)}{dz}\left(1 - \frac{\rho_c(z)}{\rho_{\rm ism}(z)}\right)$$

Other relations between the dynamical variables are derived from the equations relating to the conservation of mass, number density and energy. The complete set enables one to derive, inter alia, the run of velocity along the jet.

Turbulence also produces some dissipation and this has an effect on the surface brightness via particle acceleration (see Bicknell and Melrose, 1982). For low Mach numbers this dissipation can be neglected in the first approximation. However, it will be important to consider it when the question of reacceleration of energetic particles is further addressed. In this adiabatic model, the evolution of jet surface brightness is given by $I_\nu \propto (v_c a^2)^{-s/3}\ a\, B^{(s-1)/2}$ with $B \propto (v_c\, a)^{-1}$ when the field is perpendicular to the jet.

2.3 Fit to NGC315

In Bicknell (1986b) this turbulent adiabatic model was fit to the well studied jet in NGC 315 (Bridle et al. 1979; Willis *et al.*, 1981). This jet is well known for the intriguing behaviour of its diameter as a function of distance from the core. The surface brightness could be fit by the model for a range of values of the relevant parameters. However, all fits imply that the jet is initially light and that the ratio of virial to ISM temperature, $\beta \sim 1$. This fact alone is a good indication that the model has some merit.

Other features of the model fits are that in the case of NGC315, the velocity increases over the collimation plateau, and the rate of entrainment becomes zero. These two features indicate that buoyancy is winning out over turbulence in this region and provides an attractive explanation for the collimation behaviour of this and other similar jets. This also highlights another attractive feature of the model - the explanation of the anti-correlation between rate of surface brightness decrease and spreading rate.

2.4 Generic picture of turbulent class I jets

The generic picture that arose from the modeling of NGC 315 is the following: Initially the jet is free as it emerges from the core and the magnetic field is perpendicular to the jet. After a distance of the order of a few hundred pc to a kpc the jet starts to come into equilibrium with the atmosphere, at which point particle acceleration at the associated oblique shocks starts to reaccelerate the high energy electrons which have cooled on their travel from the core. The shocks realign the magnetic field so that it is closer to parallel to the jet direction, explaining the presence of parallel magnetic field at the luminous *bases* of class I jets. This also solves the well known problem of the magnetic field strength in jets. An initially perpendicular field is consistent with the estimates of core and large scale magnetic field, whereas a parallel field is not.

Turbulence associated with the shocks driven into the jet causes turbulent flaring and the jet expands in this way until the turbulence starts to die out. It will do so since turbulence is hard to maintain in a decreasing pressure gradient. The jet then collimates as buoyancy starts to win out over the turbulent stresses. When the jet starts to plough into the more or less constant pressure of the intergalactic medium, turbulence reappears and the jet starts to reexpand. Some jets may not go through the intermediate collimation phase depending upon the initial level of turbulence and the point at which the background medium starts to dominate the pressure.

3. PARAMETERS OF JETS IN A SAMPLE OF CLASS I SOURCES

A large sample of radio galaxies has been observed at the VLA by the group at il Istituto di Radioastronomia, Bologna (see Fanti et al.1987). In a number of these sources the jets were well enough resolved that fitting of this model could be attempted (Bicknell et al. 1990). In order of importance the parameters to be constrained are: $\beta = T_*/T_{\rm ism}$, the jet density ratio η, the core radius of the galaxy potential, the Mach number of the jet, M, and the normalizing surface brightness. The two most important parameters are β and η. The other parameters vary in a wide range but generally within the expected range. For example, usually $1 \lesssim M \lesssim 3$, $\eta \lesssim 1$. The best indication of the success of the fits can be gauged by the distribution of the parameter β. This is very similar to the distribution of this parameter derived from an independent sample of X-Ray ellipticals (Thomas et al. 1986). The velocities of the jets implied by this model fitting range from about 1,000 to 10,000 km s^{-1}.

Interestingly, in the Bologna sample there is a "high Mach number subset" comprising 8 jets for which a range of high Mach numbers are possible ($1 \lesssim M \lesssim 10$). These jets are more well-collimated and have a lower density ratio than the other jets in the sample. All of these features indicate that this high Mach number subsample represents jets in the transition from class I to class II.

Consideration of the energy budget in class I jets (Bicknell 1986a) shows that a jet tends to come into equilibrium with the radiation from the lobe in a synchrotron plus inverse Compton radiative time scale. Thus a more or less random sample of jets, such as the Bologna sample, should show values of the ratio, κ, of lobe luminosity to jet energy flux, ranging from 0 to 1. When one uses the values of the Mach number and density ratio inferred from the models, combined with values of the minimum pressure in the jet to compute the energy flux $F_E = 1.5 \times 10^{40}\,(\sigma/300\ \mathrm{km\,s^{-1}})\,(\Phi/\mathrm{kpc})^2\,(P/10^{-11}\ \mathrm{dy\,cm^{-2}})\,\beta^{-1/2}\eta^{-1/2}M\,(1+M^2/6)$ ergs s^{-1} then one indeed obtains values of κ more or less uniformly distributed over the range 0 to 1.

4. OTHER INDICATORS OF JET VELOCITY - THE CENTAURUS A FILAMENTS

Emission line filaments in the closest radio galaxy, Centaurus A, have attracted attention for more than a decade (see Blanco et al., 1975; Peterson, Cannon and Dickens, 1975; Graham and Price, 1981). However, the source of excitation of these filaments has been a mystery. Recently, Morganti et al. (1991) have proposed that the filaments are photoionized by beamed ionizing radiation from the nucleus. Here, I summarize an alternate view, namely that the filaments are locally excited by radiative shocks and that this is consistent with a velocity $\sim$ a few thousand km s^{-1} for the northern jet of Centaurus A.

There are two parts to this argument (1) The energy flux incident upon the filaments is sufficient to account for the emission line flux and (2) The observed turbulent velocities can account for the observed H_β luminosity.

The observations show that the inner filaments consist of a number of individual clouds each of which intercepts a power from the jet

$$\dot{E}_{jet} = \frac{\gamma}{\gamma - 1} P_{\rm jet}\, v_{\rm jet}\, A_{\rm cloud} \left(1 + \frac{\gamma - 1}{2} M^2\right)$$

where $A_{\rm cloud} \approx 4 \times 10^4$ pc^2 is the cross-sectional area of the cloud perpendicular to the jet flow and $v_{\rm jet} = (2\gamma k T_{ism}/\mu m_p)^{1/2} \eta^{-1/2} M = 520 \gamma^{1/2} \eta^{-1/2} M$ km s^{-1} where γ is the effective jet polytropic index. If the northern jet has entrained enough material then $\eta \approx 1$ at the location of the filaments. Taking $P_{\rm jet} \approx 10^{-11}$ dy cm^{-2} (Feigelson et al. 1981), $v_{\rm jet} \approx 1000$ km s^{-1} and $M \approx 1$ then $\dot{E}_{jet} \approx 1.3 \times 10^{39}$ ergs s^{-1}. If all of this incident energy is dissipated in line emission then, since the H_β luminosity in a photoionized region is typically 0.5-1% of the total emission line luminosity, the predicted H_β luminosity is of order $0.5 - 1.0 \times 10^{37}$ ergs s^{-1}, in agreement with what is observed for each blob of emission. Any discrepancy between the inferred and observed H_β luminosity can be accounted for by a modest increase in the assumed jet velocity.

The velocities observed by Morganti et al show substantial gradients across each blob $\gtrsim 200$ km s^{-1}. I interpret the emission line blobs as regions wherein clouds, jostled by the jet, have collided with a supersonic relative velocity $\Delta v \gtrsim 200$ km s^{-1} and produced shocks with a temperature $T_{\rm shock} \approx \mu m_p (\Delta v)^2/24k = 1.2 \times 10^5 (\Delta v/200 \text{km s}^{-1})^2$. The soft X-ray emission from the shocked gas photoionizes the surrounding dense material producing characteristically high excitation spectra (see Binette, Dopita and Tuohy, 1985). The post-shock cooling time is short compared to the flow time scale so that most of the kinetic energy of the colliding clouds is dissipated in photoionizing radiation. The end result (Bicknell 1991) is an H_β luminosity $L(H_\beta) \approx$

$1.4 \times 10^{36} (\Delta v/200 \text{ km s}^{-1})^3 (A_{\text{cloud}}/4 \times 10^4 \text{ pc}^2)$ ergs s^{-1}. For $\Delta v \approx 200\sqrt{3}$ km s^{-1} (allowing for projection) the predicted H_β luminosity is $\approx 7 \times 10^{36}$ ergs s^{-1} reasonably close to the observed value $\approx 1 - 2 \times 10^{37}$ ergs s^{-1}. Obviously the dependence upon Δv is fairly sensitive so that with only a slight change to the suggested parameters the H_β luminosity will be considerably greater.

Thus, with really no adjustable parameters, this simple theory is capable of explaining the luminosity of the filaments in Centaurus A to within an order of magnitude in a physically plausible way. This has ramifications for the narrow line emission from AGN's as outlined in the following section.

5. DECELERATING JETS AND THE CORE PROPERTIES OF RADIO GALAXIES

Radio galaxy and quasar cores are frequently classified as either steep or flat spectrum. In this section I briefly describe some of the results from Caganoff's (1989) Ph.D. thesis which relate to the notion of decelerating jets and the excitation of the narrow line region of extragalactic radio sources. Caganoff focussed on radio sources with powers just above the break. Other studies had focussed on powerful 3C sources.

The combined data of Caganoff (1989), Phillips et al.(1986) and Baum et al.(1988) show that there is a good correlation between the 1.4 GHz radio power and H_α+$[NII]$ luminosity in class II radio galaxies. Rawlings et al. (1989) have also shown that P^{tot}_{178MHz} and P_{OIII} are correlated. One of the important discoveries in Caganoff's work was that steep spectrum cores are associated with smaller source sizes. The simple explanation for this is that in some sources there is strong interaction of the jet with the ISM surrounding the core and that the consequent deceleration of the jet enhances the steep spectrum optically thin emission through the processes outlined earlier. When the jets do not interact strongly with gas in the vicinity of the core then the optically thick emission from the base of the jet dominates. The VLBI observations of the compact steep spectrum core source 3C48 (Wilkinson et al. 1991) showing a plume-like radio structure support this notion.

Rawlings and Saunders (1987) also showed that the optical [OIII] emission correlates better with the jet luminosity rather than with the total radio luminosity. In the light of the theory for the filaments in Centaurus A this suggests that it is the jet energy flux that powers the optical emission and that this occurs through the interaction of dense clouds with the jet. There is some support in Caganoff's data for the proposition that the correlation between emission line power and radio power is stronger for the steep spectrum core sources than it is for the flat spectrum core sources. However, further data are required.

REFERENCES

Baum S.A., Heckman T.M., Bridle A.H., van Breugel W., and Miley G.K., 1988, *Ap. J. Suppl.*, **68**, 833

Bicknell G.V., 1986a, *Ap. J.*, **300**, 591

Bicknell G.V., 1986b, *Ap. J.*, **305**, 109

Bicknell G.V., 1991, *Proc. A.S.A.*, in press

Bicknell G.V., de Ruiter, H.R., Fanti, R., Morganti, R., Parma, P., 1990, *Ap. J.*, **354**, 98–115

Bicknell, G.V., and Melrose D.B., 1982, *Ap. J.*, **262**, 511

Binette L., Dopita M.A., and Tuohy I.R., *Ap. J.*, **297**, 476

Blanco V.M., Graham J.A., Lasker B.M., and Osmer P.S., 1975, *Ap. J. (Letters)*, **198**, L63

Blandford R.D. and Rees M.J., 1974, *M.N.R.A.S.*, **169**, 395

Bridle A.H., 1984, *A. J.*, **89**, 979

Bridle, A.H., Davis, M.M., Fomalont, E.B., Willlis, A.G., and Strom R.G., 1979, *Ap. J. (Letters)*, **228**, L9

Blandford R.D. and Rees M.J., 1974, *M.N.R.A.S.*, **169**, 395

Caganoff S., 1989, *Ph.D. thesis*, Australian National University

Fanaroff B.L. and Riley J.M., 1974, *M.N.R.A.S.*, **167**, 31P

Fanti R., Lari C., Parma P., Bridle A.H., Ekers R.D., and Fomalont E.B., 1982, *Astr. Ap.*, **110**, 69

Fanti C., Fanti R., de Ruiter H.R., and Parma P., 1987, *Astr. Ap. Suppl.* **69**, 57

Feigelson E.D., Schreier E.J., Delvaille J.P., Gaicconi R., Grindlay J.E., Lightman A.P., 1981, *Ap. J.*, **251**, 31

Graham J.A., and Price R.M., 1981, *Ap. J.*, **247**, 813

Morganti R., Robinson A., Fosbury R.A.E., di Serego Aligheri S., Tadhunter C.N., and Malin D.F., 1991, *M.N.R.A.S.*, in press.

Peterson B.A., Dickens R.J., and Cannon R.D., 1975, *Proc. A.S.A.*, **2**, 366

Phillips M.M., Jenkins C.R., Dopita M.A., Sadler E.M., and Binette L., 1986, *A. J.*, **91**, 1062

Rawlings S.G. and Saunders R., 1987, in *Emission Lines in Active Galactic Nuclei*, ed. P.M. Gondhalekar, 160

Rawlings S., Saunders R., Eales S.A., and Mackay C.D., 1989, *M.N.R.A.S.*, **240**, 701

Scheuer P.A.G., 1974, *M.N.R.A.S.*, **166**, 513

Thomas P.A., Fabian A.C., Arnaud K.A., Forman W., Jones C., 1986, *M.N.R.A.S.*, **222**, 655

Willis, A.G., Strom, R.G., Bridle, A.H. and Fomalont, E.B., 1981, *Astr. Ap.*, **95**, 250

Wilkinson P.N., Tzioumis A.K., Benson J.M., Walker R.C., and Simon R.S., 1991, preprint

Laing–Garrington Effect: Radio Depolarization Asymmetry

R.G. CONWAY, S.T. GARRINGTON & G.F. HOLMES

University of Manchester, Jodrell Bank, U.K.

1. THE LAING–GARRINGTON EFFECT

Powerful radio sources, both quasars and FRII radio galaxies, regularly show two lobes of emission, but radio maps show only one jet, not two (Bridle & Perley, 1984). It is clearly important to know whether this "jet-asymmetry" is intrinsic, or is an illusion caused by relativistic beaming of the approaching jet. Faraday effects in single-jet sources are related to this asymmetry, being systematically weaker on the jet side than on the jet-less or counter-jet side (Laing, 1988; Garrington et al. 1991). This remark applies both to the rotation measure, and to Faraday depolarization (DP).

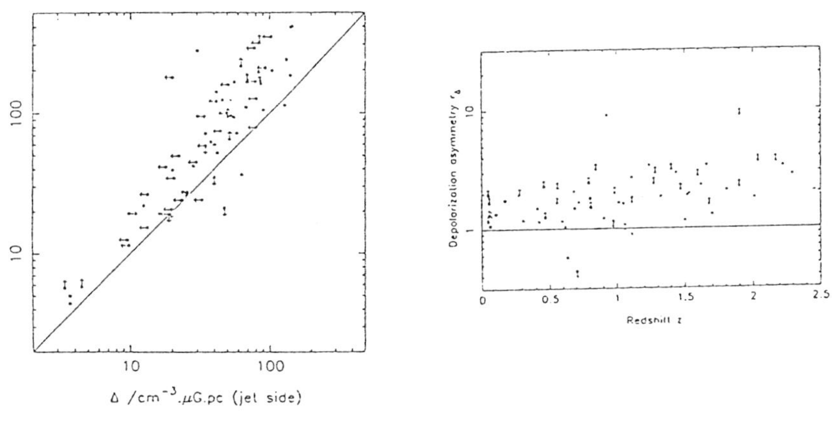

Fig.1 (a) Plot of Δ_{cj} versus Δ_j. The excess of points above the line, which have $r_\Delta > 1$, forms the DP asymmetry.
(b) The ratio r_Δ as a function of redshift z.

Weaker depolarization on the jet side is found in 49 out of 69 single-jet sources, including 53 quasars and 16 radio galaxies (Garrington & Conway, 1991). Other sources, only weakly depolarized at 20 cm, show the DP asymmetry in the same sense at 49 cm. Faraday depolarization arises because different parts of the source have different Faraday depth (Burn, 1966)

$$\phi = \int n.B.d\ell \qquad \text{in } cm^{-3}.\mu G.pc \qquad (1)$$

and can be used to calculate the Faraday Dispersion Δ, i.e. the spread of ϕ about its mean value. Typical values for Δ_j on the jet side are 10 to 100 $cm^{-3}.\mu G.pc$, while Δ_{cj} is 20 to 200 $cm^{-3}.\mu G.pc$. The ratio $r_\Delta = \Delta_{cj}/\Delta_j$ has a median value of 1.90. Both Δ_j and Δ_{cj} increases with redshift z, as does the ratio r_Δ, i.e. the DP asymmetry becomes stronger with redshift (Fig. 1).

2. QUASARS VS. GALAXIES

Fewer radio galaxies (RG) than quasars (QS) show jet-asymmetry. To compare depolarization in RG and QS we have made more observations of polarization in 3CR galaxies (Holmes et al., in preparation). Because DP depends on redshift, we have selected matched samples of RG and QS in the range $0.3 < z < 1.0$. In this range 9 out of 31 RG show jet-asymmetry, compared with 18 out of 19 quasars. Values of Δ for RG were found to be in the range 20 to 50 $cm^{-3}.\mu G.pc$, similar to Δ_j for quasars in this z-range, but less than Δ_{cj}.

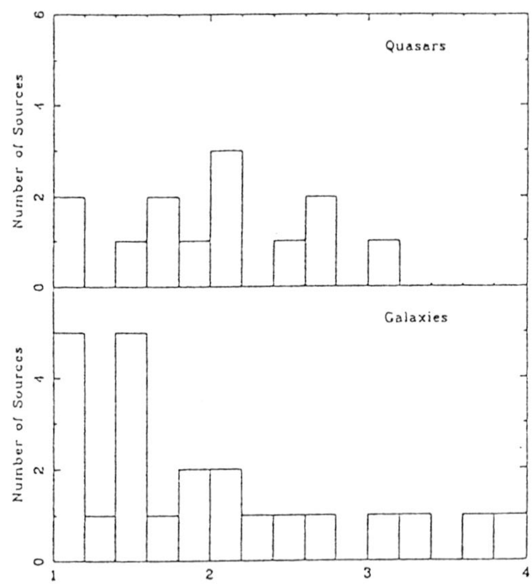

Fig.2 Histograms of the unsigned ratio r^*_Δ. Each plot forms the upper bound to the true r_Δ distribution.
(a) QS with redshift 0.3 to 1; (b) RG in same z range.

When no jet is present, one cannot identify Δ_j and Δ_{cj} to find r_Δ, but one can find the unsigned ratio r^*_Δ equal to the larger Δ divided by the smaller (hence $r^*_\Delta \geq r_\Delta$).

The histogram of r^*_Δ for RG (Fig.2) has many values close to 1.0, suggesting that the graph for r_Δ (if it could be measured) would show little or no asymmetry. Thus the Laing-Garrington effect in galaxies is less evident than in quasars both because fewer single-jet sources are found, and also because the depolarization asymmetry is weaker.

3. INTERPRETATION

The depolarizing Faraday medium may in principle lie either within the source or in front or both; the balance of evidence now suggests that the greater part is in front. Foreground depolarization requires a screen with irregularities of scale size d, less than the beamwidth of observation, which in our case corresponded to a few kpc. For such a screen the Faraday Dispersion Δ is given by

$$\Delta^2 = \int n^2 B^2 d.d\ell \qquad (2)$$

(Note in passing that any positrons in the medium will tend to cancel Faraday rotation; hence the density n deduced from Faraday effects always refers to the excess of e^- over e^+, i.e. to the proton density).

Two alternatives suggested for the magneto-ionic medium are a halo of ionised plasma at 10^7K (X-ray halo), or gas at 10^4K with visible line emission (Garrington & Conway, 1991). The nature of the jet-asymmetry will differ in the two cases; if depolarization occurs in a halo, the jet-asymmetry is probably due to beaming. However, if the depolarization is by emission-line gas (not subject to relativistic beaming), the jet-asymmetry is likely to be intrinsic.

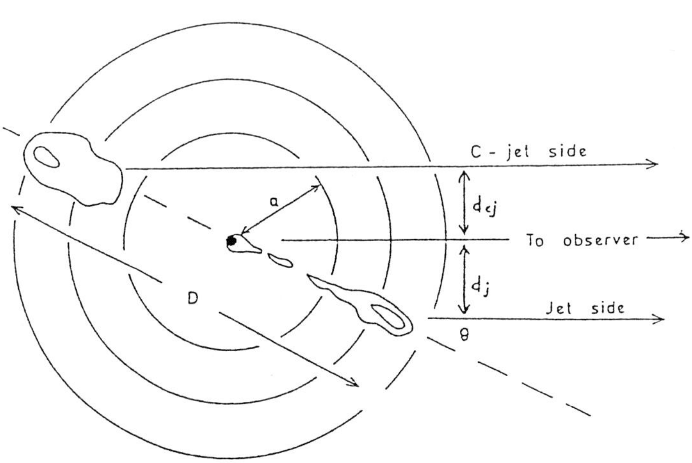

Fig.3 Sketch of radio source and halo, illustrating core radius a, angle θ, and armlength d.

3.1 X–Ray Halo

If the jet side is towards the observer (Fig.3), as required for beaming, we have shown that the depolarization asymmetry found in single-jet sources is satisfactorily explained by a hot-plasma halo model. The central density needed is typically 3 10^{-2} cm^{-3} with a core radius of 70–100 kpc; the plasma is permeated by a tangled B-field of 0.3 μG. Though there are no X-ray observations as yet of high-z sources, these estimates agree with known X-ray properties of small clusters (Jones & Forman 1984).

In this model, the jet axis is at an angle θ to the line-of-sight, which needs not be limited to very small values. From the upper limits on jet: counter-jet intensity, the values of (v/c).cosθ must be greater than about 0.5. Thus the jet-asymmetry can arise with a velocity of 0.7c at $\theta = 45°$. With higher velocities, angles of up to 60° would be possible.

Although the X-ray halo model explains the depolarization data very satisfactorily, counter-arguments exist. A second correlation (between spectral index and jet side) found by us and by Liu & Pooley (1991) has no natural explanation on the beaming hypothesis. In a few cases (e.g. the galaxies 3C337 & 3C441), Δ_j exceeds Δ_{cj}, against the general rule. Here at least, one must invoke another cause of depolarization asymmetry.

3.2 Optical emitting Gas

Depolarization may also occur in gas in Extended Emission Line Regions (EELR). (For much of the information in this section, and for references to the literature on EELR in radio galaxies, see McCarthy, 1988). Gas in RG is mostly concentrated about the central object out to a radius of order 2 kpc, but outlying gas may also occur up to 100 kpc from the centre, aligned along the radio axis. Central gas densities are of order 10^2 cm^{-3}, with filling factors of only 10^{-5}. Since the EELR are smaller than the radio source (at least at low z) by a factor 5 to 10, the EEL gas is probably not itself the depolarization agent, but may be a tracer for more extended gas (cf also van Breughel et al. 1985).

Though centrally concentrated, the centroid of the EELR may be displaced relative to the central AGN (McCarthy et al. 1991). The shifts are often small, but are physically significant. Nearby radio galaxies are rather symmetrical in armlength (arms equal to within a factor 2), but at high z many galaxies have unequal arms, and in such cases the short-arm side shows both stronger DP and stronger EEL radiation (Pedelty et al. 1989; McCarthy et al. 1991). The discrepant galaxies 3C337 and 3C441 noted above, with Δ_j greater than Δ_{cj}, are both in this category.

A correlation between armlength and DP is not itself surprising, if the gas is centrally concentrated (Strom & Conway 1985). The connection between armlength and EEL gas is more interesting, giving evidence that gas density variations actually <u>cause</u> the inequality in armlength (McCarthy et al. 1991). In sum, gas density/armlength effects seem to control depolarization behaviour in radio galaxies.

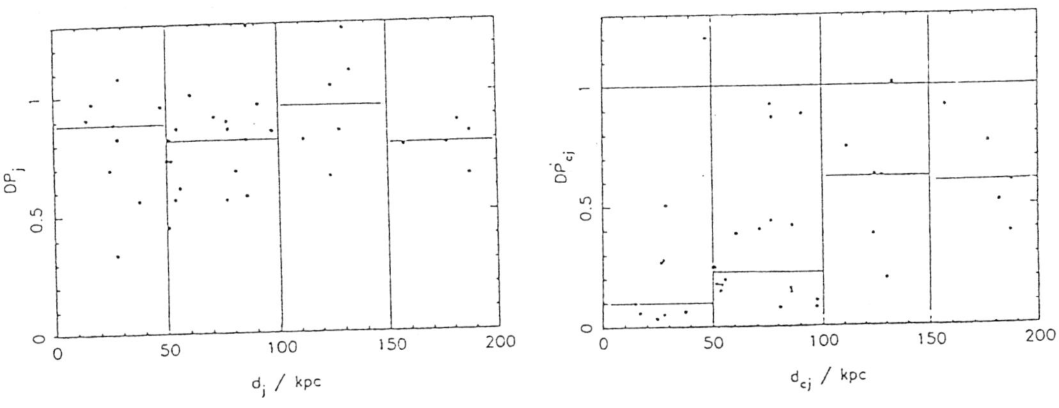

Fig.4 Depolarization ratio DP against armlength in kpc, showing median values for different ranges of armlength.
(a) Jet side; (b) counter-jet side.

These remarks specifically concern radio galaxies. There are fewer EELR observations for quasars : those by Heckman et al. (1991) report Lyα in excess on the jet-side, i.e. in the contrary sense to that required to explain DP asymmetry. In our sample (mainly QS) we can find no systematic relation between EEL gas and jet-sidedness (Garrington & Conway 1991). We find the expected correlation between DP and armlength in our data, but only on counter-jet sides (Fig.4). This is further confirmation that jet sides lie in front, between the observer and the Faraday screen.

4. SUMMARY

Asymmetric radio depolarization in quasars is definitely linked to jet-sidedness, and is almost certainly a front/back effect. DP asymmetry in galaxies is linked to armlength ratio, and to the presence of optical line emission, though the depolarizing medium is not the optical gas itself. We cannot say whether one mechanism overrides the other, since the two samples are almost entirely disjoint. By and large, one-sided jets occur in quasars but not in galaxies; EELR are typical of galaxies rather than quasars.

If, as suggested, the asymmetry properties of galaxies at high-z are the result of variations in gas density, while the asymmetry of quasars is aspect-dependent, the DP asymmetry in galaxies and quasars would be the result of entirely unconnected causes, which seems an unsatisfactory interpretation. The radio depolarization data by itself can be interpreted in terms of symmetrical gas distributions, centrally concentrated, simply by assuming quasars to lie at small angles ($\theta < 45°$) to the line-of-sight, while galaxies lie more nearly in the plane of the sky. On the major question, that of the nature of the jet-asymmetry in quasars, our evidence at present favours the beaming hypothesis.

REFERENCES

Bridle, A.H. & Perley, R.A., 1984, A. Rev. Astr. Astrophys. **22**, 319

Burn, B.J., 1966, Mon. Not. R. Astr. Soc. **133**, 67

Garrington, S.T., Leahy, J.P., Conway, R.G. & Laing, R.A., 1988, Nature, **331**, 147

Garrington, S.T. & Conway, R.G., 1991, Mon. Not. R. Astron. Soc. **250**, 198

Garrington, S.T., Conway, R.G. & Leahy, J.P., 1991, Mon. Not. R. Astron. Soc. **250**, 171

Heckman, T.M., Lehnert, M.D., van Breughel, W. & Miley, G.K., 1991, Astrophys. J. **370**, 78

Jones, C. & Forman, W., 1984, Astrophys. J. **276**, 38

Laing, R.A., 1988, Nature, **331**, 149

Liu, R. & Pooley, G., 1991, Mon. Not. R. Astron. Soc. **249**, 343

McCarthy, P.J., 1988, Ph.D. Thesis, University of California, Berkeley, USA

McCarthy, P.J., van Breughel, W. & Kapahi, V.K., 1991, Astrophys. J. **371**, 478

Pedelty, J.A., Rudnick, L., McCarthy, P.J. & Spinrad, H., 1989, Astron. J. **97**, 647

Strom, R.G. & Conway, R.G., 1985, Astron. Astrophys. Suppl. **61**, 547

van Breughel, W., Heckman, T.M., Miley, G.K., Butcher, H. & Bridle, A.H., 1985, Astrophys. J. **290**, 496

Radio and Optical Asymmetries in Powerful Radio Galaxies

R. LIU AND G.G. POOLEY

MRAO, Cavendish Laboratory, Cambridge CB3 OHE, England

In a recent study of the environment of powerful radio galaxies, we find strong correlations between asymmetric radio properties and the associated optical line emission. In the 4 sources for which we have both multi-frequency radio data and optical emission images, we find that the optical line emission is brighter on the side of the radio source for which the depolarization is stronger, the spectrum steeper and the hotspot closer to the core. The four sources (3C247, 263.1, 280 and 299) are all small ($D < 200$ kpc), powerful ($P_{178} > 10^{27}$ W Hz^{-1} sr^{-1}) and distant ($z > 0.4$). None of them shows clear signs of a jet. We adopt $H_0 = 50$ km s^{-1} Mpc^{-1} and $q_0 = 0$.

Garrington et al. (1988) and Laing (1988) find that the sources with one-sided jets have stronger depolarization on the counter-jet side. This correlation is naturally explained by Doppler effects. The jet is on the side closer to the observer and its radiation is relativistically enhanced. The lobe furthest from the observer is seen through a greater Faraday depth of the galactic envelope.

The correlated asymmetries presented here seem unrelated to that of Garrington & Laing and may have a different origin. Liu & Pooley (1991a) have shown that in order to account for the spectral differences observed in both the hotspots and lobes in terms of Doppler effects, the velocities required for the bulk motion of the emitting material are much higher than predicted from the current models of radio sources. Again, the emission-line asymmetry cannot be explained by any orientation effect in any simple way. The most probable cause of the asymmetries seems to be that the power output of the central engine or the environments of the sources are different on opposite sides of the nucleus. We speculate that the gas around the host galaxy is not symmetrical along the radio axis, which consequently leads to the asymmetries observed here. Full discussion is presented in Liu & Pooley (1991b).

REFERENCES

Garrington, S.T., Leahy, J.P., Conway, R.G. & Laing, R.A., 1988. *Nature*, **331**, 147.
Laing, R.A., 1988. *Nature*, **331**, 149.
Liu, R. & Pooley, G.G., 1991a. *Mon. Not. R. astr. Soc.*, **249**, 343.
Liu, R. & Pooley, G.G., 1991b. accepted by *Mon. Not. R. astr. Soc.*.

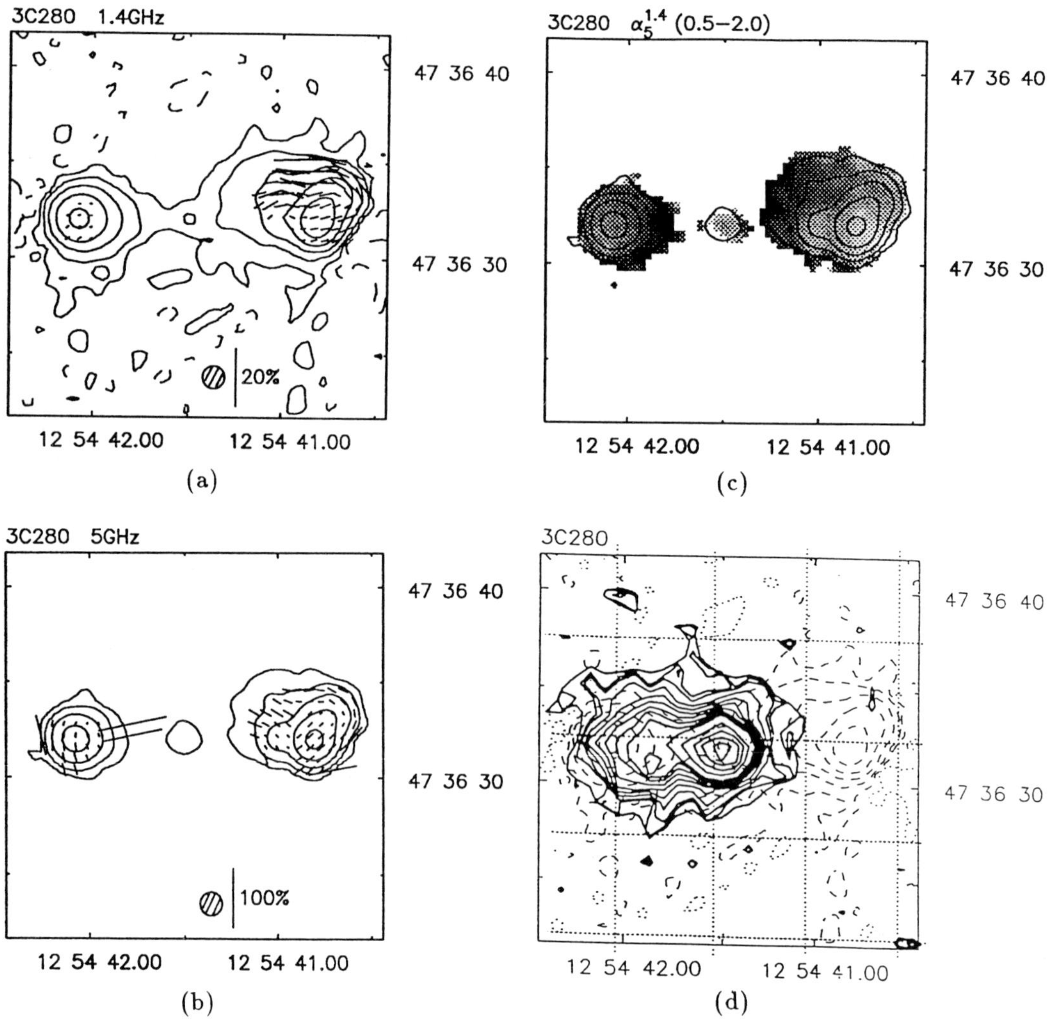

Figure 1. Maps of 3C280. (a) & (b) Contours of total intensity and the 'vectors' indicating the fractional polarizations and the position angles of the received B-field, which will represent the projected magnetic field in the source if the Faraday rotation is small. The radio observations are made with the VLA and described in Liu & Pooley (1991a). The hatched circle indicates the half-power beamsize. (c) Grey-scale map of spectral index (defined as $S_\nu \propto \nu^{-\alpha}$), overlaid with total intensity map. The range of α, which runs from light to dark, is indicated on the top of the map. (d) Contour map (solid) of the extended [OII] emission-line image (from McCarthy, van Breugel & Spinrad 1991, ApJ. Suppl., in press), overlaid with total intensity map (dashed contours). The images are aligned by assumming that the position of the radio core coincides with the peak of the optical continuum.

Mixed Shocks and the Formation of Hot Spots

R. LEHOUCQ(1) & J. ROLAND(2)

(1) CE–Saclay, Service d'Astrophysique. (2) IAP, Paris

1 INTRODUCTION

The hot spots in the extended lobes of extragalactic radio sources can be explained by the interaction of a non–relativistic supersonic jet with the intergalactic medium (see in this volume Roland & al.). Hot spots are considered to be the downstream flows behind the shocks at the end of the jets. The acceleration of the relativistic particles and the formation of power law distributions can be understood using a mixed shock model. After the early works of Axford & al. 1977, Krimsky 1977, Bell 1978 and Blandford & Ostriker 1978, a non–linear theory of the mixed shock structure has been performed by Drury & Völk 1981, Mc Kenzie & Völk 1982. Further developments of mixed shocks have been done latter and references can be found in Kang & Jones 1990 and Fraix–Burnet & Pelletier 1991.

In this article we will focus our attention on the influence of the Mach number and the Alfvenic Mach number on:
i) the acceleration of relativistic particles by mixed shocks
ii) the spectral index of the power law distribution of the relativistic particles.

2 THE INFLUENCE OF THE MACH NUMBER

Let us note upstream and downstream quantities with indices 1 and 2 respectively. Pelletier and Roland 1986, developped an analytical determination of the structure function ϕ of mixed shocks and a spectral theory parametrized by the ratio θ of the non–relativistic pressure p_c to the relativistic pressure p_r, namely $\theta = p_c/p_r$. Here, as we are interested by the influence of a finite Mach number, we will suppose that the pressure jump $\Delta = (p_{c2} + p_{r2})/(p_{c1} + p_{r1})$ is finite. Moreover, assuming that $p_{r1} \neq 0$, namely $\theta_1 \leq 10^5$, we will not investigate the question of the injection of relativistic particles at the front shock.

2.1 Basic Assumptions and Equations

We will suppose that :
i) the ratios θ_1, and θ_2 are arbitrary,

ii) the shock is adiabatic,
iii) the ratio δ of the acceleration length, δ_c, of the non-relativistic particles to the acceleration length, δ_r, of the relativistic particles is very small compare to 1, i.e. $\delta = \delta_c/\delta_r \ll 1$,
iv) the diffusion coefficient D_r of the relativistic particles in independant of the energy, thus the diffusivity κ_r is $\kappa_r \simeq D_r$,
v) the transition is monotonic.

The hydrodynamical equations can be reduced to a single differential equation, for the velocity u of the scattering centers, of the form $\kappa_r \partial u/\partial x = \phi(u)$, where ϕ is a negative function that we call the structure function of the shock. Using this equation to eliminate the variable x in the transport equation for the energy distribution $\rho_r(E)$ of the relativistic particles (Skilling 1975, Ginzburg & Ptuskin 1985), we obtain the following equation for the stationary distribution function

$$\frac{E^3}{3}\frac{\partial}{\partial E}(\rho_r/E^2) - u\frac{\partial \rho_r}{\partial u} + \frac{\partial}{\partial u}\left(\phi\frac{\partial \rho_r}{\partial u}\right) = 0 \qquad (1)$$

where E is the energy of the relativistic particles.
In this section, we will assume that the Alfvenic Mach number is infinite, thus the velocity u of the scattering centers is identical to the velocity v of the flow.
The Rankine-Hugoniot relations can be written

$$J = \rho_1 v_1 = \rho_2 v_2 \qquad (2)$$

$$F = J v_1 + p_{c1} + p_{r1} = J v_2 + p_{c2} + p_{r2} \qquad (3)$$

$$\frac{v_1^2}{2} + \left(\frac{\gamma_r p_{r1}}{\gamma_r - 1} + \frac{\gamma_c p_{c1}}{\gamma_c - 1}\right) / \rho_1 = \frac{v_2^2}{2} + \left(\frac{\gamma_r p_{r2}}{\gamma_r - 1} + \frac{\gamma_c p_{c2}}{\gamma_c - 1}\right) / \rho_2 \qquad (4)$$

where $\gamma_r = 4/3$, $\gamma_c = 5/3$ and ρ is the mass density.
The compression ratio $r = \rho_2/\rho_1 = v_1/v_2$ is related to the pressure jump Δ by

$$r = \frac{\dfrac{7+4\theta_2}{1+\theta_2} + \dfrac{1}{\Delta}}{1 + \dfrac{1}{\Delta}\dfrac{7+4\theta_1}{1+\theta_1}} \qquad (5)$$

Denoting the square of the Mach number $M^2 \equiv J v_1/(p_{c1} + p_{r1})$, the relation between Δ and M is

$$\frac{\Delta - 1}{M^2} = 1 - \frac{1}{r} \qquad (7)$$

From the definition of the relativistic pressure, namely $p_r = (1/3)\int \rho_r(E)EdE$ and from (1), we find that p_r is governed by

$$\gamma_r p_r + v\frac{dp_r}{dv} = \frac{d}{dv}\left(\phi\frac{dp_r}{dv}\right) \tag{8}$$

We will describe the evolution of the thermal component by the same kind of equation, i.e.

$$\gamma_c p_c + v\frac{dp_c}{dv} = \frac{d}{dv}\left(\delta\phi\frac{dp_c}{dv}\right) \tag{9}$$

where $\delta \ll 1$.
The description of mixed shocks, i.e. the determination of ϕ, p_c and p_r can be obtained if we add the equation $F = Jv + p_c + p_r$.

2.2 Results

The system has been solved numerically using an implicit Runge–Kutta method. The parameter δ has been assumed to be 10^{-2}, a reasonable value for mixed shocks responsible for the formation of hot spots (Roland & al. 1988) The solutions $\theta_2(\theta_1)$ for various values of Δ are presented in Figure 1.

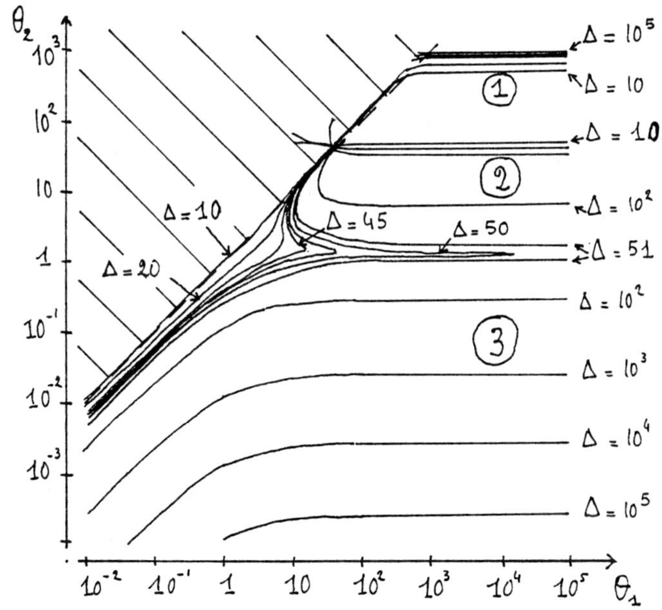

Figure 1: The solutions $\theta_2(\theta_1)$ corresponding to various values of the pressure jump Δ. Mixed shocks which accelerate efficiently relativistic particles are those with $\theta_1 > \theta_2$.

If the pressure jump is greater than a critical value $\Delta_c \simeq 50$, i.e. if the Mach number is greater than about 7.7, there exist stationary solutions even if the pressure of the relativistic particles is very small compared to the thermal pressure in the upstream flow.

For a given value of θ_1 greater than about 300, there exist 3 solutions. Denoting by $\epsilon \simeq p_{r2}/(p_{r2}+p_{c2})$ the efficiency of the solution, we obtain in the strong shock limit
i) solution (1) with $\theta_2 > 1$
when $M \rightarrow \infty$, $\theta_2(\theta_1 = 10^5) \rightarrow 961$, and $\epsilon \simeq 10^{-3}$
ii) solution (2) with $\theta_2 > 1$
when $M \rightarrow \infty$, $\theta_2(\theta_1 = 10^5) \rightarrow 48.4$, and $\epsilon \simeq 2 \times 10^{-2}$
iii) solution (3) with $\theta_2 < 1$
when $M \rightarrow \infty$, $\theta_2(\theta_1 = 10^5) \propto 1/M^2 \rightarrow 0$, and $\epsilon \simeq 0.999$

The observations of extragalactic radio sources indicate that the efficiency of the acceleration due to the bow shock in front of the hot spots is low when the Mach number is very large. Thus, in the strong shock limit, the acceleration corresponds to the low efficiency solutions, i.e. to solutions with $\theta_2 > 1$. A similar result is obtained from the observation of the acceleration efficiency in shocks due to the interaction of supernovae remnants with the interstellar medium. More generally, observations of shocks in collisionless plasmas indicate that the efficiency is always much less than unity. This result is not surprising, because in the strong shock limit, the solution $\theta_2 < 1$ supposes that the kinetic energy of the thermal protons in the jet is converted into the energy of the relativistic particles mostly without heating the thermal component. In the strong shock limit, the solution (3) corresponds to a shock without entropy increase and does not fit the observations.

3 THE INFLUENCE OF THE ALFVENIC MACH NUMBER

3.1 Basic Assumptions and Equations

In the jets powering the hot spots, the magnetic field is quasi–toroidal. Thus we assume that the magnetic field is quasi–perpendicular to the jet, but that the angle between the shock front and the magnetic lines is greater than few degrees. Thus the first order Fermi process occurs. In that case, we must take into account the magnetic pressure jump effect and again the velocity u of the scattering centers is equal to the velocity v of the flow (Pelletier & Roland 1988, Fraix-Burnet & Pelletier 1991).

To the basic equations (8) and (9), we have to add the two equations

$$F = Jv + p_c + p_r + p_m \tag{10}$$

$$Bv = Cst \tag{11}$$

where $p_{\mathrm{m}} = B^2/(8\pi)$ is the magnetic pressure and $J = \rho v$.
Let us call $V_{\mathrm{A}} = B/\sqrt{(4\pi\rho)}$ the Alfven speed, $M_{\mathrm{A}} = v_1/V_{\mathrm{A}1}$ the Alfvenic Mach number. The compression ratio is now given by

$$r = \frac{\dfrac{7+4\theta_2}{1+\theta_2} + \dfrac{1}{\Delta} + \dfrac{\Delta-1}{\Delta}\dfrac{r^2+1}{r^2-1}\dfrac{1}{\dfrac{2M_{\mathrm{A}}^2}{r(r+1)} - 1}}{1 + \dfrac{1}{\Delta}\dfrac{7+4\theta_1}{1+\theta_1} + \dfrac{\Delta-1}{\Delta}\dfrac{r^2+1}{r^2-1}\dfrac{1}{\dfrac{2M_{\mathrm{A}}^2}{r(r+1)} - 1}} \tag{12}$$

The square of the Alfvenic Mach number is $M_A^2 = Jv_1/(2p_{\mathrm{m}1})$ and the relation between Δ and M^2 is now

$$\frac{\Delta - 1}{M^2} = 1 - \frac{1}{r} - \frac{r^2-1}{2M_A^2} \tag{13}$$

In the precursor, a fraction of the incident particles of the hydrodynamical flow can be reflected, which leads to a singularity in the hydrodynamical equations. If, $\theta_2 > 1$, we must have $M_{\mathrm{A}} > 1$ and if $\theta_2 < 1$, $M_{\mathrm{A}} > 18$ (see Pelletier & Roland 1988).

3.2 Results for $\Delta = 10$

Bearing in mind the application to the formation of hot spots we want to investigate the solutions corresponding to Alfvenic Mach numbers as low as $M_{\mathrm{A}} \simeq 6$ (Pelletier & Roland 1988). The system admits solutions if the pressure jump Δ is smaller than about $2 \times M_{\mathrm{A}}$, thus we will present the solutions $\theta_2(\theta_1)$ for $\Delta = 10$ and $6 \le M_{\mathrm{A}} \le \infty$ in Figure 2.

We have shown in the previous section (see Figure 1) that, if Δ is small enough, only the solution (1) exists when $\theta_1 \simeq 10^5$. When the magnetic pressure jump effect is important, we can see that solution (1) is shifted toward $\theta_2 \simeq 1$ when M_{A} decreases, but remains greater than 1. Indeed, $\theta_2(\theta_1 = 10^5, M_{\mathrm{A}} \to \infty) \simeq 518$ and $\theta_2(\theta_1 = 10^5, M_{\mathrm{A}} = 6) \simeq 2.67$.
This result shows that inside the hot spots the pressure is dominated by a non–relativistic gas, i.e. the non–relativistic protons which mean energy per particle is $E_{\mathrm{th,p}} \simeq$ 8–10 MeV (Roland & al. 1988).

The other usefull result concerns the variation of θ_2 with both Δ and M_{A}. For instance, $\theta_2(\Delta = 18, M_A = 10) \simeq 3.9$ and $\theta_2(\Delta = 14, M_A = 14) \simeq 15.4$

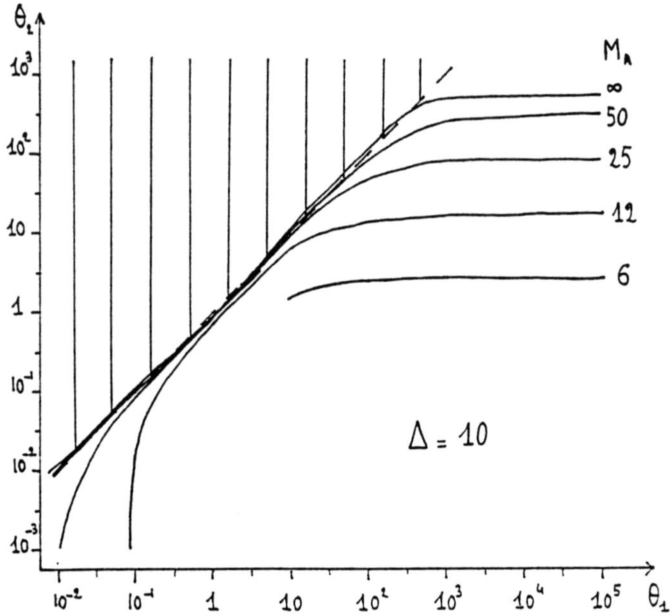

Figure 2: The solutions $\theta_2(\theta_1)$ corresponding to various values of the Alfvenic Mach number for $\Delta = 10$.

4 SPECTRAL THEORY OF MIXED SHOCKS

4.1 The Method

As indicated in the introduction, the relativistic particles spectrum is a power low distribution at high energy. The spectral index of the power law is a function of Δ, M_A and θ_1. A perturbation method to calculate the spectral index has been presented by Pelletier & Roland 1986, 1988. At high energy, the distribution of the relativistic particles is $\rho_r(E) \propto E^{-\eta}$.

The structure function ϕ can be written $\phi(u) = g(u)(u - u_1)(u - u_2)$ with $g(u) > 0$. The spectral index η is given by

$$\eta = \frac{3r}{r-1}\Big(1 + \frac{1}{a(r-1)}\Big) - 2 \ , \tag{14}$$

where

$$a = \frac{\int g(u)(u-u_2)^{\alpha}(u_1-u)^{\beta}du}{\int (u-u_2)^{\alpha}(u_1-u)^{\beta}du} \ , \tag{15}$$

with $\alpha = 1/[a(r-1)]$ and $\beta = r/[a(r-1)]$ (Pelletier & Roland 1986).

4.2 Results

In the previous sections, we have obtained numerically the functions $\phi(u)$ and $g(u)$. So, we can calculate the radio spectral index $\alpha_r = (\eta - 1)/2$ as a function of Δ and M_A, assuming $\theta_1 = 10^4$.

First, one verify that the 3 solutions obtained with $\theta_1 = 10^4$ and $M_A \to \infty$ in Section 2 have a radio spectral index $\alpha_r = 0.5$ when $\Delta = 10^5$, i.e. in the strong shock limit.

For the application to hot spots (see Roland & al. in this volume), we calculated the radio spectral index for $10 \leq \Delta \leq 20$ and $8 \leq M_A \leq 16$. The results are presented in Table 1.

M_A / Δ	8	10	12	14	16
10	0.75	0.76	0.77	0.77	0.78
12	0.67	0.70	0.71	0.72	0.72
14	0.63	0.66	0.67	0.68	0.68
16		0.62	0.64	0.65	0.65
18		0.58	0.61	0.62	0.63
20			0.58	0.60	0.61

Table 1: The radio spectral index α_r for various values of Δ and M_A.

REFERENCES

Axford, W.I., Leer, E. & Skhodron, G., 1977, 15th Inter. Conf. Cosmic Rays, Plovdiv
Bell, A.R., 1978, Monthly Notices Roy. Astron. Soc. **182**, 147
Blandford, R.D. & Ostriker, J.P., 1978, Astrophys. J. **221**, L29
Drury, L.O.C. & Völk, H.J., 1981, Astrophys. J. **248**, 344
Fraix–Burnet, D. & Pelletier, G., 1991, Astrophys. J. **367**, 86
Ginzburg, V.L. & Ptuskin, V.S., 1985, Sov. Sci. Rev. E, Astrophys. Space Phys. **4**
Kang, H. & Jones, T.W., 1990, Astrophys. J. **353**, 149
Mc Kenzie, J.F. & Völk, H.J., 1982, Astron. Astrophys. **116**, 191
Krimsky, G.F., 1977, Sov. Phys. Dokl. **23**, 327
Pelletier, G. & Roland, J., 1986, Astron. Astrophys. **163**, 9
Pelletier, G. & Roland, J., 1988, Astron. Astrophys. **196**, 71
Roland, J., Pelletier, G. & Muxlow, T.W.B., 1988, Astron. Astrophys. **207**, 16
Skilling, J., 1975, Monthly Notices Roy. Astron. Soc. **172**, 557

On the Formation of Extragalactic Radio Sources

J. ROLAND(1), R. LEHOUCQ(2) & G. PELLETIER(3)

(1) IAP, Paris (2) CE–Saclay, Service d'Astrophysique (3) Observatoire de Grenoble

1 RADIO SOURCES SIMILAR TO CYGNUS A

Cygnus A type radio sources are characterized by
i) the existence of hot spots in the extended lobes,
ii) weak radio jets powering the hot spots,
iii) a magnetic field quasi–toroïdal in the jets and the hot spots,
iv) a magnetic field parallel to the surface of the extended lobe in the cocoon,
v) an increase of the radio spectral index from the hot spots to the galaxy along the cocoon,
vi) the Mach number in the jet is $M > 1$. The interaction of the jet with the intergalactic medium produces a shock at the end of the jet. The shock is responsible for the acceleration of the relativistic e^- in the hot spots.
vii) they are generally high luminosity sources.
Let us note that using iii) to v) we can deduce that the well known arc second jet of 3C 273 is in fact a cocoon (Roland et al. 1991–a).

1.1 Main Physical Parameters of the Jets and Hot Spots

The formation of Cygnus A hot spots can be explained by the interaction of a non-relativistic supersonic jet with the intergalactic medium (Muxlow et al. 1988). A similar result has been found by Meisenheimer et al. (1989).
In the case of Cygnus A, the main results concerning the jets and the hot spots are (Roland et al. 1988)
– the speed of the jet is $v_{\rm j} \simeq 0.4\ c$,
– the ejected mass is $\dot{M} \simeq 0.5 \times M_{\odot}/y$,
– the kinetic power of the jet is $P_{\rm Kj} \simeq 2.4 \times 10^{45}\ erg/s$ and the integrated synchrotron luminosity of one hot spot is $L_{\rm HS} \simeq 0.03 \times P_{\rm Kj}$,
– the Alfvenic Mach number can be as low as $M_{\rm A} \simeq 6$, indicating the importance of magnetized mixed shocks. In that case, the parameter $\beta = p_{\rm par}/p_{\rm m} \simeq 2$–$3$, where $p_{\rm par}$ is the total pressure of the particles and $p_{\rm m}$ is the magnetic pressure.
– The dominant pressure in the hot spots is the pressure of the non-relativistic protons. The mean energies of the thermal protons and of the thermal electrons are

respectively $E_{\rm th,p} \simeq$ 8–10 MeV and $E_{\rm th,e} \simeq$ 1–2 MeV.
– Inside the hot spots, the pressure of the relativistic protons is smaller than the pressure of the relativistic electrons.
The formation of jets whose speeds are $v_{\rm j} \simeq 0.4\ c$ can be explained by the MHD extraction process (Pelletier & al. in this volume and references quoted). One important characteristic of the solution of this model is that the speed of the jet is $v_{\rm j} \simeq 0.4\ c$ if the central object is a black hole.

1.2 Formation of Asymmetries in Sources with Hot Spots

Radio sources with hot spots can be characterized by asymmetries. We will see how we can explain the formation of asymmetrical sources assuming non-relativistic jets.

a) sources similar to 3C 263
3C 263 is characterized by asymmetrical hot spots and extended lobes (Leahy et al. 1989), i.e.
– one lobe contains a strong hot spot and the other lobe without hot spot is weaker
– the weaker lobe is more distant from the galaxy than the strong lobe.
The advance speed $v_{\rm ad}$ of the hot spots in the intergalactic medium is $v_{\rm ad} \propto 1/n_{\rm ex}^{1/2}$, where $n_{\rm ex}$ is the density of the intergalactic medium.
In the case of Cygnus A, the external density is $n_{\rm ex} \simeq 5 \times 10^{-3}\ e^-/cm^3$ and the advance speed in $v_{\rm ad} \simeq 0.05\ c$. Thus, for a radio source with the same characteristics as those of Cygnus A, when the external density is $n_{\rm ex} \leq 10^{-4}\ e^-/cm^3$, the advance speed becomes $v_{\rm ad} \simeq 0.35\ c \simeq v_{\rm j}$, and the strong shock at the end of the jet disappears and we do not observe a bright hot spot.
Thus, if the two jets have the same characteristics, i.e. the same speeds, mass ejected and Mach numbers, an asymmetry of the density of the intergalactic medium produces an asymmetrical source similar to 3C 263 and one observes that the weak lobe is more distant from the galaxy than the strong lobe.

b) sources similar to 3C 68.1
3C 68.1 (Leahy et al. 1989) is an asymmetrical source characterized by
– one lobe contains a strong hot spot and the other without hot spot is weaker
– the separations between the two lobes are the same.
Since the two lobes are at the same distance from the nucleus, there is no asymmetry of the density of the intergalactic medium.
The existence of a brighter hot spot suggests that the magnetic field in the jet powering the brighter hot spot is stronger than the magnetic field in the other jet.
Let us assume that it is the case and let us investigate the consequences. If the magnetic field is greater, the Alfven speed is greater and the Alfvenic Mach number is smaller. From the MHD extraction model (Pelletier & al. in this volume), one finds that the Mach number of the jet becomes higher when the Alfvenic Mach number becomes smaller. So to put some numerical value let us suppose that on

one side we have $M_A \simeq 10$ and $\Delta \simeq 18$ and on the other side $M_A \simeq 14$ and $\Delta \simeq 14$ (Δ is the pressure jump through the mixed shock and is directly related to the Mach number, see Lehoucq & Roland in this volume). From the magnetized mixed shock model (Lehoucq & Roland in this volume) one finds that θ_2 ($M_A = 10$, $\Delta = 18$) $\simeq 3.9$ and θ_2 ($M_A = 14$, $\Delta = 14$) $\simeq 15.4$ and α ($M_A = 10$, $\Delta = 18$) $\simeq 0.58 < \alpha$ ($M_A = 14$, $\Delta = 14$) $\simeq 0.68$, where θ_2 is the ratio of the pressure of the non–relativistic particles to the pressure of the relativistic particles inside the hot spots and α is the radio spectral index of the hot spots. So, a brighter hot spot with a flatter radio spectral index will form on the side where the magnetic field is greater. Moreover, on the side where Δ is greater, the mean energy of the thermal electrons will be greater, and as they are midly relativistic they will produce a lower depolarization (Jones & O'Dell 1977).
Thus, if the two jets have the same speeds and the same mass ejected, but if an asymmetry of the magnetic field exist between the two jets, an asymmetry between the Alfvenic Mach numbers and the Mach numbers of the two jets will appear. Consequently, a brighter hot spot with a flatter spectral index and a lower depolarization will form on the side of the higher magnetic field. This effect is known as the Liu–Laing–Garrington effect (Liu & Pooley 1991 and Conway & al. in this volume).

2 RADIO SOURCES SIMILAR TO 3C 449

3C 449 type radio sources are characterized by
i) the existence of two bright radio jets,
ii) the lack of hot spots in the extended lobes,
iii) the increase of the radio spectral index from the nucleus to the lobe,
iv) the Mach number of the jet is $M \leq 1$, indicating the importance of the acceleration of the relativistic particles by the turbulence (Pelletier and Zaninetti 1984, see also Fraix–Burnet, Bicknell in this volume)
v) the speed of the jets is $v_j \simeq \text{few} \times 10^3$ km/s (Bicknell 1984, 1986 a-b). In the case of head-tail sources one has $v_j \leq \text{few} \times v_{s,ex}$, where $v_{s,ex} \simeq 10^3$ km/s is the sound speed in the intergalatic medium
vi) they are generally low luminosity sources.
3C 449 type radio sources have been observed and studied by several authors (see for instance Fanti et al. 1987, Morganti et al. 1987, Parma et al. 1987 and Bicknell in this volume).

2.1 Difference Between Cygnus A Jets and 3C 449 jets

As indicated previously, the speed of the jets of one of the solutions of the MHD extraction process (Pelletier & al. in this volume) is $v_j \simeq 0.4\ c$. The main global difference between Cygnus A and 3C 449 is their radio luminosity. 3C 449 type sources are generally at least 10^3 times less luminous than Cygnus A. Let us assume that the jets are produced by the same process, their speeds close to the nucleus are $v_j \simeq 0.4\ c$. Thus, the main difference between Cygnus A and 3C 449 is the mass ejected by the nucleus. In the case of 3C 449 type radio sources, one has $\dot{M} \leq 5 \times 10^{-4} M_\odot/y$, instead of $\dot{M} \simeq 0.5\ M_\odot/y$ for Cygnus A.

When the mass ejected becomes small enough, there exist a critical value R_c, of the radius of the jet, and if the radius R_j of the jet becomes greater than R_c, the kinetic energy density of the jet becomes comparable to the pressure of the external medium, i.e. $\rho_j v_j^2 \simeq P_{ex}$. In that case, the jet becomes unable to push the external medium, it interacts with it and dissipates its kinetic energy via turbulence. The relativistic e^- are accelerated by the turbulence in the dissipation area. A bright radio jet forms and when the dissipation of the kinetic energy is finished the speed of the flow, i.e. the bright radio jet, is smaller or equal to the sound speed of the external medium. Let us note that a detailed description of the scenario describing the dissipation of the initial jet and the acceleration of the relativistic e^- has to be done.

To illustrate our talk, let us calculate the critical value of the radius of the jet, assuming $n_{ex} \simeq 2 \times 10^{-3}\ e^-/cm^3$ and $T_{ex} \simeq 10^8\ K$ for the external medium and an ejected mass $\dot{M} \simeq 10^{-4} M_\odot/y$. We will have $\rho_j v_j^2 \simeq 2\ n_{ex} K T_{ex}$ when $R_j \simeq R_c \simeq 230\ pc$. The existence of this critical value explains the gap between the nucleus and the bright radio jet, indeed during the gap, one has $R_j < R_c$.
For Cygnus A type sources, one has always $\rho_j v_j^2 \gg P_{ex}$, and the jet pushes the external medium. A bright hot spot forms at the end of the jet and the relativistic e^- are accelerated by a shock.

2.2 Observational Evidence: the Case of M 87
Although M 87 is a source with a single bright jet, it is a good example of a n-earby source whose bright jet is produced as indicated previously. Indeed, VLBI observations (Reid et al. 1989) show:
- the speed of the jet close to the nucleus is $v_j \simeq 0.35\ c$,
- a limb brightening of the jet, indicating the interaction of the jet with the external medium.

The 14 arc second bright jet corresponds to the dissipation area of the jet with the external medium, area where the dissipation of the kinetic energy occurs and the acceleration of the relativistic e^- by the turbulence takes place (see Fraix-Burnet, Machetto in this volume).
After 14 arc second, or few kpc, the dissipation process is finished and the speed of the jet is $v_j \leq v_{s,ex} \simeq 10^3\ km/s$, i.e. the slow opening jet which powers the extended lobe is formed.

2.3 The Origin of the Fanaroff–Riley Correlation

It has been noted by Fanaroff and Riley (1974) that generally Cygnus A type sources are high luminosity sources and that 3C 449 type sources are low luminosity sources. Let us note that there exist

- low luminosity sources with hot spots, i.e. Cygnus A type sources, e.g. B2 1113+29 (Morganti & al. 1987),
- high luminosity sources with bright jets and no hot spots, i.e. 3C 449 type sources, e.g. 0800+608 (Jackson & al. 1990).

We can now easily understand the origin of the Fanaroff–Riley correlation and of its exceptions.

For a given value of the external pressure, if the mass ejected is great enough and/or if the jet does not open too much, we will have $\rho_j v_j^2 \gg P_{ex}$, and hot spots will form, now if the mass ejected becomes too small and/or if the jet opens too much, for a critical value of the radius of the jet, we will have $\rho_j v_j^2 \simeq P_{ex}$ and a slow bright radio jet will form. This provides a natural explanation to the Fanaroff–Riley correlation and is related to the fact that close to the nucleus, one has $v_j \simeq 0.4\ c$.

Moreover, we can understand why a low luminosity source like B2 1113+29 is a Cygnus A type source. If the galaxy associated with B2 1113+29 is in an intergalactic medium which pressure is low, say for instance $n_{ex} \simeq 10^{-4}\ e^-/cm^3$ and $T_{ex} \simeq 5 \times 10^7\ K$, with $\dot{M} \simeq 10^{-4} M_\odot/y$ we have $R_c \simeq 1.4\ kpc$. Thus, hot spots will be able to form and a low luminosity radio source similar to Cygnus A will be observed.

Finally, if the galaxy associated with 0800+608 is in a high pressure intergalactic medium characterized by $n_{ex} \simeq 10^{-2}\ e^-/cm^3$ and $T_{ex} \simeq 10^8\ K$, due to the large radius of the jet, i.e. $R_j \simeq 0.7\ kpc$ (Jackson & al. 1990), we will have $\rho_j v_j^2 \simeq P_{ex}$ and a bright jet without hot spot will form, even if the mass ejected and the luminosity are quite high.

3 COMPACT RADIO SOURCES

3.1 The Two–Fluid Model

VLBI observations of variable radio sources and the lack of the inverse Compton catastrophe for these sources indicate that the $e^- - e^+$ responsible for the VLBI synchrotron emission are ejected relativistically with a bulk Lorentz factor $\overline{\gamma} \simeq 10$, but observations do not imply that the thermal component of the plasma is also ejected relativistically. Thus, we arrive to the following conclusion: the nucleus ejects two fluids (Sol & al. 1989 and Pelletier & Roland 1989), i.e.

a) a non-relativistic $e^- - p$ jet whose speed is $v_j \simeq 0.4\ c$ and which mass ejected is $\dot{M} \leq 0.5 \times M_\odot/y$. It carries most of the mass and the kinetic power. It is responsible for the formation of the hot spots, the bright kpc scale radio jets and the extended

lobes
b) A relativistic $e^- - e^+$ beam which bulk Lorentz factor is $\gamma_b \leq 20$ and which is responsible for the formation of compact variable radio sources. Note that the low energy cut-off of the relativistic e^- and e^+ is $E_o \simeq$ few $\times 10^2$ MeV (Ghisellini in this volume).

3.2 Observational Evidences for the Two–Fluid Model

There are several pieces of evidence that variable radio nuclei of galaxies eject two fluids with different speed. Indeed,
i) two different speeds have been detected in the VLBI jet of 3C 84 (Maar et al. 1989)
ii) the radio source 3C 371 is a Cygnus A type radio source which nucleus shows superluminal expansions (Wrobel and Lind 1990). The Doppler beaming effect due to the relativistic ejection can be observed in the nucleus and within the first arc seconds of the large scale jet (Figure 1). After few arc seconds, the large scale jet becomes weaker suggesting that the Doppler beaming effect does not occur any more and that the speed of the large scale jet is not relativistic.

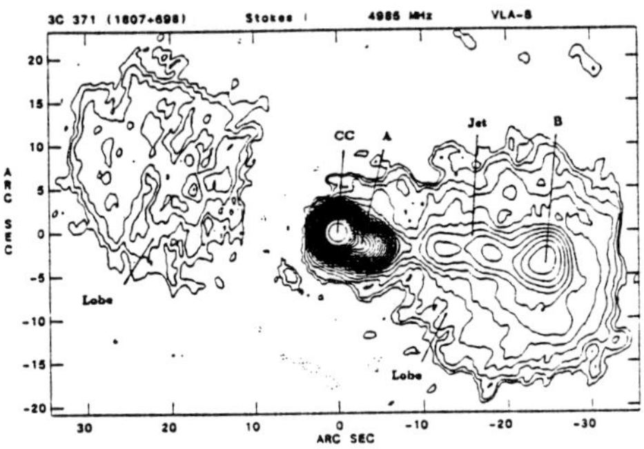

Figure 1: The radio map of 3C 371 obtained by Wrobel and Lind 1990.

iii) 1928+738 is a 3C 449 type radio source with two bright large scale jets. Its nucleus shows superluminal expansion and the Doppler beaming effect (Johnston & al. 1987). After few arc seconds, one observes the two large scale jets which suggets again that the speed of the large scale jets is non-relativistic.

3.3 Consequences of the Two–Fluid Model

a) Propagation and stability of the beam in the jet

The relativistic $e^- - e^+$ beam can propagate in the jet if the magnetic field is parallel to the direction of the propagation in the beam and in the mixing layer and if it is greater than a critical value B_c (Pelletier et al. 1988, see also Istomin, Achatz & Schliekeiser in this volume). Let us note that the beam propagates on distances smaller than few kpc in extragalactic radio sources.

b) Magnetic field configuration in the beam and in the jet

Pelletier and Roland (1990) showed that VLBI polarization observations imply that the magnetic field is helical close to the nucleus and becomes rapidly toroïdal when the jet opens. As indicated previously in the beam and in the mixing layer the magnetic field is parallel to the direction of propagation. This magnetic configuration in the beam and in the jet is produced by the MHD extraction process.

c) Different kind of variable sources

The difference between VLBI observational properties of compact sources suggests a possible classification of compact source based on the value of the magnetic field close to the nucleus compared to the critical value B_c (Roland & al. 1991–b)

d) Further consequences of the beam dissipation in the jet

When the beam dissipates in the jet, one can have coherent emission (Baker et al. 1988, see also Benford, Lesch in this volume).

e) Correlation between VLBI jets and bright hot spots

As indicated in §1.2, when a source similar to 3C 68.1 has a bright hot spot, the VLBI jet is observed on the same side. As shown previously, the magnetic field is stronger on the side of the bright hot spot. The formation of the $e^- - e^+$ beam is easier when the magnetic field in the jet is stronger (Henri & Pelletier in this volume). Thus, the two-fluid model and the MHD extraction process produce a natural explanation to the VLBI jet – bright hot spot correlation and to its relation with the Liu–Laing–Garrington effect.

4 CONCLUSION

There are no observational evidences that after few kpc, the jets are relativistic, MERLIN observations of 3C 120 (Muxlow & Wilkinson 1991) do not confirm the relativistic motion of the 4 arc second radio knot observed previously with the VLA (Walker & al. 1988). The properties of hot spots, bright radio jets and extended lobes can be understood if the jets produced in nuclei of galaxies are non-relativistic $e^- - p$ jets whose speed are $v_j \simeq 0.4\ c$. A possible model for the formation of these jets is provided by the MHD extraction process. Depending on the ejected mass, on the pressure of the external medium and on the radius of the jet, one can explain either the formation of Cygnus A type radio sources or the formation of 3C 449 type radio source. Thus, this gives a natural explanation to the Fanaroff–Riley correlation and to its exceptions. The observational properties of compact variable sources imply that the relativistic $e^- - e^+$ are ejected relativistically from the nucleus. Consequently, the general properties of extragalactic radio sources can be understood if their nuclei eject two fluids, i.e. a non-relativistic $e^- - p$ jet and a relativistic $e^- - e^+$ beam.

REFERENCES

Baker, D.N., Borovsky, J.E., Benford, G. & Eleik, J.A., 1988, Astrophys. J. **326**, 110
Bicknell, G.V., 1984, Astrophys. J. **286**, 68
Bicknell, G.V., 1986–a, Astrophys. J. **300**, 591

Bicknell, G.V., 1986–b, Astrophys. J. **305**, 109

Fanaroff, B.L. & Riley, J.M., 1974, Monthly Notices Roy. Astron. Soc. **167**, 31P

Fanti, C., Fanti, R., de Ruiter, H.R. & Parma, P., 1987, Astron. Astrophys. Suppl. **69**, 57

Jackson, N., Browne, I.W.A., Shone, D.L. & Lind, K.R., 1990, Monthly Notices Roy. Astron. Soc. **244**, 750

Johnston, J.P., Simon, R.S., Eckart, A., Biermann, P., Schalinski, C., Witzel, A. & Strom, R.G., 1987, Astrophys. J. **313**, L85

Jones, T.W. & O'Dell, S.L., 1977, Astron. Astrophys. **61**, 291

Leahy, J.P., Muxlow, T.W.B. & Stephens, P.W., 1989, Monthly Notices Roy. Astron. Soc. **239**, 401

Liu, R. & Pooley, G., 1991, Monthly Notices Roy. Astron. Soc. **249**, 343

Maar, J.M., Backer, D.C., Wright, M.C.H. Readhead, A.C.S. & Moore, R., 1989, Astrophys. J. **337**, 671

Meisenheimer, K., Roeser, H.J., Hiltner, P.R., Yates, M.G., Longair, M.S., Chini, R. & Perley, R.A., 1989, Astron. Astrophys. **219**, 63

Morganti, R., Fanti, C., Fanti, R., Parma, P. & de Ruiter, H.R., 1987, Astron. Astrophys. **183**, 203

Muxlow, T.W.B., Pelletier G. & Roland, J., 1988, Astron. Astrophys. **206**, 237

Muxlow, T.W.B. & Wilkinson, P.N., 1991, Monthly Notices Roy. Astron. Soc. **251**, 54

Parma, P., Fanti, C., Fanti, R., Morganti, R. & de Ruiter, H.R., 1987, Astron. Astrophys. **181**, 244

Pelletier, G. & Zaninetti, L., 1984, Astron. Astrophys. **136**, 313

Pelletier, G., Sol, H. & Asseo, E., 1988, Phys. Rev. A. **38**, 2552

Pelletier, G. & Roland, J., 1989, Astron. Astrophys. **224**, 24

Pelletier, G. & Roland, J., 1990, in Parsec–scale Radio Jets, Edited by A.J. Zensus & T.J. Pearson, Cambridge University Press, Cambridge, p 323

Reid, M.J., Biretta, J.A., Junor, W., Muxlow, T.W.B. & Spencer, R.E., 1989, Astrophys. J. **336**, 112

Roland, J., Pelletier, G. & Muxlow, T.W.B., 1988, Astron. Astrophys. **207**, 16

Roland, J., Lehoucq, R. & Pelletier, G., 1991–a, in Gamma–Ray Line Astrophysics, Edited by Durouchoux P. & Prantzos N., American Institute of Physics, New–York, p 453

Roland, J., Pelletier, G. & Sol, H., 1991–b, in Variability of Active Galaxies, Edited by Duschl W.J., Wagner S.J. & Camenzind M., Spinger–Verlag, Berlin, p 225

Sol, H., Pelletier, G. & Asseo, E., 1989, Monthly Notices Roy. Astron. Soc. **237**, 411

Walker, R.C., Walker, M.A. & Benson, J.M., 1988, Astrophys. J. **335**, 668

Wrobel, J.M. & Lind, K.R., 1990, Astrophys. J. **348**, 135

Sub-kpc detail of nearby FRII radio galaxies

A.R.S. BLACK

M.R.A.O., Cavendish Laboratory, Cambridge CBE OHE, England

1 THE PROJECT

Multi-configuration VLA observations at $\lambda 3.6\ cm$ have been made for 30 nearby ($z < 0.15$) FRII ($P_{178} > 2 \times 10^{25}\ WHz^{-1}sr^{-1}$, for $H_0 = 50\ km\,s^{-1}Mpc^{-1}$) extragalactic radio sources. A sub-sample of 8 sources have so far been mapped – high sensitivity, detailed images of the hotspots (at .25″ and .75″ resolution and with average noise levels of $\bar{\sigma}_I \simeq 30\mu\ Jy$) are supplemented by high sensitivity, lower resolution images of each source: jets are detected in 4 (perhaps 5) of these.

2 REVIEW

Detailed astrophysical interpretation of the maps in this project and consideration of trends and correlations is deferred until the whole sample is available. Nevertheless, some striking features of the currently mapped sub-sample of 8 (sometimes augmented by existing maps of other sample members) invite comment.

a). Long narrow features like those generally called jets appear in 5 of the 8 sources (and perhaps also in 3C223.1 Nf lobe – see fig. 1): 17 have so far been detected in the total sample of 30. It is well-known that large-scale jets are common in quasars, but such a high incidence in radio galaxies is more surprising.

b). The jet features detected in these observations are generally faint (often ~ 5 times the surface-brightness of the background lobe), knotty and non-continuous and quite broad (of width $\sim 5\ kpc$). If these reflect the more general properties of jets in FRII radio galaxies it would explain why so few have been detected. The jet width and detail in the hotspots are comparable in size, though little more can so far be said of their relationship.

c). No standard hotspot structure is prevalent. Even the two ends of the same source are often very different in detail (e.g. 3C223.1, see figs 2 & 3 and 3C382, see fig. 4), though the lobe structure and radio luminosity are generally rather similar (with obvious and striking exceptions). This suggests that hotspots are not steady patterns but transient features, changing over time–scales at least ten times and perhaps 100 times less than the age of the source. We should therefore expect observational differences due to dynamical evolution as well as those arising due to differing environments.

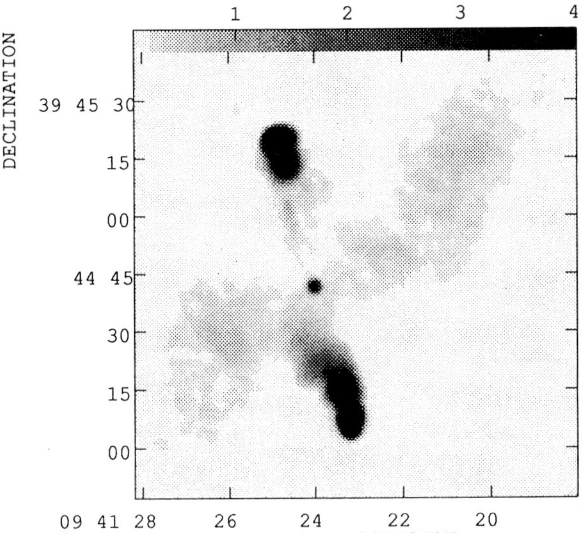

Fig 1: $3C223.1$ at $2.5''$ resolution. Greyscales are shown for $.12\ mJy/beam$ (white) to $4\ mJy/beam$ (black).

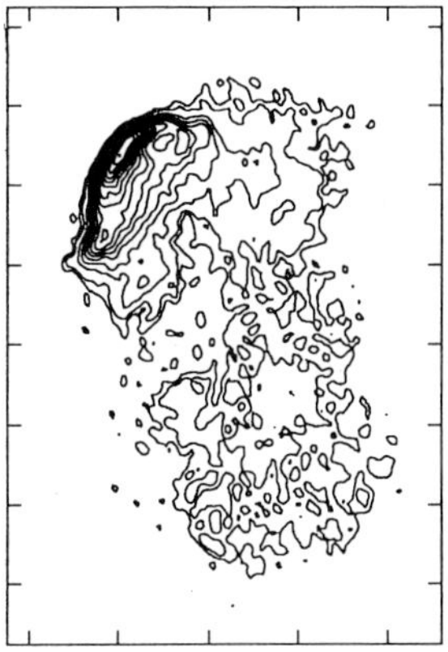

Fig 2: Detail of hotspot in Nf lobe of 3C223.1 at $.25''$ resolution. Contours are shown at $(3, 5, 10, 15, 20, 25, 30, 35, 40, 45, 50, 55, 60, 65, 70, 75, 80, 85, 90, 95, 100) \times 20\ \mu Jy/bm$.

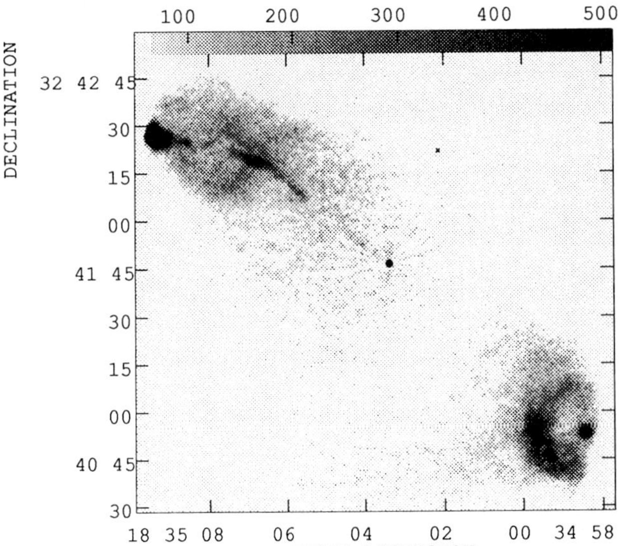

Fig 4. 3C382 at $.75''$ resolution. Greyscales for $50\ \mu Jy/bm$ (white) to $500\ \mu Jy/bm$ (black).

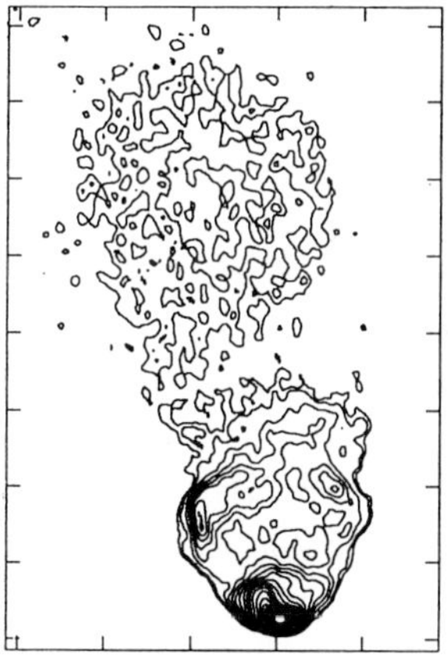

Fig 3: Detail of hotspot in Sp lobe of 3C223.1 at $.25''$ resolution. Contours as for fig 2.

d). In particular, at high resolution many sources do not show the canonical pattern of a compact ('primary') accompanied by a less compact ('secondary') hotspot: such simplistic hotspot definitions become meaningless – indeed some sources show three or more hotspots in at least one end.
e). There is a very large difference in scale between the hotspot structures and the size of the whole source, exceeding 1000:1 in some cases.
f). Several sources show large wings of low surface–brightness (e.g. 3C223.1, see fig. 1), at a position angle quite different from that of the main source. The fact that the wings are longer then the main source seems to indicate that the central engine has undergone a drastic change in orientation at some time during the past 10^8 years; it is difficult to imagine how backflow such as that evisaged by Leahy & Williams 1984 (where it is channelled back into the old cavity of a former orientation) could become longer than the main source. It is possible that less dramatic 'wing' sources demonstrate a related phenomenon, though with a more recent change of direction.
g). A trend that appears in the sample as a whole is that the more distorted and strikingly asymmetric sources are the smallest (often no bigger than the associated galaxy) while the simpler and more symmetric are the largest, when examined at the same linear resolution.
h). Some sources have lobes that cut-off sharply and widen at straight boundaries on either side of the central cores; in such sources the core then lies in a region devoid of emission. This morphology is suggestive of central, dense, cold gas enveloping the central galaxy.

It is not clear at this stage whether these points reflect the fact that most of our sample are not far above the FRI/FRII division in radio power, or whether these trends are more general and reflect the greater detail offered by the new maps.

REFERENCES

Leahy, J.P., and Williams, A.G., 1984, *MNRAS*, **210**, 929.

ACKNOWLEDGEMENTS

I thank Julia Riley, Peter Scheuer, Patrick Leahy and Rick Perley for their time, patience and stimulating discussion. I also thank N.R.A.L. and the V.L.A. for their hospitality to me. I acknowledge Clare College and M.R.A.O. who have financially supported my attendance at this conference. Finally, I thank my dear parents.

The VLA is part of the National Radio Astronomy Observatory and is operated by Associated Universities, Inc., under co-operative agreement with the National Science Foundation.

VLA Observations of Head-Tail Radio Sources in Poor Clusters of Galaxies

D. J. BATUSKI(1), R. J. HANISCH(2) & J. O. BURNS(3)

(1) Univ. of Maine, (2) Space Tel. Sc. Inst., (3) New Mexico State Univ.

In a continuing study of radio sources in poor clusters of galaxies, we have obtained VLA observations of four head-tail sources as probes of the intra-cluster environments: NGC 742, NGC 1044, NGC 4061, and NGC 7503. These sources represent an exceptional opportunity for the study of radio sources (and radio jets in particular). They are very nearby ($z \sim 0.02$ to 0.04) and thus can be examined in great detail, much like NGC 1265, which is the 'Rosetta Stone' of rich cluster HTs, largely because of its proximity to the Earth ($z \sim 0.02$). The four sources contain some remarkable structures. NGC 742 apparently has its lobes wrapped around a companion, NGC 741. NGC 7503 (Figure 1) has a horseshoe shape very similar to the NGC 1265 head-tail. The NGC 4061 source maintains a very linear structure out to equal distances on either side of the galaxy center, where the lobes are sharply 'swept back.' These structures are remarkable because the sources are found in poor groups, where both the density of the ICM and the velocities of the galaxies (thus the ram pressures) are supposedly much lower than in the rich clusters. Yet these poor groups have narrow-angle-tail sources (NATs) with the same general characteristics as those in rich clusters. Finding NATs at all in poor groups is rather surprising, but then finding that they have morphologies very similar to rich cluster NATs is especially interesting.

From the opening angles, bending, luminosities, and spectral index variations within the jets and lobes of these and other poor cluster HTs, we plan to calculate pressure and energy limits within the intra-cluster gas for comparison with similar calculations for rich cluster head-tail sources. For example, Figure 2 shows the results of preliminary measurements of the widths of the jets in NGC 7503, compared against data on NGC 1265 from O'Dea and Owen (1986, ApJ **301**, 814). While the error bars are large on the NGC 7503 data points (as much as $\pm 50\%$ for some points), there is a strong indication that the poor cluster sources have roughly twice the opening angle of NGC 1265. This result suggests that the poor cluster jets are considerably less pressure-confined than the jets of NGC 1265.

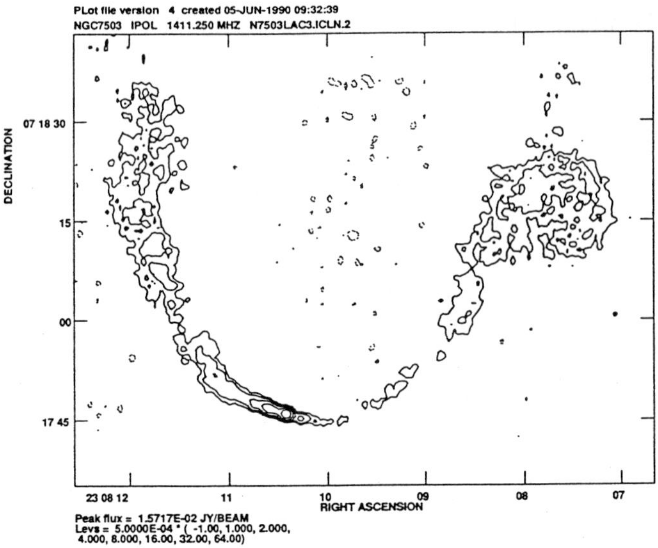

Figure 1. VLA 'A' array 40 minute observation of NGC 7503 at 20 cm. The source is **very** similar in shape to the well-studied rich cluster source, NGC 1265. The right hand jet is weak, while the jet on the left is very prominent and maintains a straight path for about 5 h^{-1} kpc before gradually bending upward. Resolution 1.2 arcseconds.

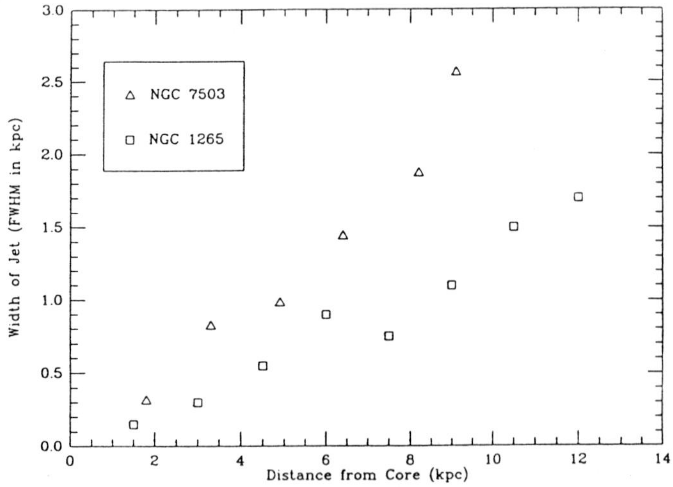

Figure 2. The opening of the eastern jet of NGC 7503 as a function of distance from the nucleus of the galaxy. Plotted with triangles is the projected, deconvolved FWHM for a gaussian fit to a slice through the jet at the given projected distance from the core source. Also plotted here are similar data for NGC 1265 adapted from O'Dea and Owen (1986).

Multiple Outbursts in Radio Galaxies

J.P. LEAHY[1] & P. PARMA[2]

[1]Nuffield Radio Astronomy Laboratories, Jodrell Bank, Macclesfield, Cheshire SK11 9DL, England

[2]Istituto di Radio Astronomia, Via Irnerio 46, I-40126, Bologna, Italy

ABSTRACT

Recent observations of faint structure in radio galaxies show a number of features which imply that the nuclear activity is highly intermittent on periods from 1 to 100 Myr. Most dramatic are 'wings', a pair of secondary radio lobes at an angle to the main source axis, giving the total source an 'X' shape.

We only consider cases where the wings are at least 80% of the length of the main lobes. Winged sources occupy a narrow range of radio luminosities near the Fanaroff & Riley (1974) division ($L_{1.4GHz} \approx 10^{25}$ W Hz^{-1}). Wings are found in about 7% of sources with $0.3 < L_{25} < 30$. The absence of wings at low luminosities is statistically significant, but at the highest luminosities there are too few adequate images to be sure there is a real lack. Of the eleven winged sources known, seven are clearly FRIIs with well-defined hotspots, 4C12.03 contains both faint hotspots and weak twin jets, NGC326 is a typical FRI with prominent jets and no hotspots, 3C315 lacks any compact structure except for a kiloparsec-scale core, and 3C433 has a bizarre structure with the luminosity of an FRII.

By comparing the structure of winged sources with that of the rest of the population, we can show that wings are neither small distortions enhanced by projection effects, nor the result of one particular viewing angle on a population of steadily precessing jets (see Fig. 1). We can also show that it is highly implausible that wings are created by backflow, or by buoyancy. Distortions produced by these two processes cannot grow faster than the external sound speed, while the advance speed of the main lobes is substantially supersonic; such distortions must therefore be much shorter than the main lobes, unlike true wings. Instead we propose that wings were created in an earlier outburst, some tens of Myrs previous to the current renewal of nuclear activity, during which time the ejection axis has precessed. This is consistent with the low surface brightness, steep radio spectrum and high polarization of wings, all features of the oldest components of radio sources. Material does seem to flow from the main lobes into the wings, which may re-energize the old cavities.

The rarity of wings but the presence of S-type distortion in FRIs suggests that their lifetimes are similar to precession timescales. Long lives for FRIs have also been suggested by Owen & Laing (1989).

Multiple events are only obvious when precession (or, rarely, motion through the IGM) occurs, but there is no reason to suppose that the two phenomena are physically linked. There is more common but also more ambiguous evidence for multiple outbursts along the *same* axis in 'bottleneck' structures. Here hotspots are found at the end of a narrow tail projecting out of a much broader and lower surface-brightness bridge, suggesting that the jet thrust has dramatically increased. Finally, many bridges contain well-defined regions with surface-brightness several times higher than the background, which may represent the expanding remnants of previous pulses of jet emission.

REFERENCES

R.D. Ekers, R. Fanti, C. Lari & P. Parma, 1978. *Nature*, **276**, 588.
B.L. Fanaroff & J.M. Riley, 1974. *MNRAS*, **167**, 31P.
F.N. Owen & R.A. Laing, 1989. *MNRAS*, **238**, 257.

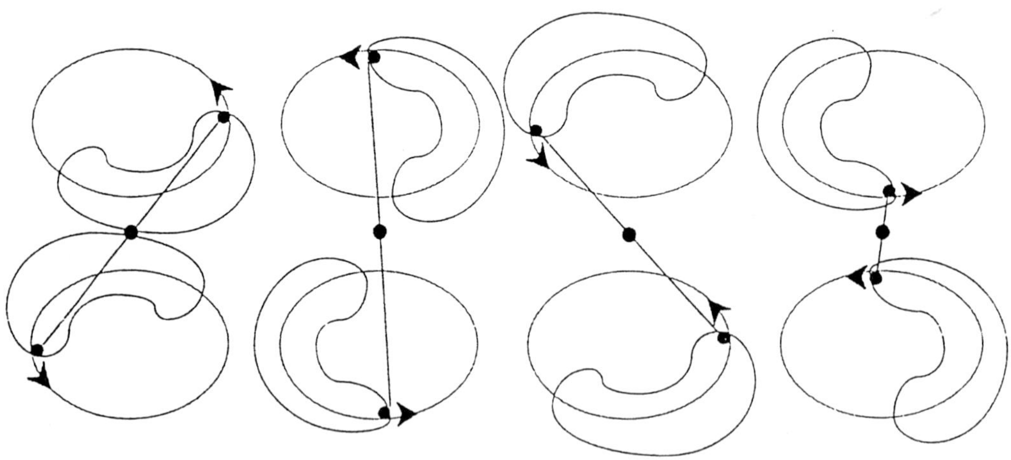

Figure 1. Sketch of a precessing source (after Ekers *et al.* 1978) in four stages of its cycle. It is only X-shaped when close to the first stage; objects resembling the other stages are much rarer than winged sources.

Observations of Jets with the Faint Object Camera

F. MACCHETTO [1]

Space Telescope Science Institute

1 INTRODUCTION

The study of the optical counterparts to radio jets will be the subject of an intensive investigation program with the European Space Agency's Faint Object Camera first proposed 10 years ago (Macchetto 1981, Miley 1981).

Since the optical emission appears to be the continuation of the radio synchrotron spectrum, optical observations are essential to define the physical parameters and constrain the jet emission models.

High spatial resolution observations in the visible and ultraviolet, offer the possibility to determine the precise location where particle acceleration occurs, since the electron lifetime for optical emitting electrons is extremely short. Comparison of radio and optical morphologies obtained with similar resolution will allow the investigation of confinement mechanisms and diffusion processes within the relativistic plasma.

The following sections will describe the results obtained for PKS 0521-36, (Macchetto et al, 1991a) for 3C 66B (Macchetto et al, 1991b) and for M87.

2 PKS 0521-36

One of the most prominent radio and optical jets is that found in the elliptical galaxy PKS 0521-36, a relatively isolated radio galaxy at a redshift $z = 0.055$ which also harbors a bright $V = 16$ BL Lac nucleus and extended optical line emission. Detailed spectroscopic and morphological studies have been carried out by Danziger et al. (1979, 1983); Cayatte & Sol (1987); and Boisson, Cayatte, & Sol (1989). Recently Sparks, Miley, & Macchetto (1990) reported optical polarization measurements of the jet and nucleus, which confirmed the expected high polarization if the emission is due to synchrotron radiation.

Images of PKS 0521-36 were obtained on 1990 August 25 with the FOC using the

[1]On assigment from the Space Science Department of ESA

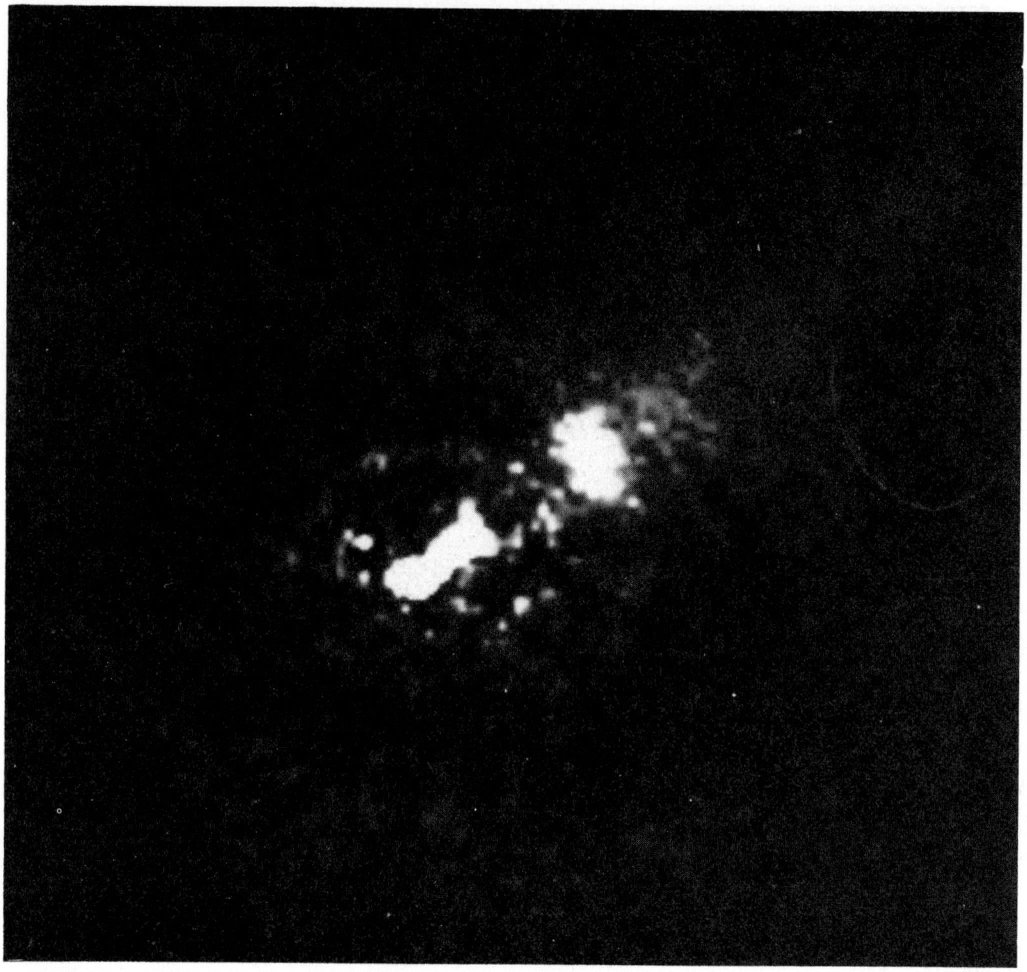

1. The jet in PKS 0521-36 as observed by the FOC. An elliptical model of the galaxy has been subtracted. The jet is approximately 6.″5 long.

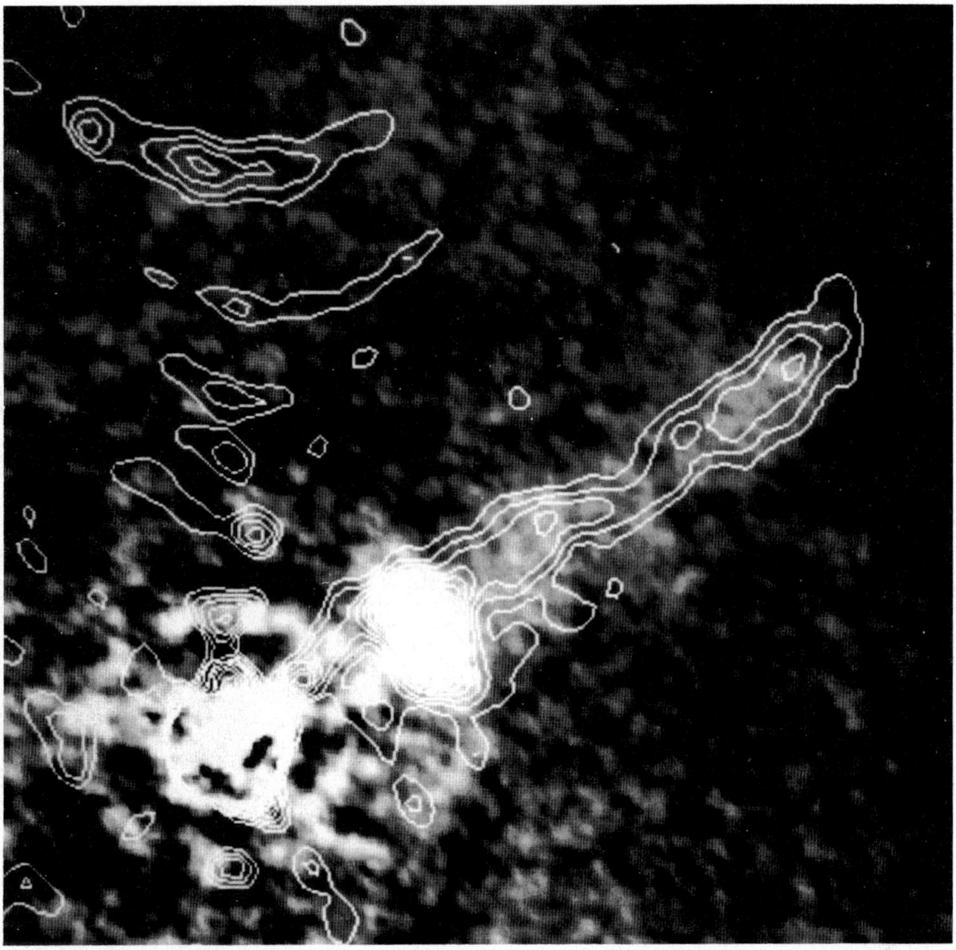

2. The VLA contour data of Keel (1986) is shown superposed on the FOC data. Note the excellent agreement of the two data sets.

F430W and F320W filters and f/96, 512 × 512 mode (Paresce 1990) with a corresponding pixel size approximately 0".022. A galaxy subtracted deconvolved image is shown in Figure 1.

The VLA contour data as published by Keel (1986) is shown superposed on the FOC data in Figure 2. It is immediately obvious that the FOC data has a resolution very similar to that of the VLA data, but shows considerably more morphological information that the ground-based optical data.

The FOC image shows a bright knot located $\approx$ 1."8 to the NE and clearly resolved as in the VLA data. The width of the knot is $\approx$ 0."8. Beyond this bright knot, the jet has approximately constant surface brightness and a morphology similar to the VLA image with a total length of 6."5. The jet is also resolved in width, 0."6 wide in the fainter regions of the jet, with little or no evidence of structure on a scale of $\leq$ 0."1. The FOC data appears to show more flux than the VLA data in the region at slightly larger radius from the nucleus but close to the southern tip of the knot.

There is a second component seen in the deconvolution at 0."36 from the nucleus, Figure 2. It is visible in the raw data, although by no means as clearly as in the deconvolved image. Such a source would be completely within the core of the highest resolution VLA contour map published by Keel. We believe we have detected a previously unseen inner jet structure at a distance corresponding to 300 pc, $H_o = 75$ km s^{-1} Mpc^{-1}. Presumably, the optical polarization measured by Sparks, Miley & Macchetto (1990) comes from both these components.

The large bright knot further along the jet is a clear counterpart to the radio knot. The radio and optical polarization position angles (Keel 1986; Sparks et al. 1990) suggest a magnetic field aligned along the jet direction. This knot is unresolved in the optical polarimetric measurements, but the general sense of the magnetic field is still along the jet direction at that position. The bright knot is clearly an important site where particle acceleration is occurring. The general optical morphology of the jet does not exhibit any significant degree of clumpiness even at FOC resolution.

Using standard formulae and values for the relevant parameters, we derive a mean lifetime for the electrons of $t_{1/2} \approx 600$ yrs. This implies that there must be continuous acceleration along the jet of the electrons responsible for the optical emission, since electron diffusion from the bright knot could not account for the observed optical extent.

3 3C 66B

3C 66B is a relatively nearby bright radio source associated with a 13th magnitude galaxy at a redshift of 0.0215 (Matthews, Morgan, & Schmidt 1964). The galaxy lies in a small group close to the cluster Abell 347, at a distance of 86 Mpc. At this distance an angular scale of 0."1 corresponds to a projected linear size of 41 pc.

The radio source has been the subject of several comprehensive studies (e.g., Northover 1973; Miley & van der Laan 1973; van Breugel & Jägers 1982; Leahy, Jägers, & Pooley 1986). An optical counterpart of the jet in 3C 66B was detected by Butcher et al. (1980) and further studied by Fraix-Burnet et al. (1989b).

Images of 3C 66B were obtained with the Faint Object Camera (FOC) on 1990 September 23 using the F320W filter in the f/96 512 × 512 mode (Paresce 1990) and reduced as in Macchetto et al. (1991a). The filter bandpass is about 900 Å FWHM centered at 3360 Å. The pixel size after correction for geometric distortion is approximately 0."022, ≈ 10 pc, and the field size approximately 11" ≈ 4 kpc. Two exposures of 1500 s duration were made with this filter, both in fine lock, resulting in a tracking accuracy of 0."007 rms.

The galaxy was clearly visible in the raw data, but in order to investigate the jet alone, a model of the underlying galaxy light was subtracted from the deconvolved image (Fig. 3). In order to compare our HST image with the best VLA map of 3C 66B, we smoothed the deconvolved and galaxy-subtracted optical image with a circular Gaussian function having $\sigma = 3.1$ binned pixels (FWHM ≈ 0."35; Fig. 4a). Figure 4b shows a contour map of the "A-configuration" VLA map of 3C 66B obtained at 6 cm wavelength by Leahy et al. (1986).

Several conclusions can be drawn from these figures.

1. The similarities between the smoothed optical image and the radio map imply that on a scale of 0."3 the radio and optical emission have very much the same radio-to-optical spectral index all along the jet.

2. On the scale of the HST resolution, the jet of 3C 66B is filamentary. Filamentary structure has not previously been seen for optical jets, although it has been observed on a scale of 0."1 for the M87 jet in the radio (Owen et al. 1989), (and now with the FOC, see Section 4)

3. Two distinct "strands" appear in the HST image. These can be traced from ≥ 3."7 (1.5 kpc) from the nucleus out to a distance of 7."6 (3 kpc), where they disappear into the noise. The separation between the strands varies between about 0."3 and 0."4,

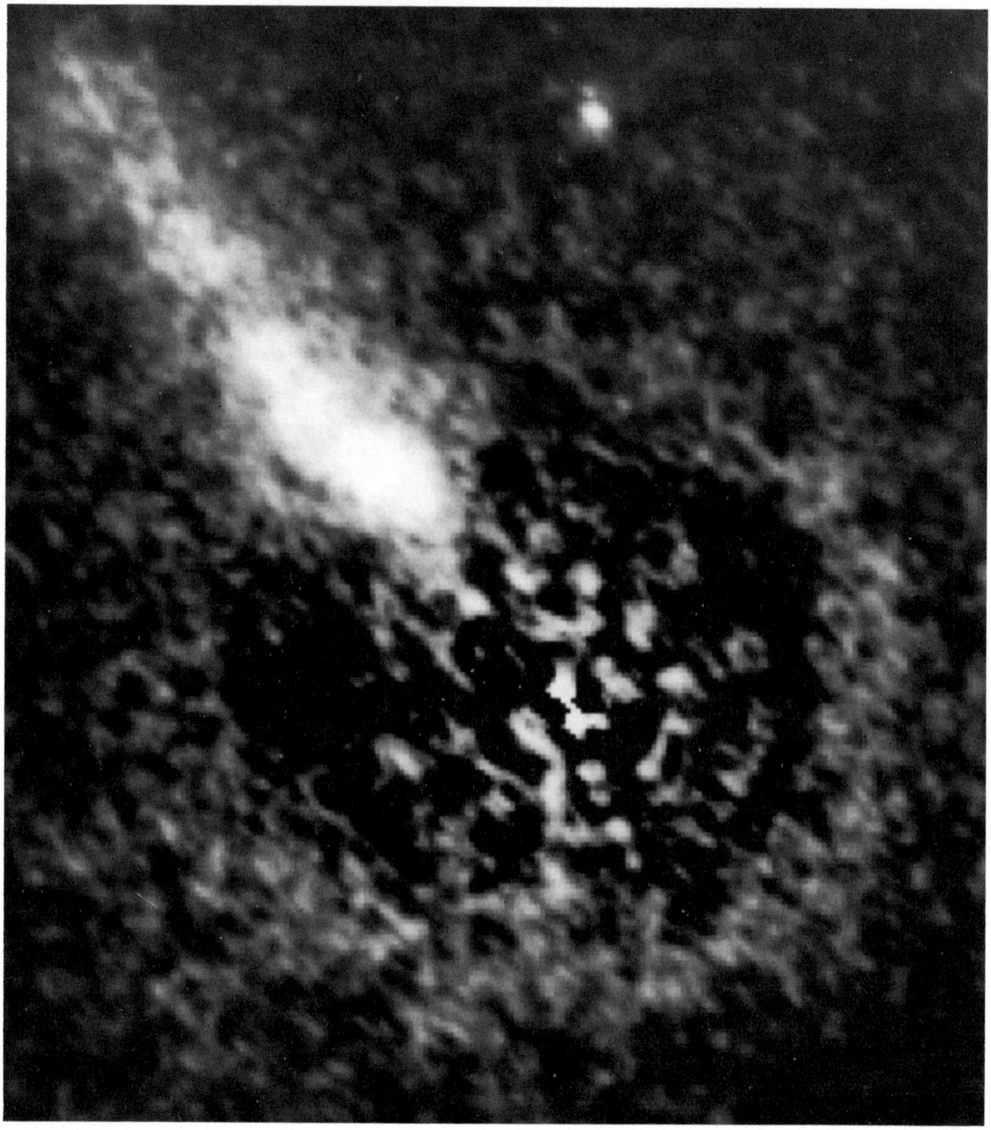

3. Deconvolved and galaxy-subtracted image of the jet of 3C 66B. The jet is 7.″6 (3 kpc) long. The FOC resolution of 0.″1 corresponds to 40 pc at the distance of the galaxy.

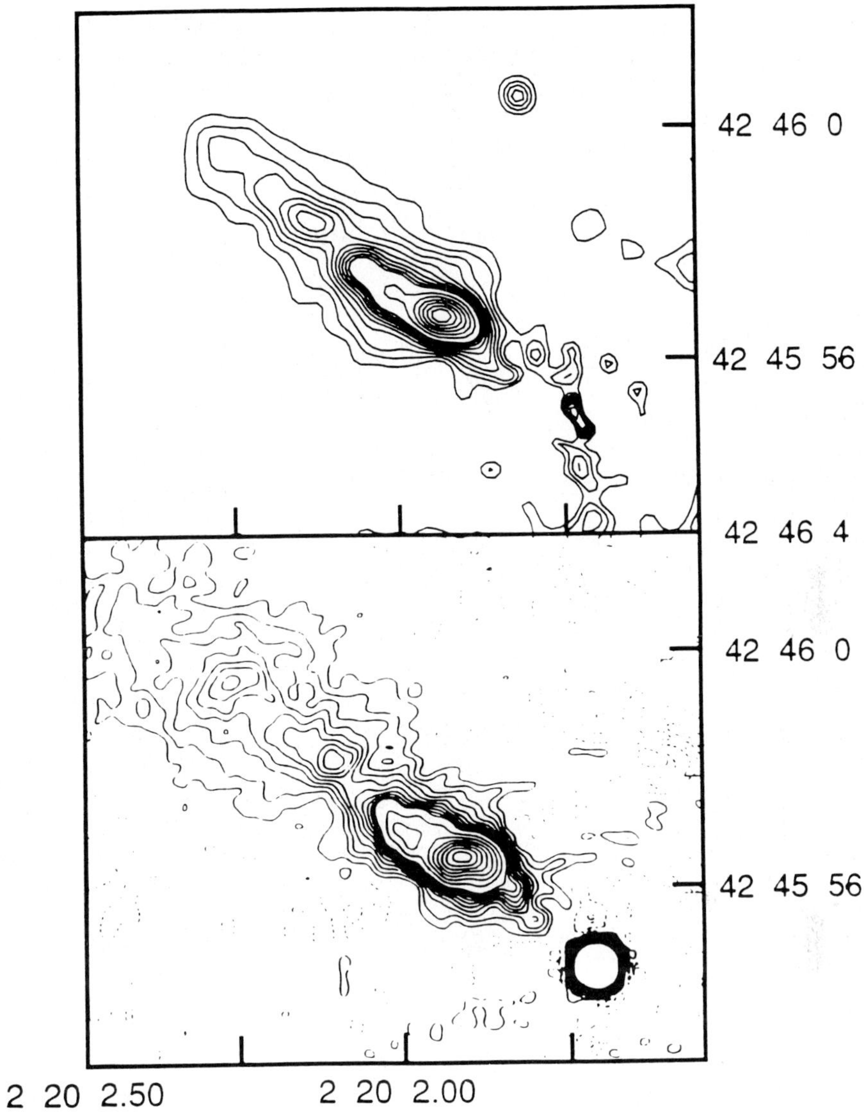

4. Contour map of 3C 66B jet smoothed with a Gaussian "beam" with σ = 3.1 pixels (0″.44 each). The contours are at intervals of 1.12 counts from 1.12 to 11.2 counts and intervals of 3.73 counts from 14.92 to 37.3 counts, corresponding to the contours of Leahy et al. 1986 for a nominal flux conversion and spectral index α = 1.1. Note that the spatial registration and scale are only approximate for the HST data. The coordinates' R.A. and Decl. axes are from Leahy et al. *(bottom)* The 5 GHz VLA map at comparable resolution to the optical image of *(top)* (from Leahy et al. 1986).

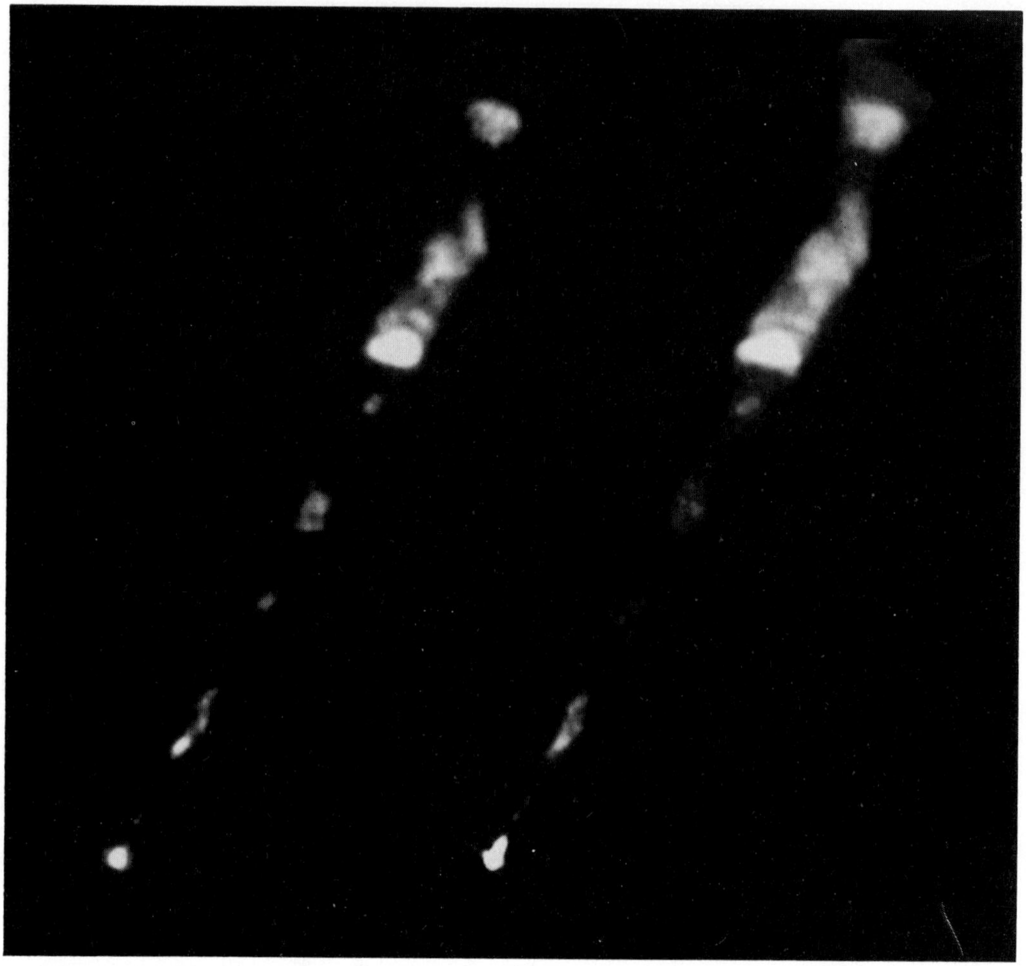

5. Left. Deconvolved FOC image of the M87 jet. Two FOC images are needed to cover the length of the jet (≃ 20"), hence the small gap in the spatial coverage. Right. Deconvolved VLA image kindly supplied by F. Owen and J. Biretta. The images show the remarkable agreement of the ultraviolet and radio observations.

that is, about 150 pc and they appear to undergo sharp "kinks" at distances of 2.″5 (1.0 kpc) and 6.″2 (2.5 kpc) from the nucleus.

A very basic question posed by these observations concerns constraints on the regions where the synchrotron-radiating electrons are accelerated. Does the presence of optical radiation along the jet imply that localized particle acceleration is required, or could the electrons be accelerated in the nucleus and transported to the sites of the radiation?

The average age of the radiating synchrotron electrons, with typical values for the magnetic field strength and cut-off frequency, is less than 1000 yr for the optical electrons, while the light travel time from the nucleus to the end of the observed optical jet is about 10^4 yr. Hence, at first sight, localized particle acceleration would be required, however, as discussed later for M87, in situ acceleration is not needed in a two fluid model, (Pelletier and Roland 1989).

The other new features present in our observations are the sharp bends or "kinks" in the double-stranded filaments. The origin of these kinks is unclear. The fact that they are mimicked in more than one filament suggests that they are not due to an instability mode in an individual filament. They may well trace out irregularities in the ISM of the galaxy and/or be due to time-dependent variations in the power of the nuclear machine responsible for producing the jet.

4 M87

The EO galaxy M87 harbours the prototypical and most studied example of an optical jet. The jet has been observed at radio (Owen, Hardee & Cornwell, 1989, Biretta, Stern & Harris 1991), optical (de Vaucouleurs & Nieto 1979, Keel 1988, Fraix-Burnet, Le Borgne & Nieto 1989) and X-rays wavelengths (Schreier, Gorenstein & Feigelson, 1980).

Images of M87 centred on the nucleus and positions along the jet were obtained with the FOC, utilizing the following filters in the f/96 512 × 512 modes: F120M, F140W, F220W (direct and with polarizers POL0,POL60,POL120), F430W (with polarizers only) and F501N. The exposures were made in fine lock, with an expected tracking accuracy of 0.″007.

The F220W exposures were corrected using the Lucy (1974) deconvolution technique to produce the final deconvolved images with minimum beam size 0.″2 FWHM. The results of this restoration are shown in Figure 5. This figure shows side-by-side the deconvolved FOC and VLA data. As is easily observed, two FOC frames (each of

11" × 11") are needed to cover the jet which extends for over 20" in length.

The complex structure and the wide range of intensities make it very difficult to display all the features in a single picture. Furthermore, the value of the relative intensity of the ultraviolet and radio data is arbitrary and was chosen only to show the most prominent features. Detailed intensity comparisons and determination of spectral indices as a function of position along the jet will be carried out in the near future.

The FOC observations demonstrate for the first time that the radio and ultraviolet brightness distribution is generally the same over a scale of about $0\farcs1$ or about 10 pc (M87 presents 78 pc arcsec^{-1} at an assumed distance of 16 Mpc).

The FOC data show that all the prominent optical knots (A,B,C, etc) have now been fully resolved and show the same remarkable structure as the radio data. (See Owen, Hardee and Cornwell, 1989, also for the knot notation). The jet shows very well-defined edges along its conical structure with an opening angle of $\sim 6.5^o$ and has a tight filamentary structure. The appearance of this structure indicates that the filaments are wrapped around the jet with pitch varying angles. Slices taken across the jet at different locations show prominent limb brightening similar, but not identical, to the radio data. Detailed comparisons are beyond the scope of this paper, but this general agreement provides strong evidence in favor of the two fluid model for this jet. (Pelletier & Roland 1989, Owen, Hardee & Cornwell 1989).

In this scenario, the jet consists of a cone around which one or more bright filaments are wrapped. The optical and radio emission comes mostly from a surface layer in which these filaments are embedded. The synchrotron lifetimes in the ultraviolet are typically only of the order of 100 yr, corresponding to light travel times of about 30pc. This is comparable to the width and presumably the thickness of the optical and radio strands.

The low emissivity in the jet's core indicates that the energetic particles are not suffering significant synchrotron losses. This is also compatible with the model in which the jet's core has a relatively low magnetic field. In this case, the high-energy particles can be produced in the central black-hole and propagate along the jet's interior in a low magnetic field region, thereby suffering only modest synchrotron losses. As they diffuse across the jet and into the high magnetic field boundary layer, they can produce the optical emission without the need for in situ acceleration.

Several mechanisms can be invoked to explain how the emission from this boundary layer is produced. Non-linear evolution of synchrotron instabilities has been proposed by Bodo et al, 1991, to explain the formation of filaments in jets out of equipartition, such as that in M87, where the energy of the relativistic electrons exceeds that of the magnetic field. They find that in a plasma subject to constant heating, after an initial phase in which the instability growth rate follows the linear model, the instability reaches a quasi equilibrium state on timescales of the order of several synchrotron timescales. This mechanism can explain the formation of filaments of enhanced emission observed in the lobes and jet of M87.

Optical synchrotron emission can also be produced through a diffusive shock acceleration mechanism (Fraix-Burnet, 1991). This process is so efficient that requires the magnetic field turbulence to be quite low. The source of energy of this turbulence could be the kinetic energy of the jet which can be transferred to the magnetic field (or to the plasma) through the interaction of the jet with the interstellar medium.

The answer to which of the competing mechanisms and scenarios are at work and the determination of the relevant physical parameters will have to wait further detailed analysis of the FOC optical, ultraviolet and polarization data.

5 CONCLUSIONS

The study of the optical counterparts to radio jets with the Faint Object Camera on board the Hubble Space Telescope has already produced new and unexpected results.

The jet in PKS 0521-36, which is the most distant, has been fully resolved. Because of its length, magnetic field configuration and optical morphology, it seems to require reacceleration sites for the optical electrons. These could well be provided by shocks at the site of the brighter knots observed.

For both 3C 66B and M87, we have observed, for the first time at optical wavelengths, a filamentary structure which is similar to the radio data. In this case, we conclude that the emission comes from a boundary layer where the filaments and the strong magnetic field are located.

One of the most puzzling, but fundamental, results that must be explained by models of particle acceleration is why within an object, over more than five decades of frequency and a large variation of physical conditions, the old and young electrons have similar spatial distributions, although large differences are observed from object to object.

Our FOC observations of PKS 0521-36, 3C 66B and M87 show that, even with its present aberration problems, the HST is a uniquely important instrument for studying synchrotron jets. Future observations with the HST of other extragalactic jets should provide fundamental information about the nature of collimated activity in galaxy nuclei.

6 ACKNOWLEDGEMENTS

The FOC/IDT members are: Rudolf Albrecht, Cesare Barbieri, David Baxter, J. Chris Blades, Alec Boksenberg, Phil Crane, Jean Michel Deharveng, Michael J. Disney, Perry Greenfield, Peter Jakobsen, Robert Jedrzejewski, Theo M. Kamperman, Ivan R. King, F. Duccio Macchetto, Craig D. Mackay, Antonella Nota, Francesco Paresce, William Sparks, and Gerd Weigelt.

We wish to thank Drs. F. Owen and J. Biretta for making available to us the VLA radio data for M87, and Dr. Keel for the PKS 0521-36 data.

7 REFERENCES

Biretta, J.A., Stern, C.P. & Harris, D.E. 1991, *A. J.,* **101**, 1632

Bodo, G., Ferrari, A., Massaglia, S., Rossi, P., Shibata, K., & Uchida, Y. 1991, *Astr. Ap.,* in press.

Boisson, C., Cayatte, V., & Sol, H. 1989, *Astr. Ap.*, **211**, 275

Bridle, C., Hough, J.H. Bailey, J., Axon, D.J., & Hyland, A.R. 1986, *M.N.R.A.S.,* **221**, 739

Butcher, H.R., van Breugel, W.J.M., & Miley, G.K. 1980, *Ap. J.,* **235**, 749

Cayatte, V., & Sol, H. 1987, *Astr. Ap.*, **171**, 25

Danziger, I.J., Ekers, R.D., Goss, W.M., & Shaver, P.A. 1983, in *Astrophysical Jets*, ed. A. Ferrari & A.G. Pacholczyk (Dordrecht:Reidel), 131

Danziger, I.J., Fosbury, R.A.E., Goss, W.M. & Ekers, R.D. 1979, *M.N.R.A.S.,* **188**, 415

de Vaucouleurs, G., & Nieto, J.L. 1979, *Ap. J.,* **231**, 364

Fraix-Burnet, D., Golombek, D., Macchetto, F., Nieto, J.L., Lelievre, G., Perryman, M.A.C., & di Sergo Alighieri, S. 1991, *A. J.,* **101**, (1), 88

Fraix-Burnet, D., Nieto, J.L., Lelievre, G., Macchetto, F., Perryman, M.A.C., & di Serego Alighieri, S. 1989b, *Ap. J.,* **336**, 121

Fraix-Burnet, D., Le Borgne, J.F., & Nieto, J.L., 1989, *Astr. Ap.*, **224**, 17

Fraix-Burnet 1991, Proceedings 7th IAP Meeting Extragalactic Radio Sources: from Beams to Jets, Paris, France, 2-5 July, 1991

Keel, W.C. 1986, *Ap. J.,* **302**, 296

Keel, W.C. 1988, *Ap. J.,* **329**, 532

Leahy, J.P., Jägers, W., & Pooley, G.G. 1986, *Astr. Ap.*, **156**, 251

Lucy, L.B. 1974, *A. J.,* **79**, 745

Macchetto, F. 1981, in *Proc. of ESO/ESA Workshop Optical Jets in Galaxies*, ed. F. Macchetto, G. Miley, & M. Tarenghi (Noordwijk:ESA), 15

Macchetto, F., Albrecht, R., Barbieri, C., Blades, J.C., Boksenberg, A., Crane, P., Deharveng, J.M., Disney, M.J., Jakobsen, P., Kamperman, T.M., King, I.R., Mackay, C.D., Paresce, F., Weigelt, G., Baxter, D., Greenfield, P., Jedrzejewski, R., Nota, A., & Sparks, W.B. 1991a, *Ap. J.,* **369**, L55

Macchetto, F., Albrecht, R., Barbieri, C., Blades, J.C., Boksenberg, A., Crane, P., Deharveng, J.M., Disney, M.J., Jakobsen, P., Kamperman, T.M., King, I.R., Mackay, C.D., Paresce, F., Weigelt, G., Baxter, D., Greenfield, P., Jedrzejewski, R., Nota, A., Sparks, W.B. & Miley, G.M. 1991b, *Ap. J.,* **373**, L55

Matthews, T.A., Morgan, W.W., & Schmidt, M. 1964, *Ap. J.,* **140**, 35

Miley, G.K. 1981, in *Proc. of ESO/ESA Workshop Optical Jets in Galaxies*, ed. F. Macchetto, G. Miley, & M. Tarenghi (Noordwijk:ESA), 9

Miley, G.K., & van der Laan, H. 1973, *Astr. Ap.*, **28**, 359

Northover, K.J.E.D. 1973, *M.N.R.A.S.,* **165**, 369

Owen, F.N., Hardee, P.E., & Cornwell, T.J. 1989, *Ap. J.,* **340**, 698

Paresce, F. 1990, FOC Instrument Handbook, Space Telescope Science Institute

Pelletier, G. & Roland, J., 1989, Astron. Astrophys. **224**, 24

Schreier, E.J., Gorenstein, P., & Feigelson, E.D. 1982, *Ap. J.,* **261**, 42

Sparks, W., Miley, G., & Macchetto, F. 1990, *Ap. J.,* **361**, L41

van Breugel, W.J.M., & Jägers, W. 1982, *Astr. Ap. Suppl.,* **49**, 529

Optical Synchrotron Emission and Turbulences in Extragalactic Jets

D. FRAIX-BURNET [1,2]

Space Telescope Science Institute
3700 San Martin Drive
Baltimore, MD 21218, USA

ABSTRACT
The optical synchrotron emission from extragalactic jets can be explained by the diffusive shock acceleration. But this process seems to be so efficient that the magnetic turbulences must be quite low. As to the origin of these turbulences, the interaction of the jet with the ISM may provide a common explanation for optical emission and one-sidedness of extragalactic jets.

1 INTRODUCTION

Two important questions concerning extragalactic jets remain unsolved: why a few radio jets have a synchrotron spectrum which extends to optical wavelengths? and why some radio sources show two jets, others one and others none? Even if relativistic beaming could explain the difference between one- and two-sided jets, it is not satisfactory because, besides its own intrinsic difficulties (the huge sizes sometimes implied and inconsistency with the best studied of all jets, i.e. M87, Reid *et al.*, 1989), it does not help to understand optical jets and sources without jet.

The synchrotron radiation, especially at optical wavelengths, provides strong constraints on the physics of the jet because the relativistic particles need to be reaccelerated *in situ*. This can be done with Fermi-type mechanism, the most efficient being the diffusive shock acceleration. It has thus been possible to fit the synchrotron spectrum for optical jets (see for example 3C 273 in Heavens and Meisenheimer, 1987). To understand why only four or five extragalactic jets have optical emission it is useful to derive the important physical parameters (Section 2). These parameters probably depend either on conditions internal to the jet (speed, magnetic field, density) and/or external conditions. I propose in Sections 3 and 4 to use large scale properties of radio sources to investigate the influence of external conditions on the jet luminosity. In Section 5 a "unified" model based on the characteristics of the

[1] Affiliated with the Astrophysics Division, Space Science Department, European Agency.
[2] Now at Observatoire Midi-Pyrénées, Observatoire de Toulouse, URA 285, 14 Avenue Edouard Belin, F-31400 Toulouse, France.

interaction between the jet and the ISM is proposed and may explain both optical emission and one-sidedness.

2 DIFFUSIVE SHOCK ACCELERATION WITH OBLIQUE SHOCKS

2.1 The analytical calculations

The relativistic particles responsible for the optical synchrotron emission in extragalactic jets must have Lorentz factors of about 10^6. Numerical calculations have shown that this is possible with diffusive shock acceleration. But to determine the important parameters of the mechanism, an analytical theory is essential. Pelletier and Roland (1988) developed analytical calculations of the particle energy spectrum obtained with the diffusive shock acceleration at work in magnetized mixed strong shocks. Fraix-Burnet and Pelletier (1991) subsequently generalized these calculations for oblique shocks which can accelerate particles to higher energies. The results are briefly presented below for both the spectral index at radio frequencies and the cut-off frequency which determines the visibility of the jet at optical wavelengths.

2.2 The spectral index

The synchrotron radiation spectral index (α) is related to the energy spectrum index (η) by $\alpha = (\eta - 1)/2$ at very low frequencies. In the case where the ratio between thermal and relativistic plasma pressures is greater than 0.1 ($\theta = p_g/p_r > 10^{-1}$), η is related to the shock compression ratio r by:

$$\eta = \frac{3r_s}{r_s - 1} - 2 \qquad \text{with} \qquad r_s = r\left(1 - \frac{\cos\beta^u}{M_A}\right)$$

r_s is the compression ratio experienced by the scattering centers of the relativistic electrons, β^u is the angle between the upstream magnetic field and the normal to the

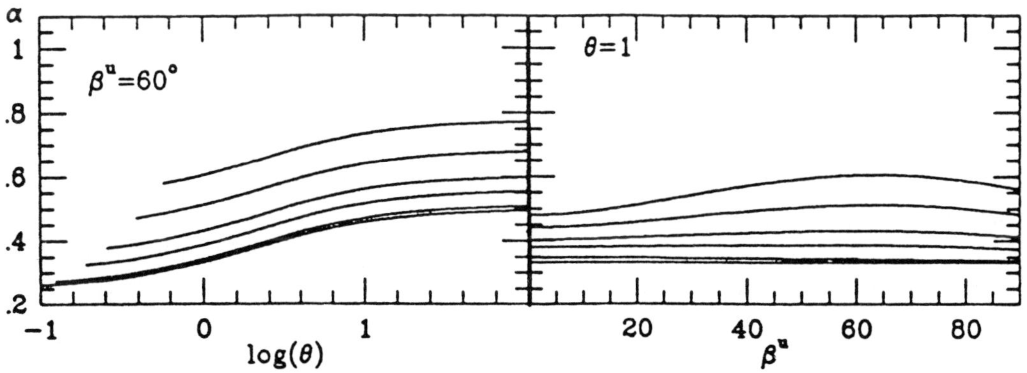

shock front and M_A is the Alfvenic Mach number upstream.

Fraix-Burnet and Pelletier (1991) find that r is given to the first order in M_A^{-1} by:

$$r = r_0 / \left(1 + \frac{r(r_0 - 3) + r_0 + 1}{2} \frac{\sin^2 \beta^u}{M_A^2}\right) \qquad \text{with} \qquad r_0 = \frac{7 + 4\theta}{1 + \theta}$$

The spectral index α as a function of the obliquity β^u and θ is plotted above for different values of M_A (4, 5, 7, 10, 30, and ∞ downward). It can be seen that α depends mainly on M_A and θ. The influence of β^u is weak.

2.3 The cutoff frequency

The cut-off frequency is defined by the balance of energy gain through the acceleration process with the synchrotron losses. Fraix-Burnet and Pelletier (1991) establish the relation between the cut-off frequency (given by the synchrotron formula: $\nu_c = 1.2\,10^6 B_{\mathrm{Gauss}} \gamma^2$) and the obliquity of the shock:

$$\cos^2 \beta^u = \left(\frac{18\pi m}{\sigma_T c} \frac{K}{\gamma(\gamma^2 - 1)^{1-\nu_A/2}} \frac{u^{u2}}{B^2} - H\right) / (1 - H)$$

where

$$H = \left(\frac{Kmc}{eB}\right)^2 (\gamma^2 - 1)^{\nu_A - 1} \qquad \text{and} \qquad K = \eta_{turb} \frac{\nu_A - 1}{\nu_A(\nu_A + 2)} \frac{3\pi}{2} \frac{c}{l_0} \left(\frac{mc^2}{l_0 e B}\right)^{\nu_A - 2} .$$

η_{turb} is the degree of magnetic turbulence $< (\delta B)^2 > / < B >^2$, l_0 is the long spatial wavelength cut-off of the spectrum, ν_A is the power law of the spectrum of the Alfvén waves ($\omega^{-\nu_A}$), $r_{gr} = mc^2/eB$, σ_T is the Thomson cross-section and u^u the upstream speed of the scattering centers (Alfvén waves).

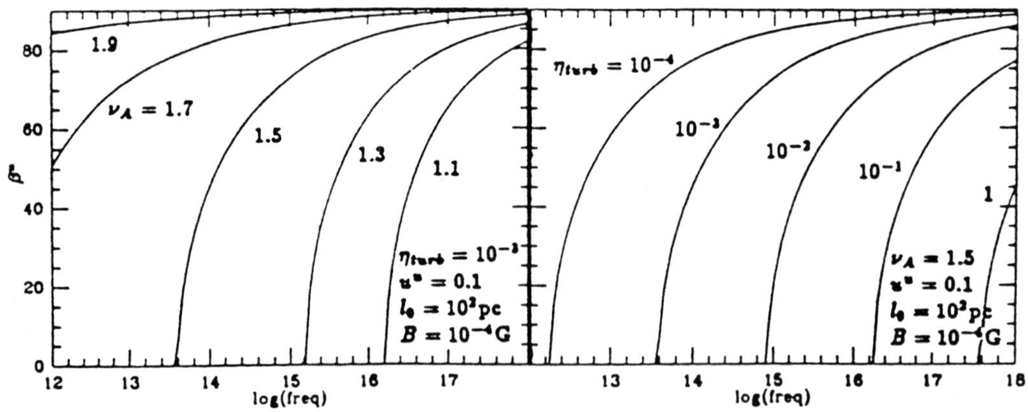

This study clearly shows that the most important parameter is the magnetic turbulence. This is illustrated on the figure above showing the variations of ν_c with β^u for different values of ν_A and η_{turb}. Depending on the assumption about the turbulence spectrum ($\nu_A = 5/3$ for Kolmogorov or 3/2 for Kraichnan), the degree would have to be quite high (> 0.15) or only $10^{-3} - 10^{-2}$ (respectively) to account for optical emission.

2.4 Application to the jet of 3C 273

Fraix-Burnet and Pelletier (1991) applied their calculations to the jet of 3C 273 in order to constrain the parameters of the model. It immediately appears that the obliquity of the shock cannot be much constrained due to its low influence on the synchrotron spectrum. More interesting is the result obtained on the turbulences because the turbulent conditions inside a radio source are totally unknown. Fraix-Burnet and Pelletier favor a Kraichnan spectrum with a rather low degree of turbulences ($3\,10^{-3} < \eta_{turb} < 4\,10^{-2}$) for the optical jet of 3C 273. Note that the speed of the jet is found to be only mildly relativistic within the range 0.2–0.6 c.
Assuming that the turbulence spectrum has the same slope in all radio sources, this result implies that degrees of turbulence have to be very low in general ($\lesssim 10^{-3}$) . This is in agreement with Fritz (1989) and Roland *et al.* (1988), as well as with the high level of polarization observed in jets and lobes.

3 THE ORIGIN OF THE TURBULENCES

3.1 The 3-D structure of jets

Now that the efficiency of the diffusive shock particle acceleration mechanims is clearly established, what is the observational evidence for shock waves in jets? If this process is really at work, the shock front itself is not expected to be seen because of the synchrotron lifetime of the electrons: a few hundreds of parsecs for optical emission and a few tens of kpc for radio emission. It is therefore very surprising to see filaments with thickness of the order of 10 pc on radio maps of M87 (Owen *et al.*, 1989) and optical data from HST on M87 and 3C 66B (Macchetto, this conference). These observations also show that the synchrotron emission is not distributed in separated knots, but is rather continuous with strong indications of being localized in the boundary layer between the jet and the ISM.
Two difficulties arise for the diffusive shock acceleration: no shock front is detected yet and the degree of turbulences has to be quite low. This raises indeed the possibility that a stochastic acceleration mechanism is at work, and because it would be much more difficult to obtain optical emission (see for example Eilek and Henriksen, 1984), this could explain the very small number of optical jets. In this case the jet luminosity would depend even more on magnetic turbulences.
Because of their importance in the synchrotron emission process, we thus have to assume that the magnetic turbulences are very probably localized in the boundary

layer.

3.2 The interaction between the jet and the ISM

The only source of energy available to feed the turbulences is the kinetic energy of the jet which can be transferred to the magnetic field or to the jet plasma through the interaction of the jet with the interstellar medium. This process has been proposed and studied by Eilek (1979) and Ferrari *et al.* (1979) using the stochastic acceleration of electrons.

Ferrari *et al.* (1982) studied the influence of the geometry of the boundary layer between the jet and the ISM and showed that the growth rates of the turbulence depends significantly on this geometry, especially the smaller scalelengths which depend on the layer width. These small scalelengths of turbulences are the most important for the particle acceleration. Thus one can reasonably expect the synchrotron spectrum to depend on the energy in the turbulences (see Bicknell and Melrose, 1982) and on the geometry of the boundary layer that are imposed by the characteristics of the interaction between the jet and the ISM.

4 THE EFFECT OF THE ENVIRONMENT ON RADIO SOURCES

The propagation of radio jets is certainly influenced by the insterstellar medium. Very few direct evidences for the interaction between the jet and the ISM have been found. But one strong support for this is the correlation beween asymmetries of the ISM and of the radio sources.

4.1 The asymmetry of the ISM

The correlation between emission-line and radio characteristics found by Pedelty *et al.* (1989) and McCarthy *et al.* (1991) clearly shows that the asymmetry of radio sources is intrinsic. They note the importance of environmental effects on the structural and dynamical properties of radio sources. These results also show that the ISM is by no means symmetric in a galaxy so that the jets certainly encounter different propagation conditions on the two opposite sides of the radio source.

4.2 One-sidedness of extragalactic jets: intrinsic or apparent?

One of the most important observation concerning large-scale one-sidedness of jets is the Laing-Garrington effect (see Garrington and Conway, 1991, and Garrington *et al.*, 1991): in one-sided radio sources, the visible jet is in general on the same side as the less-depolarized lobe. By itself this result does not rule out either interpretation (intrinsic or apparent). However, Liu and Pooley (1991) have recently confirmed another important correlation, simply mentioned in Garrington *et al.* (1991), between the spectral index of the lobe and its depolarization from a sample of high-luminosity radio sources which have not been selected for their jet structure. I reproduce below this correlation using data from Garrington *et al.* (1991; black dots represent jet side and open circles counterjet side) and Liu and Pooley (1991; $\times$), DPM being the depolarizing parameter defined by the last authors (DPM increases with the depolarization).

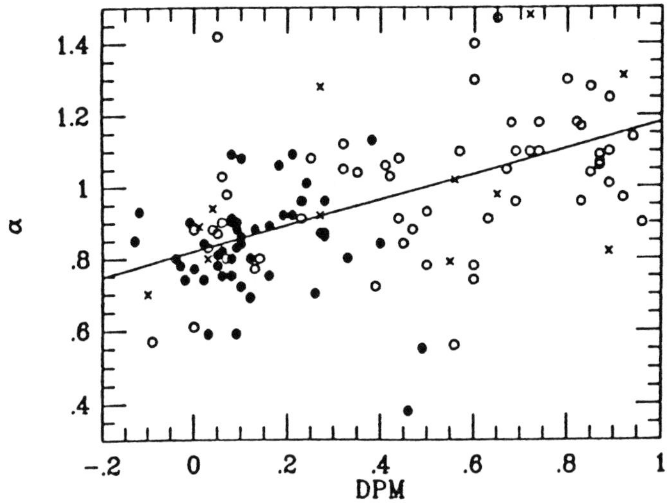

Note that the less depolarized lobe has the flatter spectrum, and that for one-sided jets the range of DPM values is narrower on the jet side. We conclude that the three observables (jet visibility, depolarization and spectral index of lobe) are correlated. Since the spectral index can be hardly affected by any aberration (see Liu and Pooley, 1991), *the asymmetry of jets and lobes is intrinsic.*

The confirmation of this result is found in M87 where Owen *et al.* (1990) find that the lobe on the jet-side is less depolarized, but also that the jet itself is much less depolarized than both lobes. This shows that the jet is in front of the Faraday medium which is inside or more probably around each lobe rather than around the whole radio source as supposed in the relativistic beaming (apparent asymmetry) interpretation.

5 A UNIQUE MODEL FOR ONE-SIDEDNESS AND OPTICAL SYNCHROTRON EMISSION

We have seen that the interaction between the jet and the ISM are different for the jet and the counterjet. One consistent way to explain the correlations between jet visibility, spectral index and depolarization of the lobe, is to assume that *a denser ISM decreases the synchrotron emissivity of a jet.* A denser ISM, revealed by a more depolarized lobe, creates a larger turbulent boundary layer so that the small turbulence scalelength cutoff is higher and the particle acceleration is less efficient. Of course, there should be a limit because a jet propagating in the void has probably

no turbulent edge. But within the adequate range, the visible jet looses less energy in the interaction with the ISM so that more energy is available for the particles in the corresponding lobe, explaining the flatter spectrum in the lobe on the side of the visible jet.
Globally this model seems to be able to explain with a minimum number of parameters and in a unified frame, radio sources with no jet, one jet or two jets, the small number of optical jets and why they are one-sided, gaps in jets and also why powerful sources tend to be one-sided. This model and its implications will be studied in detail in a forthcoming paper (Fraix-Burnet, in preparation).

Acknowledgements
I acknowledge support by the European Space Agency through an ESA postdoctoral fellowship.

References
Bicknell, G.V., Melrose, D.B. 1982, Astrophys. J, 262, 511
Eilek, J.A. 1979, Astrophys. J, 230, 373
Eilek, J.A., Henriksen, R.N. 1984, Astrophys. J, 277, 820
Ferrari, A., Massaglia, S., Trussoni, E. 1982, Month. Not. Roy. Astron. Soc., 198, 1065
Ferrari, A., Trussoni, E., Zaninetti, L. 1979, Astron. Astrophys., 79, 190
Fraix-Burnet, D., Pelletier, G. 1991, Astrophys. J, 367, 86
Fritz, K.-D. 1989, Astrophys. J, 347, 692
Garrington, S.T., Conway, R.G. 1991, Month. Not. Roy. Astron. Soc., 250, 198
Garrington, S.T., Conway, R.G., Leahy, J.P. 1991, Month. Not. Roy. Astron. Soc., 250, 171
Heavens, A.F., Meisenheimer, K. 1987, Month. Not. Roy. Astron. Soc., 225, 335
Liu, R., Pooley, G. 1991, Month. Not. Roy. Astron. Soc., 249, 343
McCarthy, P.J., van Breugel, W., Kapahi, V.K. 1991, Astrophys. J, 371, 478
Owen, F.N., Eilek, J.A., Keel, W.C. 1990, Astrophys. J, 362, 449
Owen, F.N., Hardee, P.E., Cornwell, T.J. 1989, Astrophys. J, 340, 698
Pedelty, J.A., Rudnick, L., McCarthy, P.J., Spinrad, H. 1989, Astron. J., 97, 647
Pelletier, G., Roland, J. 1988, Astron. Astrophys., 196, 71
Reid, M.J., Biretta, J.A., Junor, W., Muxlow, T.W.B., Spencer, R.E. 1989, Astrophys. J, 336, 112
Roland, J., Pelletier, G., Muxlow, T.W.B. 1988, Astron. Astrophys., 207, 16

On the Formation of Emission Filaments in Jets

SILVANO MASSAGLIA

Istituto di Fisica Generale dell'Università, Torino (Italy)

GIANLUIGI BODO AND ATTILIO FERRARI

Osservatorio Astronomico di Torino, Pino Torinese (Italy)

PAOLA ROSSI

Harvard–Smithsonian Center for Astrophysics, Cambridge (USA)

1 INTRODUCTION

High resolution observations of synchrotron extragalactic sources show the presence of emission filaments. They have been observed, at radio wavelengths, with the VLA, in the lobes of Cygnus A (Dreher, Carilli and Perley 1987) and in the jet of M87 (Hines, Owen and Eilek 1989); very recently, optical observations performed by HST, have shown filaments in the jet of 3C66B (Macchetto *et al.* 1991). Relying upon the fact that structuring is observed in an increasing number of objects as the resolution and sensitivity of the observations increase, we suggest that these phenomena are a common feature of synchrotron extended sources.

A possible mechanism which can explain the formation of these filaments is a radiative instability driven by the synchrotron emission itself (Simon and Axford 1967; Bodo *et al.* 1990; Gouveia dal Pino and Opher 1990). The analyses of the instability presented in those papers suffer, however, of the very limiting assumption of a maxwellian distribution of the relativistic synchrotron emitting particles. Here we present an analysis of the instability process including an equation for the evolution of the distribution function which is assumed to be initially a power law. We then obtain growth rates which depend on the spectral index and our results show a faster development of the instability for flatter spectra. If we consider the cases of 3C66B and M87 jets, for example, we can predict that the instability sets in much closer to the jet origin in the source which has a lower spectral index, which is M87. Therefore in this source filaments should be visible at a smaller distance from the nucleus.

2 MODEL AND RESULTS

We consider a two-component medium: a thermal part which contributes to the inertia of the system and a component of relativistic electrons which provides the main contribution to the pressure given by

$$P(\mathbf{r}, t) = \frac{1}{3}\int_0^\infty f(E, \mathbf{r}, t) E dE \quad , \tag{1}$$

where $f(E, \mathbf{r}, t)$ is the electron distribution function which must obey the evolution equation (Kardashev 1962)

$$\frac{\partial f}{\partial t} + \nabla(f\mathbf{v}) = -\frac{\partial}{\partial E}(f\dot{E}) + Q \quad , \tag{2}$$

where we have included synchrotron losses and adiabatic losses or gains according to Kardashev (1962).

We study the stability of a homogeneous infinite system, with a uniform background magnetic field along the z axis, for perturbations that fulfill the conditions typical of the classical condensation mode (Field 1965, Simon and Axford 1967). We therefore restrict ourselves to perturbations which keep constant total pressure, namely

$$\frac{\delta B}{B} = -\frac{\beta}{2}\frac{\delta P}{P} \quad , \tag{3}$$

where $\beta = 8\pi P_0/B_0^2$. The dispersion relation is obtained from the linearization of (2) under the assumption that there is no variation in the particle injection rate, i.e. we assume that the perturbation of the flux $F = \dot{E}f$ at $E = E_{max}$ is zero (see Rossi *et al.* 1991). E_{max} is the high energy cut-off of the equilibrium distribution function.

The dispersion relation has been solved adopting for the initial distribution function

$$f_0 = \exp(-E_0/E)E^{-\alpha} \quad E < E_{max} \quad ; \qquad f_0 = 0 \quad E > E_{max} \quad .$$

Fig. 1a shows the growth rate as a function of the spectral index for different values of β and for $E_0 = 0.01E_{max}$. We can see that, for $\beta = 10$, the system is unstable for every value of α, the growth rate, however, increases, decreasing α, *i.e.* going from steep to flat spectra. The behaviour of the growth rate as a function of β is represented in fig. 1b. A cut-off is found for $\beta = 1$ independently of the value of α.

The important result is that the system is unstable for $\beta > 1$, i.e. if it is out of equipartition with particle pressure prevailing over magnetic field pressure, and for every value of the spectral index. We can apply our results to the interpretation of the formation of filaments in the jets of two extragalactic sources, namely 3C66B and M87. The two sources have different spectral indexes: the spectrum of 3C66B is steeper ($\alpha \sim 3.2$), while that of M87 is flatter ($\alpha \sim 2.2$). On the bases of the present analysis we can predict a faster development of the instability in this last case, and therefore, for M87, we expect the formation of filaments much closer to the jet origin. We find in fact, in the case of 3C66B, that the minimum distance from the origin of the jet is (see Rossi *et al.* 1991 for details) $d \sim 1$ kpc, in agreement with observations (Macchetto *et al.* 1991). In the case of the jet of M87 we estimate a minimum distance for filaments formation $d \sim 3$ pc. We recall that the emission knot closest to the nucleus (feature G) is located at about 1".3, corresponding to $\sim$ 100 pc.

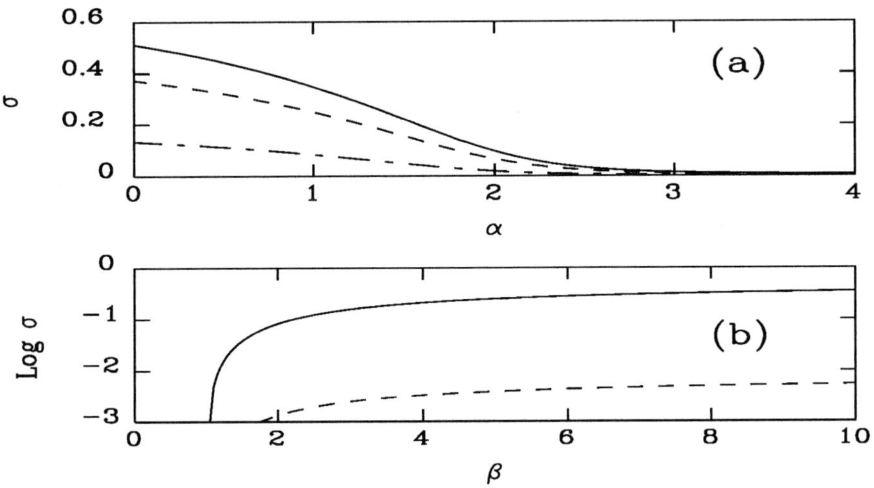

Figure 1 – Panel (a): Growth rate σ as a function of α for $E_0 = 0.01E_{max}$ and for $\beta = 10$ (solid line), $\beta = 5$ (dashed line) and $\beta = 2$ (dashed–dotted line). Panel (b): σ vs β for $\alpha = 1$ (solid line) and $\alpha = 4$ (dashed line).

REFERENCES

Bodo, G., Ferrari, A., Massaglia, S., and Trussoni, E., (1990) *Mon. Not. R. astr. Soc.,* **244**, 530.
Dreher, J.W., Carilli, C.L., and Perley, R.A., (1987) *Astrophys. J.,* **316**, 611.
Field, G., (1965) *Astrophys. J.,* **162**, 531.
Gouveia Dal Pino, E.M., and Opher, R., (1989) *Astrophys. J.,* **342**, 686.
Hines, D.C., Owen, F.N. and Eilek, J.A., (1989) *Astrophys. J.,* **347**, 713.
Kardashev, N.S., (1962) *Sov. Astr.*, **6**, 317.
Macchetto *et al.*, (1991) *Astrophys. J. Lett.,* in press.
Rossi P., Bodo G., Massaglia S. and Ferrari A., (1991) *Astron. Astrophys.*, submitted.
Simon, M. and Axford, W.I., (1967) *Astrophys. J.,* **150**, 105.

Correlations between the Radio and Optical Properties of Extragalactic Radiosources

STEVE RAWLINGS

Astrophysics, Department of Physics, Oxford University, Keble Road, Oxford OX1 3RH, UK, and Mullard Radio Astronomy Observatory, Cavendish Laboratory, Madingley Road, Cambridge, CB3 0HE, UK

1. CORRELATION BETWEEN RADIO AND OPTICAL NARROW-LINE LUMINOSITIES

Recent studies of extragalactic radiosources at both low redshift ($z \lesssim 0.5$; e.g. Baum & Heckman 1989, Rawlings *et al.* 1989) and high z ($z \gtrsim 0.5$; McCarthy & van Breugel 1989) have confirmed the correlation (Fig. 1) between low-frequency radio luminosity L_{151} and narrow-line luminosity L_{NLR} first suspected by Hine & Longair (1979). Objects from unbiased radio-bright samples (e.g. Rawlings *et al.* 1989) and from fainter samples (e.g. Allington-Smith *et al.* 1991) obey the same relation, which is thus caused neither by a selection effect nor by a correlation with z.

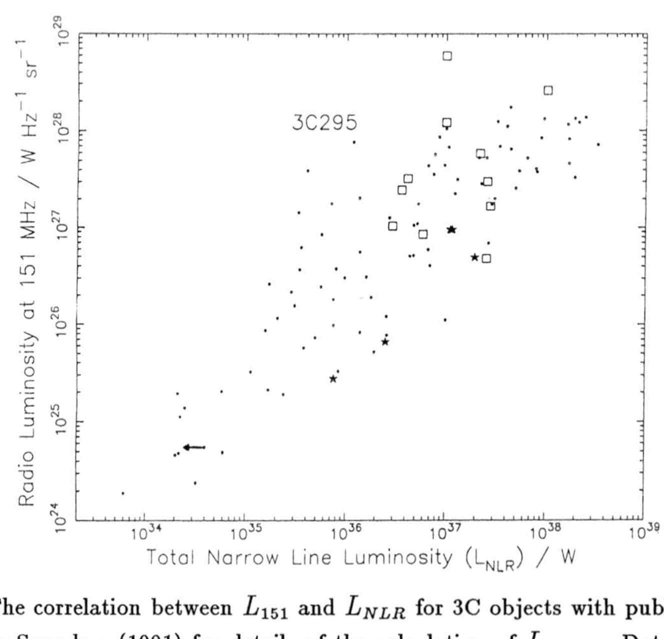

Figure 1. The correlation between L_{151} and L_{NLR} for 3C objects with published optical data; see Rawlings & Saunders (1991) for details of the calculation of L_{NLR} . Dots represent narrow-line radiogalaxies (NLRGs); asterisks, broad-line radiogalaxies (BLRGs); and squares, radioquasars (RQs). Note only one object lacks any emission lines.

Fig. 1 is a classic 'brighter is brighter' relation and as such probably reflects a strong underlying physical link. The basic features of narrow-line emission seem firmly established, with a consensus (e.g. Baum & Heckman 1989) that the availability of photoionizing photons from the active nucleus controls L_{NLR} (but see Sec. 4.1). The implication is of a link between the photoionizing UV/X-ray luminosity $L_{\rm phot}$ and L_{151}, which is initially surprising as these luminosities result from very different processes, and arise from very different parts of the radiosource.

2. THE UNDERLYING LINK BETWEEN JET POWER AND PHOTOIONIZING LUMINOSITY

Recall that L_{151} arises from synchrotron-emitting relativistic particles stored, with magnetic field, in the radiosource lobes. Using basic physics, L_{151} can be related (Fig. 2a) to the bulk-kinetic power Q provided by the jets, and measurable properties of the radiosource and its gaseous environment (Rawlings & Saunders 1991).

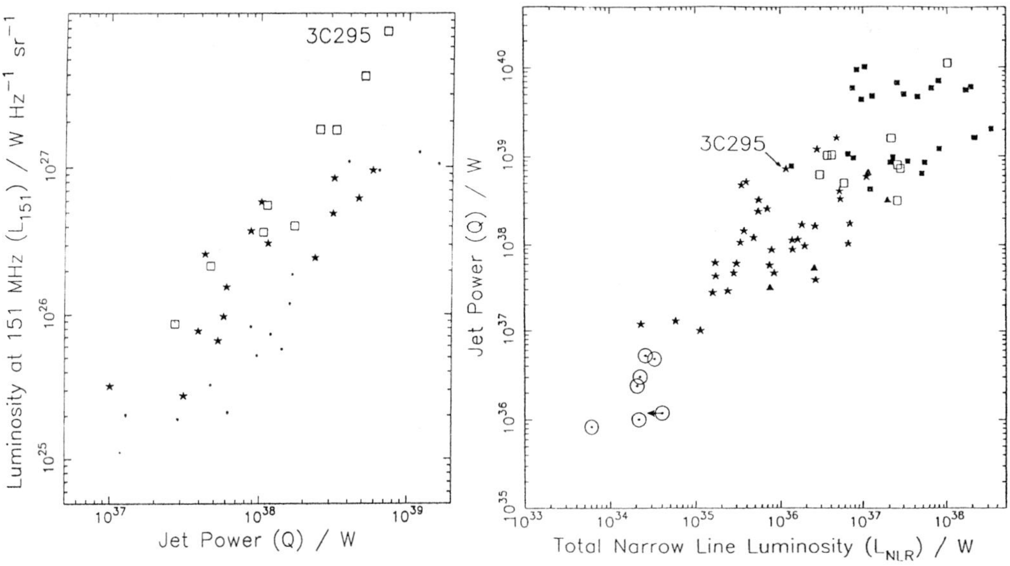

Figure 2. (a) The relation between L_{151} and Q for $z < 0.5$ FRII NLRGs; those confined by cluster gas ($n_e \gtrsim 10^3$ m^{-3}) are represented by squares, those in groups by asterisks, and isolated radiosources ($n_e \lesssim 10$ m $^{-3}$) by dots. (b) The Q–L_{NLR} relation from Rawlings & Saunders (1991); FRIs are represented by circles, $z < 0.5$ FRII NLRGs by asterisks, BLRGS by triangles, $0.5 < z < 1$ NLRGs by filled squares, and $z < 1$ RQs by open squares.

L_{151} is thus primarily an indicator of Q but also reflects the confining electron number density n_e. At fixed Q, an object has significantly higher L_{151} if n_e is high: 3C295 for example is in a very rich cluster. Bolometric radio luminosity is always a small fraction (1-10%) of Q, a fact often forgotten because Q does not itself produce photons.

Converting from L_{151} to Q produces Fig. 2b. The relation is tighter than that of Fig. 1: see e.g. the relative positions of 3C295, implying that it is the very strong link between $L_{\rm phot}$ and Q that underpins Fig. 1. Implications and applications of this result are discussed by Saunders (these proceedings). I will pursue just one aspect of Fig. 2b: that RQs and BLRGs follow the same trend defined by the NLRGs.

3. THE CONE-AND-BEAM UNIFIED SCHEME

Orientation-based unified schemes (see e.g. Barthel 1989; Antonucci 1989) propose that the angle between the radio axis and the line-of-sight (LOS) influences whether a given object is classified as a RQ, BLRG, NLRG or blazar. In such schemes objects should obey similar correlations between orientation-independent quantities, but become dispersed on plots involving orientation-dependent ones. I shall argue that Figs. 1 and 2b are examples of the former, and Fig. 3 of the latter.

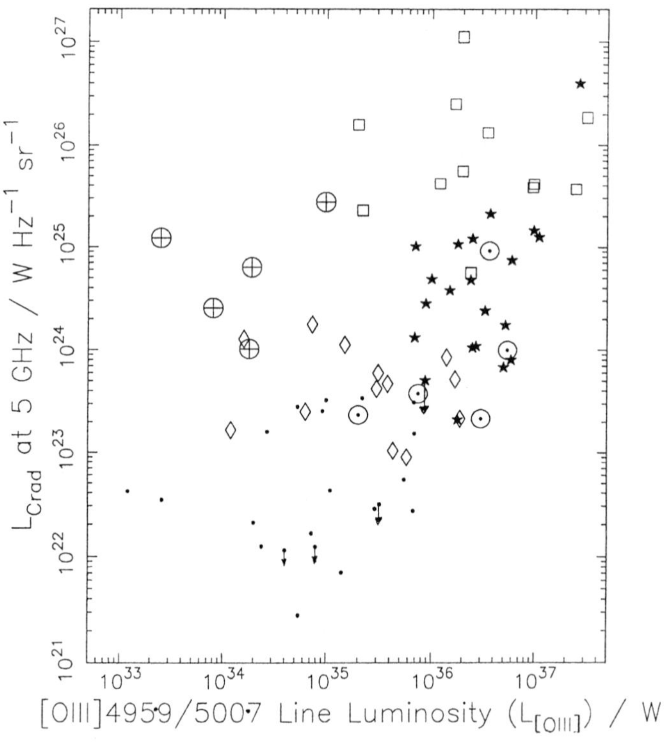

Figure 3. Separation of different classes of radiosources in the radio central component luminosity L_{Crad} – [OIII] luminosity plane: diamonds represent FRII NLRGs with luminous host galaxies ($M_K \approx -26.3$); dots, other less luminous $z < 0.5$ FRII NLRGs; circles, BLRGs; asterisks, RQs; squares, core-dominated quasars; and crossed-circles, BL Lacs.

3.1 What causes orientation dependence?

One cause of anisotropic radio/optical emission, the Doppler-beaming of synchrotron radiation from the base of a relativistic jet, is discussed elsewhere (e.g. Fosbury, these proceedings). In samples selected at low radio frequency, object properties are rarely influenced by this process because the beam has a low opening half-angle, $\theta_b \sim 10°$.

But as well as this beam there is second cause of orientation-dependent optical emission involving a cone with a much larger half-angle θ_c. Observations of the archetypal NLRG Cyg A by Tadhunter *et al.* (1991), Ward *et al.* (1991) and Djorgovski *et al.* (1991) reveal a hidden quasar which illuminates the narrow-line gas and scattering media within a cone whose axis is aligned with the jet. This quasar is hidden by a LOS-extinction of ~50 mags in V, but since we still see nuclear X-rays and radio emission, obscuration is probably by a dusty torus. Do all NLRGs have these tori? For objects of similar Q to Cyg A, the answer is almost certainly yes: a common feature of these generally high-z objects is a close alignment of the narrow-line and radio structures, with the narrow-line gas sometimes beyond the lobes, implying a broad cone of hidden photoionizing luminosity (McCarthy *et al.* 1988) as in Cyg A. At lower Q, evidence for obscuring tori is less direct but still strong: some objects have both their optical continua and broad lines highly-polarized, at a position angle (PA) perpendicular to the radio axis (Antonucci 1989); and also many objects obey Fig. 2b while appearing to lack a non-stellar source bright enough to provide the necessary photoionizing photons. The dusty torus should, of course, radiate in the far-infrared, but any meaningful comparison of RQs and NLRGs is at present too difficult due both to the inadequate sensitivity of IRAS, and to the uncertain contribution of dust heated by non-nuclear processes such as starbursts.

3.2 Do all radioquasars have these tori?

A convincing argument that they do comes from the relation between $L_{151}$and $L_{C_{rad}}$, which is anisotropic due to Doppler boosting. If any RQs lacked tori, then RQs and NLRGs would overlap in the $L_{C_{rad}}$– L_{151} plane, which they don't. Similar arguments apply to other dichotomies in the radio properties of RQs and NLRGs, namely the presence/absence of large-scale one-sided jets and associated differences in radio depolarization (e.g. Garrington & Conway 1991; Pedelty *et al.* 1989).

3.3 What is θ_c?

By considering the relative numbers of RQs and NLRGs in 3CR with redshifts $0.5 \leq z \leq 1$, Barthel (1989) proposed that $\theta_c \sim 45°$, a value broadly consistent with the comparative linear sizes and arm-length asymmetries of RQ and NLRG (Barthel 1989; Kapahi 1991). Significantly, a recent search for objects larger than the 1.6-Mpc RQ 4C74.26 (Riley *et al.* 1988), found several NLRGs but no RQs (Dingley 1990).

3.4 When is a radiosource a radioquasar?

If one accepts the premise that all radiosources contain central quasars, while allowing for a spread in $L_{\rm phot}$, the important distinction is whether the observed non-stellar continuum is dominated by direct (largely unattenuated) quasar light (a naked radioquasar; NRQ) or by light that reaches us only after scattering. Although this distinction is difficult (all NRQs have similar narrow-line equivalent widths, but for other RQs, spectropolarimetry is required to differentiate between scattered and attenuated light), it is clear that some BLRGs are naked (e.g. 3C382) but others (e.g. 3C234) are not, and the latter need not have their cones pointing at us. The light of some low-z NRQs may be swamped by the starlight of their host galaxies: several low-z 3C objects have non-stellar light (Yee & Oke 1978) but no firm evidence that it is scattered (c.f. Cyg A). Some $z > 0.5$ 3C RQs may be seen only in scattered light since, assuming a scaling between $L_{\rm phot}$ and Q and similar scattering efficiency, starlight is more easily swamped as Q increases. It should also be emphasized that the probability that a given object is classified as a RQ/BLRG rather than an NLRG is positively correlated with both θ_c and $L_{\rm phot}$ and, given ineluctable object-to-object variation in θ_c and $L_{\rm phot}$ at fixed Q (scatter in Fig. 2b), insidious biases are possible.

It is thus hard to assess the significances of (i) the positioning of the BLRGs (which lie on one edge of the correlation) in Fig. 2b; and (ii) the small apparent fraction of NRQs at low Q and z. Both effects could be used to argue against the simple single-θ_c model, but may have other interpretations. For example, Jackson & Browne (1990), who find lower [OIII] luminosities for NLRGs than for similar-L_{151} RQs, suggest some anisotropy of [OIII] due to partial obscuration by the dusty torus. I discuss the apparently low fraction of low-Q, low-z NRQs in Sec. 4.2.

4. THE EFFECTS OF ENVIRONMENT AND COSMIC EPOCH

4.1 Host galaxies at low redshift

Radiosource models by Bicknell (these proceedings) predict a slowing of the jet, and an accompanying enhancement of the steep-spectrum component of L_{Crad} over the synchrotron-opaque component if there is a strong interaction with the ISM of the host galaxy. Some, but not all, nearby FRII NLRGs have lower stellar luminosities than the D/cD galaxies typically associated with FRIs (Owen & Laing 1989). Thus FRIs and those FRIIs associated with D/cD galaxies may have L_{Crad} boosted because of properties of the host galaxy. Referring back to Fig. 3, we see some evidence for this effect from the similar, high values of L_{Crad} for those NLRGs with luminous host galaxies.

Bicknell, however, extends this model to explain Fig. 2b, employing photoionization via radiative shocks due to jet/ISM interaction. The case for photoionization by a nuclear source unassociated with the radio jet is made most strongly by the remarkable homogeneity of the narrow-line/continuum properties of radio-quiet and radio-loud quasars (e.g. Miller, these proceedings), and Rawlings *et al.* (1989) present further arguments. The tightness of Fig. 2b, in which most objects plotted show no signs of jet/ISM interaction, argues that shock-related ionization always plays a minor rôle.

4.2 High-redshift environments

Liu (these proceedings) and McCarthy *et al.* (1991) demonstrate correlated radio/optical asymmetries in luminous high-z NLRGs with more prominent extranuclear narrow-line gas lying on the side of the closer, steeper (spectral index), more-depolarized hotspot. The patchy, asymmetric environments implied by these results are in contrast with the smooth X-ray profiles and symmetric radio structures at low z. Radiosource environments are therefore different at $z \sim 0.5$ than locally (see also Hill & Lilly 1991). Systematic changes of θ_c with z seem plausible, and may perhaps be required to explain the lack of low-z NRQs. At $z \gtrsim 2$ there is already some evidence that θ_c must be lower than 45°, since the objects found at these redshifts by low-frequency radio-selected samples (e.g. Chambers *et al.* 1987) are NLRGs. Supporting this view are the quasar-like radio properties of some high-z NLRGs: e.g. the $z = 3.4$ NLRG B2 0902+34 has a strong core and a prominent one-sided jet (van Breugel & McCarthy 1990).

One remaining puzzle is why environments at high z appear patchy when $z \sim 0.5$ RQs lie in clusters of galaxies (e.g. Yee & Green 1987), and when the interpretation of the Laing-Garrington effect (Garrington & Conway 1990; Conway, these proceedings) postulates Faraday depolarization by uniform X-ray haloes. Recent results of VLB polarimetry (e.g. Cawthorne, these proceedings), suggest a new approach to this problem. If $z \sim 0.5$ RQs have X-ray haloes similar to those of nearby clusters, then there will be a large Faraday depth toward the radio central component. Mapping this component with VLBI should reveal jet blobs as well as the synchrotron-opaque core; there should be a rotation of the PAs of polarization of all components without any necessity for depolarization. No correlation would then be evident between the radio axis and the PAs of polarization in the blobs. Such a correlation *is* seen in BL Lac objects (Gabuzda, these proceedings) and its existence may constrain the properties of their magnetoionic haloes and their unification with FRI NLRGs.

Acknowledgements

It is my great pleasure to thank the organizers for arranging financial support, and for hosting a superb meeting. I acknowledge the support of a Research Fellowship at St John's College, Cambridge. I also thank Philip Miller and Richard Saunders.

REFERENCES

Allington-Smith, J.R., Peacock, J.A. & Dunlop, J.S., 1991. *Mon. Not. R. astr. Soc.*, in press.

Antonucci, R., 1989. In: *Fourteenth Texas Symposium on Relativistic Astrophysics. Proceedings.*, p. 180, ed Fenyves, E.J., N.Y. Academy of Sciences, N.Y., N.Y.

Barthel, P.D., 1989. *Astrophys. J.*, **336**, 606.

Baum, S.A. & Heckman T., 1989. *Astrophys. J.*, **336**, 681.

Chambers K.C., Miley G.K. & van Breugel W., 1987. *Nature*, **329**, 604.

Dingley, S.J., 1990. PhD thesis, University of Cambridge.

Djorgovski, S., Weir, N., Matthews, K. & Graham, J.R., 1991. *Astrophys. J.*, **372**, L67.

Garrington, S.T. & Conway, R.G., 1991. *Mon. Not. R. astr. Soc.*, **250**, 198.

Hill, G.J. & Lilly, S.J., 1991. *Astrophys. J.*, **367**, 1.

Hine, R.G. & Longair, M.S., 1979. *Mon. Not. R. astr. Soc.*, **188**, 111.

Jackson, N. & Browne, I.W.A., 1990. *Nature*, **343**, 43.

Kapahi, V.K., 1990. In: *Parsec scale jets*, p. 304, eds Zensus, J.A. & Pearson, T.J., Cambridge University Press.

McCarthy, P.J., & van Breugel, W., 1989. In: *Extranuclear Activity in Galaxies, ESO Workshop*, p. 55, eds Meurs, R.A.E. & Fosbury, R.A.E., ESO Conference and Workshop Proceedings No. 32.

McCarthy, P.J., van Breugel, W., & Kapahi, V.K., 1991. *Astrophys. J.*, **371**, 478.

McCarthy P.J., Spinrad, H., van Breugel W., Djorgovski, S., Strauss M.A. & Dickinson M., 1988. In: *Cooling Flows in Clusters and Galaxies*, p. 325, ed Fabian A.C., Cambridge University Press.

Owen, F.N. & Laing, R.A., 1989. *Mon. Not. R. astr. Soc.*, **238**, 357.

Pedelty, J.A., Rudnick, L., McCarthy, P.J. & Spinrad, H., 1989. *Astr. J.*, **97**, 647.

Rawlings, S. & Saunders, R., 1991. *Nature*, **349**, 138.

Rawlings, S., Saunders, R., Eales, S.A. & Mackay, C. D., 1989. *Mon. Not. R. astr. Soc.*, **240**, 701.

Riley, J.M., Warner, P.J., Rawlings, S., Saunders, R., Pooley, G.G. & Eales, S.A., 1988. *Mon. Not. R. astr. Soc.*, **236**, 13p.

Tadhunter, C.N., Scarrott, S.M. & Rolph, C.D., 1990. *Mon. Not. R. astr. Soc.*, **246**, 163.

van Breugel, W.J.M. & McCarthy, P.J., 1990. In: *The Evolution of Galaxies: The Hubble Centennial Symposium*, eds Kron, R.G., Publ. Astron. Soc. Pac. Conf. Series

Ward, M.J., Blanco, P.R., Wilson, A.S. & Nishida, 1991. *Astrophys. J.*, in press.

Yee, H.K.C. & Green, R.F., 1987. *Astrophys. J.*, **280**, 79.

Yee, H.K.C. & Oke, J.B., 1978. *Astrophys. J.*, **226**, 753.

Radio emission in radio-loud and radio-quiet quasars of the BQS.

PHILIP MILLER

Mullard Radio Astronomy Observatory, Cavendish Laboratory, Madingley Road, Cambridge, CB3 0HE, UK

INTRODUCTION

In radiogalaxies and radioquasars there exists a correlation between the luminosity of narrow lines and the power emitted from the extended radiosource — a correlation which occurs because of the more fundamental relation between the narrow-line luminosity and the bulk kinetic power in the jets (Rawlings & Saunders 1991). We have since been extending our study of such correlations to optically-selected active galactic nuclei in an attempt to determine the physical basis for the enormous range in radio properties which optically-similar AGN display.

SAMPLE AND SOURCES OF DATA

Our chosen sample for this work is the low-redshift ($z < 0.5$) subsample of the Bright Quasar Survey (BQS; Schmidt & Green 1983, Green, Schmidt & Liebert 1986). As outlined in Miller (1991), we have made new observations of, and gathered data from the literature on, the narrow [O III]5007 and broad Hβ lines of our 89 objects; and have obtained structural information from our own radio maps of the VLA observations of Kellermann *et al.* (1989). IRAS fluxes of many of the objects are available in Sanders *et al.* (1989).

DISCUSSION

The sample displays a remarkable homogeneity in optical properties (Miller *et al.* 1991) while containing objects of very different radio structures and luminosities. In the preliminary report of Miller (1991), the separation of the radio-loud and radio-quiet quasars in the plot of total 5 GHz radio luminosity, P, versus [O III] luminosity, $L_{[\mathrm{OIII}]}$ — although bridged by a small number of intermediate-luminosity objects — could be clearly seen. A first interpretation of the tightness of the relation between P and $L_{[\mathrm{OIII}]}$ for the radio-quiet ($P \lesssim 10^{23}\,\mathrm{W\,Hz^{-1}\,sr^{-1}}$) quasars (RQQs) was that it pointed towards the radio-emission being of AGN origin (since we expect $L_{[\mathrm{OIII}]}$ to be an orientation-independent measure of the photoionizing capacity of the central engine). However, the apparent non-linearity in the relation for these objects ($P \propto L_{[\mathrm{OIII}]}^{\sim 1.3}$; orthogonal regression) left this open to some doubt.

The more-closely linear relation for RQQs in fig. 1 — though badly affected by upper limits — has suggested to Sopp & Alexander (1991) the alternative hypothesis that the radio-emission in many of these quasars comes principally from supernovæ in the host galaxy. Indeed, our analysis has revealed that most of those RQQs which show evidence of small jet-like features at 5 GHz and/or have the highest core-to-extended radio luminosity ratios lie on the upper left side of the majority of RQQs on the $P - L_{\mathrm{FIR}}$ plot; due, presumably, to an AGN component of the radio emission which many of the RQQs lack.

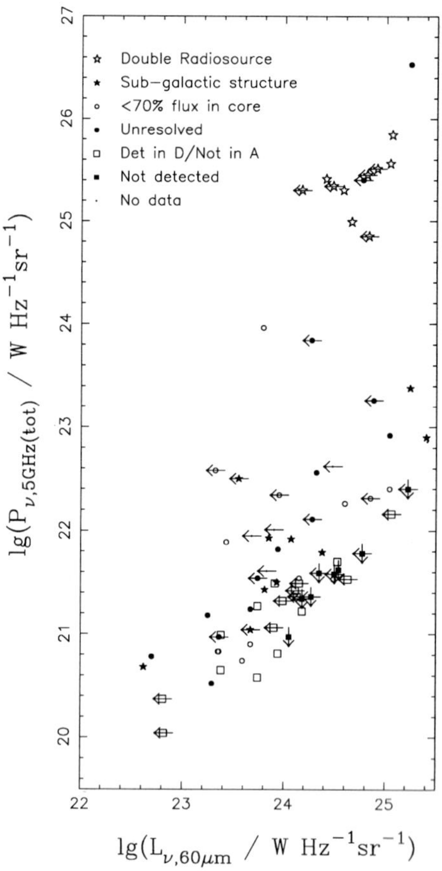

Figure 1. Monochromatic radio luminosity at 5 GHz (~18″ FWHM beam) versus monochromatic far-infrared luminosity at 60 μm with symbols for high-resolution (~0.4″ FWHM beam) radio-structure.

REFERENCES

Green,R.F., Schmidt,M. & Liebert,J., 1986. *Astrophys. J. Suppl.*, **61**, 305.

Kellermann,K.I. *et al.* , 1989. *Astr. J.*, **98**, 1195.

Miller,P.G., 1991. In: *The Physics of Active Galactic Nuclei*, eds Duschl,W.J. & Wagner,S.J., Springer-Verlag.

Miller,P.G., Rawlings,S., Saunders,R. & Eales,S.A., 1991. *M.N.R.A.S.*, in press.

Rawlings,S. & Saunders,R., 1991. *Nature*, **349**, 138.

Sanders,D.B. *et al.* , 1989. *Astrophys. J.*, **347**, 29.

Schmidt,M. & Green,R.F., 1983. *Astrophys. J.*, **269**, 352.

Sopp,H.M. & Alexander,P., 1991. *M.N.R.A.S.*, **251**, 14P.

ACKNOWLEDGEMENTS

I wish to thank my collaborators Steve Rawlings and Richard Saunders. We are grateful for the award of observing time on the University of Hawaii 2.2m telescope, to Dick Sramek for permission to use the calibrated VLA data of the BQS observations, and to Paddy Leahy for advice and assistance with AIPS reduction at the VLA.

OPTICAL FILAMENTS IN CENTAURUS A

D. TAYLOR[1], R. MORGANTI[2], R.A.E. FOSBURY[3†]

[1] *European Southern Observatory, Karl Schwarzschild Strasse-2, D-8046 Garching-bei-München, Germany*
[2] *Instituto di Radioastronomia, via Irnerio 46, Bologna, Italy*
[3] *Space Telescope - European Coordinating Facility, Karl Schwarzschild Strasse-2, D-8046 Garching-bei-München, Germany*

1 THE OPTICAL FILAMENTS

The inner optical filaments in the nuclear region of the radio galaxy Centaurus A lie along the same axis as the radio jet but at a distance of ~10 kpc from the nucleus, which is beyond the inner radio lobe (see Fig. 1 of Morganti *et al.* 1991). They are composed of a number of compact knots, with dimensions of ~100 pc, along with some more diffuse structure. The strength of the radiation field required to account for the ionisation state of the gas, in the 2–6 keV X-ray band, is around 200 times greater than that observed directly from the nucleus. It has therefore been proposed (Morganti *et al.* 1991) that the nuclear continuum is relativistically beamed towards the filaments, with a power similar to that of BL Lac. Newly obtained [OIII]/Hα images (Morganti *et al.* in prep., and also Fig.4 of Fosbury, these proceedings) provide direct evidence that the filaments are indeed ionised by the nuclear radiation field. They show that the inner knots have highest excitation on the edges directly facing the nucleus. The more diffuse filaments show a more uniform excitation, which is consistent with the interpretation that they are optically thin.

The knots have velocities which are a few hundred km s^{-1} below the systemic velocity of the central region, and significant velocity gradients ($\lesssim 300\,\mathrm{km\,s^{-1}}$) are observed over individual knots (Morganti *et al.* 1991). We now consider whether the gas can be accelerated to such velocities by either the radiation beam or by a particle jet.

2 ACCELERATION BY RADIATION PRESSURE

Acceleration of the clouds by the photon beam will occur principally by the photoionisation of hydrogen. Each ionisation imparts $\frac{h\nu}{c}$ momentum to the cloud, where $h\nu$ is the typical photon energy, which we take to be 13.6 eV. The acceleration of an optically thick cloud of radius R and density n in a beam of L ionising photons

† Affiliated to the Astrophysics Division, Space Science Department, European Space Agency

$\mathrm{cm}^{-2}\,\mathrm{s}^{-1}$ is thus

$$\frac{dv}{dt} = \frac{h\nu}{mc}\frac{L}{Rn},$$

where m is the mass of the hydrogen atom. For the Cen A filaments, we have typical values of $R \sim 10^2$ pc, $n \sim 10^2\mathrm{cm}^{-3}$ and $L \sim 10^9\,\mathrm{cm}^{-2}\,\mathrm{s}^{-1}$. The timescale required to accelerate the clouds to the observed velocities of $\sim 10^2\,\mathrm{km\,s}^{-1}$ is thus $\sim 10^9$ yrs, which is unrealistically long. Furthermore, this neglects the disruption of the cloud by the photon beam. Assuming the neutral gas to be in pressure equilibrium with its surroundings, the ionised gas, with a temperature of $\sim 10^4$ K, will be substantially over-pressured and will expand with a velocity of the order of its sound speed, $\sim 10\,\mathrm{km\,s}^{-1}$. We can estimate the maximum velocity attainable by an optically thick cloud by considering the acceleration of a 2D slab of gas under the assumption that the density of the ionised layer is uniform at any instant. We derive the following expressions for the time t, velocity v and density n at which the slab becomes optically thin:

$$t \sim 80\,(n_o R_o)^2 \text{ yrs}; \qquad v \sim 4\times 10^{-3}\, n_o R_o\,\mathrm{km\,s}^{-1}; \qquad n \sim 10^3\,(n_o R_o)^{-1}\,\mathrm{cm}^{-3},$$

where R_o is the initial thickness of the slab in pc and n_o its initial density in cm^{-3}.

As the clouds have a density $\sim 10^2\,\mathrm{cm}^{-3}$ and are still optically thick, this gives an upper limit of $n_o R_o \sim 10$. The velocity attained by the clouds as a result of radiative acceleration is thus $\lesssim 0.1\,\mathrm{km\,s}^{-1}$. Even this value is probably an overestimate as it neglects the disruption of the clouds by the Lyα photons produced by the recombinations within the cloud. We conclude that radiation pressure cannot be responsible for the acceleration of the filaments.

3 ACCELERATION BY A PARTICLE JET

It has been shown that a cloud cannot be accelerated by ram-pressure to more than about its internal sound speed before it suffers severe disruption (e.g. Nittmann *et al.* 1982, Allen 1984). Here we consider the possibility that the $\sim 10^2\,\mathrm{km\,s}^{-1}$ velocities occur in the filaments as they are being destroyed by a particle jet.

The equation of motion of a gas cloud of density n and radius R, in a jet of density n_j and velocity v_j is $\frac{dv}{dt} \approx \frac{3}{4}\frac{n_j}{n}\frac{v_j^2}{r}$. The ram pressure will produce a pressure difference across the cloud of the order $n_j v_j^2$, resulting in an expansion of the cloud at the associated sound speed, $\frac{dR}{dt} \approx \gamma \left(\frac{n_j}{n}\right)^{\frac{1}{2}} v_j$, (Nittmann *et al.* 1982). The cloud will not remain spherically symmetric and **homogeneous** throughout its expansion, so n and R must represent some average density and radius respectively. If nR^3 is constant, integration of the equation of motion yields

$$v(n) \approx \frac{1}{2\gamma^{1/2}} v_j \left(\left(\frac{n_j}{n}\right)^{\frac{1}{2}} - \left(\frac{n_j}{n_o}\right)^{\frac{1}{2}} \right) \quad \text{and} \quad t(r) \approx \frac{2}{\gamma^{1/2}} \frac{R_o}{v_j} \left(\frac{n_o}{n_j}\right)^{\frac{1}{2}} \left(1 - \left(\frac{R_o}{r}\right)^{\frac{1}{2}}\right) ,$$

where R_o and n_o are the initial values of the radius and density. Thus, assuming $R >> R_o$ and $n << n_o$, we get

$$v(n) \sim v_j \left(\frac{n_j}{n}\right)^{\frac{1}{2}} \quad \text{and} \quad t \sim \frac{R_o}{v_j} \left(\frac{n_o}{n_j}\right)^{\frac{1}{2}} .$$

Adopting the 'best guess' values $n_j \sim 10^{-2}\,\mathrm{cm}^{-3}$ and $v_j \sim 5\,000\,\mathrm{km\,s}^{-1}$ from Burns *et al.* (1983), gives

$v \sim 50\,\mathrm{km\,s}^{-1}$; $n = 10^2\,\mathrm{cm}^{-3}$ $v \sim 150\,\mathrm{km\,s}^{-1}$; $n = 10\,\mathrm{cm}^{-3}$ $v \sim 500\,\mathrm{km\,s}^{-1}$; $n = 1\,\mathrm{cm}^{-3}$.

It is thus possible that the velocities observed could be due to ram pressure accelerated gas with a density of $\sim 10\,\mathrm{cm}^{-3}$. In this case we would expect the velocity gradients across each knot to be of the same order as their bulk velocities, and that the highest velocities should occur at the edges of the knots. The knots in Cen A appear to display both phenomena.

This does, of course, raise the problem that no radio emission is detected at the position of the filaments. We suggest that the NE radio jet was at some time in the past cut off by an arm of the dust band, in which the filaments were embedded. In this picture, the portion of the jet which accelerated the clouds is now observed as the outer radio lobe. Some evidence that the radio jet has indeed been deflected is the sharp edge of the inner radio lobe, on the side facing the optical filaments. There may also be a radio 'hot-spot' on this edge (Burns *et al.* 1983).

REFERENCES

Allen, A. J., 1984. *Mon. Not. R. astr. Soc.*, **210**, 147.

Burns, J.O., Feigelson, E.D. & Schreier, E.J., 1983. *Astrophys. J.*, **273**, 128.

Morganti, R., Robinson, A., Fosbury, R.A.E., di Serego Alighieri, S., Tadhunter, C.N. & Malin, D.F., 1991. *Mon. Not. R. astr. Soc.*, **249**, 91.

Nittmann, J., Falle, S.A.E.G. & Gaskell, P.H., 1982. *Mon. Not. R. astr. Soc.*, **201**, 833.

Jet kinetic power and waste heat in radiogalaxies and quasars.

RICHARD SAUNDERS

Mullard Radio Astronomy Observatory, Cavendish Laboratory, Madingley Road, Cambridge, CB3 0HE, UK

1. INTRODUCTION.

Rawlings (these proceedings) has discussed the finding (Rawlings & Saunders 1991) of a tight correlation between jet bulk kinetic power and narrow-emission-line luminosity in double radiosources. I will comment on this further, but I first emphasize the basic result, and some cogent points that are not widely appreciated.

2. A BASIC RESULT CONCERNING JET KINETIC POWER.

You can calculate the bulk kinetic power Q in jets (which may themselves be invisible) by looking at the products of the jets, the large-scale lobes, and by then using

$$Q = \frac{kU}{\tau}$$

Here U is the stored energy in the lobes, for which the equipartition energy U_{eq} is an easy-to-obtain estimate; k, which is ~ 2, allows for factors such as $p.dV$ work expended by the jet in pushing back and warming up the extended medium; and τ is the age of the radiosource. There are two ways of estimating τ. The first is to use the radio spectral age, which for FR-IIs is both well-defined observationally and closely related to the true age — you simply make an allowance for backflow. There are, as yet, few sources for which the requisite multifrequency maps are available, however, so that you often have to use the second method. This involves combining data you can get, such as radio luminosity, lobe linear size and extended density (e.g. from X-ray studies) with a source model which relates these parameters to τ. Models we have used for FR-IIs are the analytic model A of Scheuer (1974) with a treatment for a more realistic confining density profile, and the numerical simulations of Williams (1985) with simple treatments to allow estimation of radio observables from the calculated fluid densities and pressures. There is good agreement (with a scatter of a factor of two) in the τ-predictions based on the two kinds of model, and indeed between these and the estimates based on spectral age in some ten 3C objects for which there are sufficient data, giving us confidence in the overall method.

Material radiatively excited by an AGN cools by line emission. Radiogalaxies are usually narrow-line emitters and so for these, using the established close relationships between two lines (e.g. [O II]3727 and [O III]5007) of different ionization and the total narrow-line luminosity $L_{\rm NLR}$ in all narrow lines, you can estimate $L_{\rm NLR}$ by an unambiguous procedure. Then $L_{\rm NLR}$ is related to the photoionizing luminosity $L_{\rm phot}$ by $L_{\rm NLR} = \kappa L_{\rm phot}$, where κ is the covering factor.

So we obtained radio and optical spectrophotometric data for an unbiased sample of 39 FR-II 3C radiogalaxies with $z < 0.5$ and estimated Q and $L_{\rm NLR}$ for them, and plotted these values. (We have *both* values for *all* 39 objects — no missing points or upper limits). We did the same for the low-power FR-Is (using radio spectral data and the Laycock (1987) model for twin-jets) and for the FR-IIs (galaxies and quasars) at $z < 1$ in 3C: data were available for 40 of the 54 objects. The plot is fig. 2(b) of Rawlings (these proceedings). We emphasize again that Q is the bulk power of the jets — it is not the radio luminosity of the jet or of the lobes or of anything else. We find that

- Q is proportional to $L_{\rm NLR}$ (and so perhaps Q is proportional to the accretion rate $\dot{M}$); this applies to FR-I and FR-II radiogalaxies and to radioquasars.
- $Q \sim 100\ L_{\rm NLR}$.

To get $L_{\rm phot} > Q$ would need either a covering factor of < 0.01, smaller than seen in quasars, or a narrow-line gas density very uncomfortably high given the low ionization parameters observed for the bulk of our objects. *Evidently, the powering of jets is very efficient*, for even if $L_{\rm phot}$ equals Q, tightly collimated jets are produced with remarkably little extraneous fuss.

3. POSSIBLE PROBLEMS WITH ENERGY BUDGETING.

These conclusions could be faulty if we've missed a significant energy term. I next consider the possible extra components we have thought of.

1. There is more energy in the radio lobes or the shocked environment. Probably this is true. But since we used equipartition energy (essentially the minimum value) and assumed no fast protons, almost anything we change increases Q and jet formation has to be even more efficient. The only exception is if we decrease the lobe filling factor (which we take as 1), but then, as pointed out by Leahy (1990), U and spectral age change in direct proportion, so Q is unchanged.
2. We have accounted for all the X-ray and UV radiation (because $L_{\rm phot}$ represents all the luminosity reaching the NL-region that is capable of ionizing anything) but perhaps we have missed radiation coming off in the IR (e.g. reprocessed γ-rays). This would appear not to be the case: IRAS observations of some 3CR radiogalaxies (Yates & Longair 1989) show IR excesses, but even if these are nuclear in origin, they represent IR luminosities very small compared to Q.

3. Given the current, successful orientation-models, what about the broad-line (BL) and non-stellar continuum emission hidden from us? My first thoughts about this were that some geometries may allow BL gas to see more ionizing flux than the NL gas, making analysis problematical. But this problem can be bypassed by studying *quasars*, from say the BQS, and comparing bolometric luminosity $L_{\rm Bol}$ values for radiogalaxies and BQS quasars at constant $L_{\rm [OIII]}$; $L_{\rm [OIII]}$ being orientation-independent (Rawlings & Saunders 1989). We use the notion, implicit in orientation-dependent models, that $L_{\rm Bol}$ for radiogalaxies $\leq L_{\rm Bol}$ for quasars (loud or quiet) at constant $L_{\rm [OIII]}$; this notion is supported by homogeneity of narrow-line/continuum properties of radio-loud and -quiet quasars (Miller, these proceedings). Miller *et al.* (1991) have suitable spectra of the BQS for $z < 0.5$. And you find that if you take $L_{\rm [OIII]}$ for any particular radiogalaxy, and you then take the *maximum* continuum luminosity in the BQS for this $L_{\rm [OIII]}$ value, and add to this the BL contributions and indeed all the luminosity components from X-ray to FIR (using the data in Sanders *et al.* 1989), the resultant luminosity is *about the same* as $L_{\rm phot}$ deduced from $L_{\rm NLR}$ and a covering factor of 0.01. So the fact that we do not see broad-lines in most radiogalaxies does not change the argument about jet production efficiency.
4. Radio-quiet quasars obviously do not lie on the Q-$L_{\rm NLR}$ correlation. Does this cast doubt on our assumption of photoionization? I think not, since the evidence for photoionization is very strong (e.g. Rawlings *et al.* 1989, Baum & Heckman 1989, Miller, these proceedings): for the BQS, for both radio-loud and radio-quiet objects, we find $L_{\rm H\beta} \propto L_{\rm [OIII]} \propto L_{\rm cont}$ which implies photoionization, probably with $L_{\rm phot} \propto \dot{M}$. So the radio-loud objects do have photoionizers, and for these, the facts that $Q \propto L_{\rm phot}$ *and* $Q \gtrsim L_{\rm phot}$ tell us something extra, namely that the jets are driven by a device that *controls* $\dot{M}$.
5. We've looked at the energy that doesn't go into the jets and is lost as radiation, but what about some other energy loss channel? The only plausible one I can think of is gas kinetic energy. The NL gas must have the most energy (the mass of the BL gas is insignificant in comparison); the relation between NL gas and "jets" in Seyferts (Whittle *et al.* 1988) also suggests NL gas may be important. However, taking numbers for the nucleus of a moderate-power FR-II ($n \sim 10^{10}\,m^{-3}$, filling factor ~ 0.001, $v_{\rm rms} \sim 500$ km s^{-1}, and $r \sim 1$ kpc) gives a KE of 2×10^{50} J, smaller then $U_{\rm eq}$ by a factor of at least 30. For the extended NLR, Baum and Heckman (1989) find a mass of 10^{39} kg at most, so that the KE is only 10^{50} J. Even if the nuclear and extended NL gas were continuously removed from the scene at 500 km s^{-1} for the whole radiosource lifetime, the NL gas KE still only approaches $U_{\rm eq}$.

So none of these possibilities changes our result that jet powering is very efficient.

4. APPLICATIONS.

4.1 Accretion rates.

(a) In low-Q radiosources: $U_{eq} \sim 10^{53}$ J $\sim 10^6$ $M_\odot$, so that $M_{central} > 10^7$ $M_\odot$ for a plausible efficiency, implying an Eddington luminosity $L_{Edd} \sim 10^{38}$ W. Given the observed L_{NLR}, $L_{phot} < L_{Edd}$ and the accretion is *sub-Eddington.*

(b) In high-Q radiosources: these could be super-Eddington but then we would have to contrive the continuity in the Q-L_{NLR} plot; so in high-Q radiosources the accretion is *probably sub-Eddington* too.

(c) In no-Q objects, i.e. the radio-quiet quasars: assuming the BL region is bound to the central mass, simple (and believable) physics enables $M_{central}$ to be estimated from parameters such as FWHM and L_{opt} (Joly 1987). Applying this to the BQS, we find (Miller *et al.* 1991) that $M_{central} \propto L_{opt}$ and that $L_{Bol} \leq 0.1 L_{Edd}$, implying *sub-Eddington accretion.*

So the jetless objects — the spirals — have $M_{central} \propto L \propto \dot{M}$, while the objects with jets — the ellipticals — have $\dot{M}$ not obviously dependent on $M_{central}$; it may be relevant that all the ellipticals have similar masses.

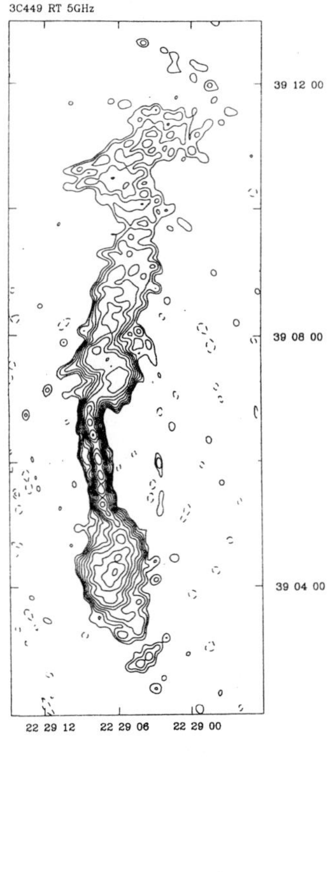

Figure 1. This meeting is about jets — here are two big ones: new Ryle Telescope 5 GHz map of 3C449, showing low-surface-brightness features on scales larger than visible with the VLA. Resolution is 6″; contours are logarithmic, factor 1.5, starting from ±0.5 mJy.

4.2 Jet speeds in twin-jet sources.

Knowing Q from $L_{\rm NLR}$ helps a lot in source physics. For example, twin-jets (e.g. 3C449, fig. 1) are though to be transonic with Mach number $M_{\rm j} \sim 1$ (e.g. Williams & Gull 1984, Bicknell 1984,1986). Combining this notion with Q estimated for 3C31 gives a jet velocity at 30 kpc from the nucleus of c/6 if you take $Q = \frac{1}{2}\rho_{\rm jet}V^3\times$ cross-sectional area, or some 6 times lower if enthalpy and/or Poynting flux are important (see Bicknell, Lovelace, these proceedings). However, the point is that according to most people's expectations of twin jets, these speeds are high.

4.3 Estimating external density in distant FR-IIs.

An estimate of Q from measurement of $L_{\rm NLR}$ is similarly useful here — combined with radio measurements and a source model, you have a good way of measuring environmental densities at redshifts beyond those reachable even by ROSAT.

4.4 Making one-sided jets by smothering a counter-jet.

Depolarization asymmetry in the lobes of radio-sources is proving very interesting (see Conway, these proceedings). One particular result is that of Liu & Pooley (1991; see also Liu, these proceedings) indicating that the shorter lobe has a steeper spectrum, is more depolarized and has more extended [O III] emission. This is (in my view) best interpreted as occurring because the shorter lobe has a higher confining density (Liu & Pooley outline the argument) — such density asymmetries are likely to arise in merger products. But another interpretation is that one jet is off. The timescales are tight for this, and the idea does not explain the [O III] asymmetry; but ignoring these difficulties, and remembering that our beloved central engine models produce the jets, can we smother one of the jets in the nucleus? Our Q-measurements and nuclear luminosities constrain this. If the smothered jet heats something up, the nuclear luminosity is expected to be $\sim Q$; the observations, especially in the IR, tend to rule this out. If the smothered jet power is turned into some other form of bulk power, we might expect the NLR KE to be $\sim Q$. We do not see this. It thus seems jets are not smothered — a one-sided jet has to be *really* one-sided ... all the way to the engine.

REFERENCES

Baum,S.A. & Heckman,T., 1989. *Astrophys. J.*, **336**, 681.

Bicknell,G.V., 1984. *Astrophys. J.*, **286**, 68.

Bicknell,G.V., 1986. *Astrophys. J.*, **305**, 591.

Joly,M., 1987. In: *Emission Lines in Active Galactic Nuclei*, p. 166, ed. Gondhalekar,P.M., Rutherford Appleton Laboratory, Didcot.

Laycock,S., 1987 *PhD thesis*, University of Cambridge

Leahy,J.P., 1990. In: *Parsec scale Jets*, p. 174,
eds Zensus,J.A. & Pearson,T.J., Cambridge University Press, Cambridge.
Liu,R. & Pooley,G.G., 1991. *Mon. Not. R. astr. Soc.*, in press.
Miller,P.G., Rawlings,S.G., Saunders,R.D.E. & Eales,S.A., 1991.
Mon. Not. R. astr. Soc., in press.
Rawlings,S.G. & Saunders,R.D.E., 1989. In: *Active Galactic Nuclei*, p. 305,
eds Osterbrock,D.E. & Miller,J.S., Reidel, Dordrecht.
Rawlings,S.G., Saunders,R.D.E., Eales,S.A. & Mackay,C.D., 1989.
Mon. Not. R. astr. Soc., **240**, 701.
Rawlings,S.G. & Saunders,R.D.E., 1991. *Nature*, **349**, 138.
Sanders,D.B., Phinney,E.S., Neugebauer,G., Soifer,B.T. & Matthews,K. , 1989.
Astrophys. J., **347**, 29.
Scheuer,P.A.G., 1974. *Mon. Not. R. astr. Soc.*, **166**, 513.
Whittle,M., Pedlar,A., Meurs,E.J.A., Unger,S.W., Axon,D.J. & Ward,M.J., 1988.
Astrophys. J., **326**, 125.
Williams,A.G., 1985. *PhD thesis*, University of Cambridge
Williams,A.G. & Gull,S.F., 1984. *Nature*, **310**, 33.
Yates,M.G. & Longair,M.S., 1989. *Mon. Not. R. astr. Soc.*, **241**, 29.

ACKNOWLEDGEMENTS

I wish to thank the organisers and their helpers for arranging an excellent meeting. I am also especially grateful to Philip Miller and Steve Rawlings for sorting out the final text of this paper while I went to Cyprus.

Anisotropic Continuum Radiation In Extragalactic Radio Sources

R. A. E. FOSBURY [1‡], R. MORGANTI [2], A. ROBINSON [3], Z. TSVETANOV [4]

[1] Space Telescope – European Coordinating Facility, Karl Schwarzschild Str. 2, D–8046 Garching bei München, Germany

[2] Instituto di Radioastronomia, via Irnerio 46, Bologna, Italy

[3] Institute of Astronomy, Madingley Road, Cambridge, CB3 0HA, England

[4] Astronomy Program, University of Maryland, College Park, MD 20742, USA

ABSTRACT

Extended emission line regions in Seyfert galaxies often exhibit evidence of broad 'radiation cones' which are usually explained as the result of an *extrinsic* process — selective obscuration of an ionizing radiation field by a nuclear disk. In the two radio galaxies PKS2152–69 and Centaurus A we measure extranuclear emission which must result from a 'misdirected' *intrinsic* beam similar to those hypothesized to emanate from blazar and BL Lac object nuclei respectively. The detection of such beams adds direct observational support to schemes which attempt to unify blazars with FR II radio galaxies – quasars and BL Lac objects with FR Is. It also carries implications for the interpretation of the optical properties of high redshift radio galaxies which are often elongated along their radio axes.

1 INTRODUCTION

There are several pieces of evidence which demonstrate, with varying degrees of directness, that Seyfert and radio galaxies have active nuclei whose output of ionizing radiation cannot easily be inferred from direct optical, UV or X-ray observations. The existence of a 'photon deficit' (see figure 1) in some objects has long been known and can be explained by two distinct hypotheses.

The first of these is *geometric:* the active nucleus is thought to radiate anisotropically so that the extended emission line gas 'sees' a radiation field which is brighter than that viewed by an earthbound observer. The origin of this is most likely to be a direction-dependent

‡Affiliated to the Astrophysics Division, Space Science Department, European Space Agency

obscuration produced by an optically thick nuclear disk; although some level of intrinsic anisotropy may be present as a result of the shape of the radiating 'photosphere'.

The second is *spectral:* the strength of the ionizing part of the radiation field — shortward of the Lyman limit — may be difficult to infer from optical or ultraviolet continuum observations. This is the case if the ionization is produced by a 'thermal bump' in the EUV. To distinguish between photoionization by a hot thermal source and 'non-thermal' power law spectrum, especially if the mean ionizing photon energy is similar, is notoriously difficult (Robinson *et al.* 1987).

Startlingly direct evidence for anisotropy is provided by the 'radiation cones' seen in narrow-band line images, particularly of some Seyfert galaxies (Pogge 1989, Tsvetanov *et al.* 1990, figure 2). Broad line spectra seen in polarized light from some Seyfert 2 galaxies also provide strong support for the unification of the classes 1 and 2 (e.g., Antonucci & Miller 1985, Antonucci 1984).

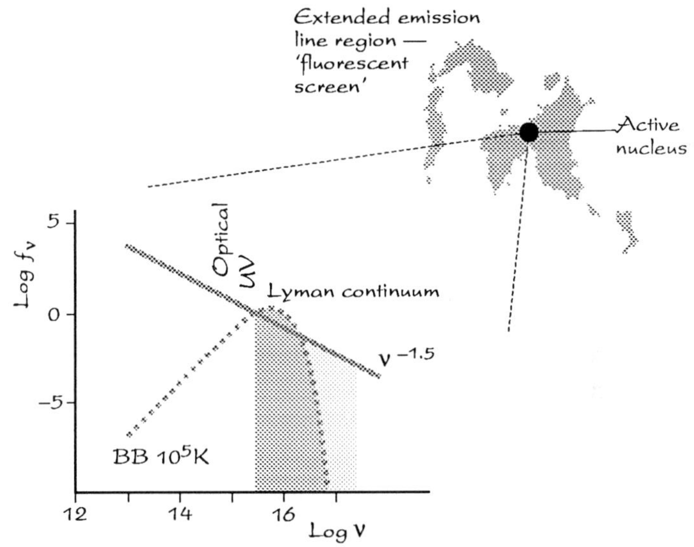

Figure 1 *By considering an extended emission line region as a 'fluorescent screen', spectrophotometric measurements can be used to infer the luminosity of the ionizing spectrum. This often leads to a requirement for photons which is larger than that inferred from direct observations of the active nucleus. Such a 'photon deficit' can be explained by hiding the nucleus in space — a radiation anisotropy — or in energy — producing photons in an EUV bump.*

Evidence for the photon beams, which when directed towards the observer appear as blazars or BL Lac objects, is harder to find. Attempts to identify the parent populations of these apparently luminous, violently variable — in flux and polarization — and radio-superluminal objects have resulted in convincing associations between the blazars and radio quasars – Fanaroff-Riley class II radio galaxies on the one hand and BL Lacertæ objects and FR Is on the other (Orr & Browne 1982, Barthel 1989, Padovani & Urry 1990, 1991, in prep.). If these unifying schemes are correct the beams should be ubiquitous, at least in the radio-loud objects. Although they may radiate only a percent or so of the total luminosity of the AGN — a fraction which appears to be similar over the wide range of luminosity spanned by the FR I through the FR II objects — the photon flux seen by an observer *within* the beam can be very high. This gives us an opportunity to find the misdirected objects by looking for the effect of the beam shining through the surrounding medium. We have to be careful to distinguish between this and local effects produced by the particle jet which powers the extended radio source. Although beams and jets may originate in the same process close to the nucleus, they may not be precisely spatially coincident after travelling some kiloparsecs: jets can be bent, beams cannot.

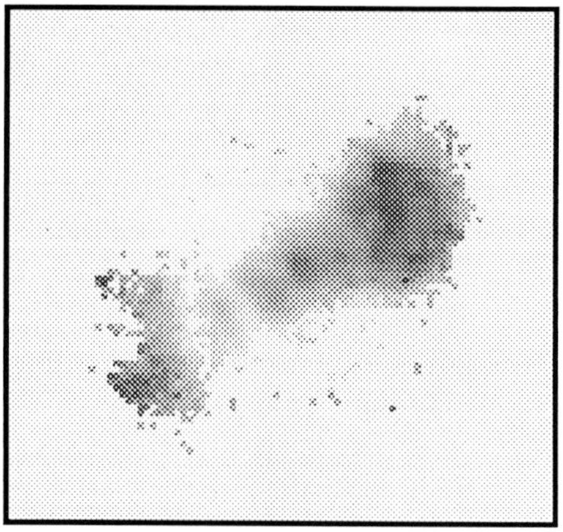

Figure 2 *An [OIII]/Ha ratio image of the Seyfert galaxy NGC5728 (Tsvetanov* et al. *in prep.) showing the highly excited biconical structure, some 20 arcsec in total extent, emerging from a heavily obscured nucleus.*

2 MISDIRECTED PHOTON BEAMS

We have now studied two cases of an interaction between a photon beam and extranuclear material. In the first of these, PKS2152–69 (Tadhunter *et al.* 1987, 1988, di Serego Alighieri *et al.* 1988, Fosbury *et al.* 1990), we see both line fluorescence from highly ionized species (up to Fe^{9+}) and polarized continuum scattering from dust associated with the emission line cloud some 8kpc or so from the nucleus of this FR II radio galaxy. Given the observed luminosities of known blazars, the requirements are not severe on the conditions and geometric configuration of the material illuminated by the beam. In PKS2152–69, only a few thousand solar masses of gas with normal gas/dust ratio are required to scatter the flux we measure.

The second case is the nearby (we assume 3.7Mpc) FR I radio galaxy Centaurus A. The optical filaments along the north-eastern radio lobes have been known for a long time (Blanco *et al.* 1975) and have been studied extensively as part of a complex of features — including X-ray (Feigelson *et al.* 1981), infrared (Joy *et al.* 1991) and radio (Burns *et al.* 1983) jets — associated in some way with the active nucleus. We noticed during investigations of the general phenomenon of extended emission line regions (EELR) around AGN (see Robinson *et al.* 1987) that the emission line spectra of the Cen A filaments fall within the same region of the diagnostic diagrams as other radio galaxy EELR. Since it is widely agreed that the EELR are photoionized by the active nucleus, it is natural to hypothesize that the Cen A filaments are being illuminated by a radiation field which is obscured from our direct view by the dust band and perhaps other absorbing circumnuclear material. This idea was tested by Morganti *et al.* (1991) who carried out detailed spectrophotometric measurements and compared these with multi-cloud photoionization models. By estimating the ionization parameter from these model fits and constraining the electron density from line ratio measurements and filling factor arguments, it was possible to estimate the ionizing photon luminosity viewed by the gas $q \geq 10^{54\pm1}$ photon $s^{-1}sr^{-1}$, with the equality holding if the nucleus and filaments lie in the plane of the sky. This is in excellent agreement with the lower limit of 10^{53} photon $s^{-1}sr^{-1}$ derived from photon counting arguments assuming that the filaments offer a similar spatial cross-section to the nucleus that they do to the earth. That this luminosity cannot be radiated isotropically from the nucleus is clear from far-infrared measurements (Joy *et al.* 1988) which show a dust re-radiation between two and three orders of magnitude less than the absorption of such a source would produce.

To measure the degree of anisotropy, we can compare the luminosity seen by the filaments with the brightness of the nucleus seen from earth. Figure 4 shows a collection of measurements of the nucleus of Cen A at radio, infrared and X-ray frequencies. Two uncertainties complicate the interpretation of the nuclear spectrum: time variability — the nucleus varies by about an order of magnitude in X-rays — and correction for absorption. In the X-ray band we have used power law fits to the high energy measurements from the Einstein and EXOSAT satellites. The J, H and K measurements of the nucleus have been corrected assuming A_V = 27mag (Israel *et al.* 1990). The broken power-law spectrum used for the photoionization modelling of the filaments by Morganti *et al.* (1991) lies between two and three orders of magnitude above these nuclear measurements. Before we draw any conclusions from this, it must be pointed out that the photoionization modelling really only measures the *ionizing* part of the spectrum and is rather insensitive to the level of the X-ray component. Our comparison will be based on the *assumption* that the Cen A spectrum — in the direction of the filaments — is similar to the average radio-selected blazar (Ghisellini *et al.* 1986). With this proviso, we can state with confidence that the Cen A nuclear radiation field is beamed strongly in the direction of the filaments in the north-eastern radio lobe with a luminosity similar to that of BL Lac itself. There is marginal evidence (illustrated by the dashed line in figure 4) that the beam becomes nar-

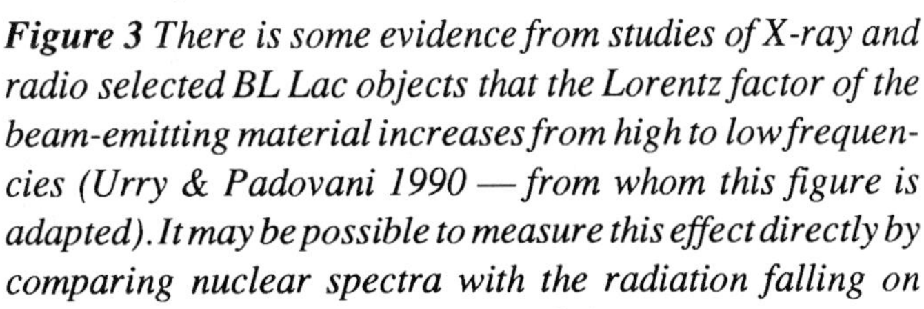

Figure 3 *There is some evidence from studies of X-ray and radio selected BL Lac objects that the Lorentz factor of the beam-emitting material increases from high to low frequencies (Urry & Padovani 1990 — from whom this figure is adapted). It may be possible to measure this effect directly by comparing nuclear spectra with the radiation falling on material in other directions around the source.*

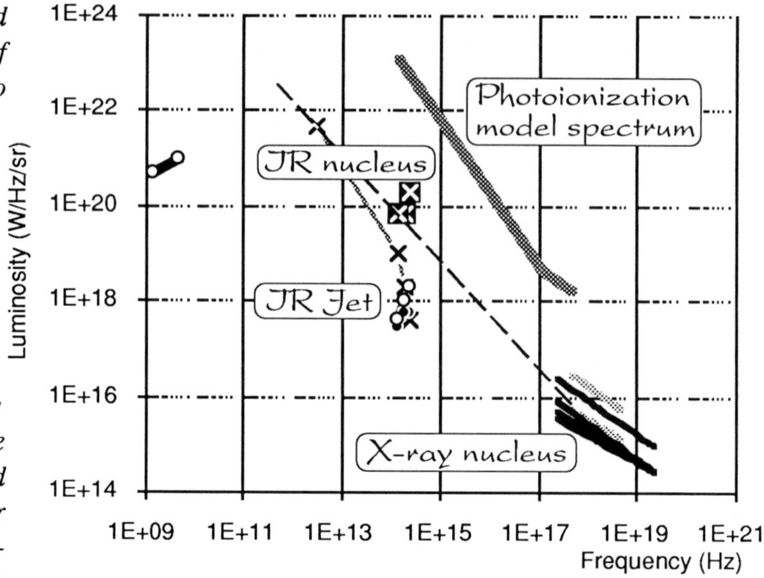

***Figure 4** Measured nuclear luminosity of Centaurus A from radio to X-ray frequencies. The 20 and 6 cm radio points are from Burns* et al. *(1983). The 100μm point is the source ≤ 16arcsec in size reported by Joy* et al. *(1988). The J, H and K measurements marked with a 'x' are the raw and de-reddened (A_V = 27mag) nuclear measurements from Bailey* et al. *(1986). The X-ray spectra are de-absorbed model fits to the multi-epoch Einstein MPC measurements (Feigelson* et al. *1981) – grey – and the EXOSAT measurements (Morini* et al. *1989) – black. We also show, at J, H and K, the infrared 'jet' reported by Joy* et al. *(1991) — marked with '•' (raw) and 'o' (de-reddened A_V = 6mag). The upper solid grey line represents the spectrum derived from the photoionization modelling of the filaments by Morganti* et al. *(1991). The dashed line is discussed in the text.*

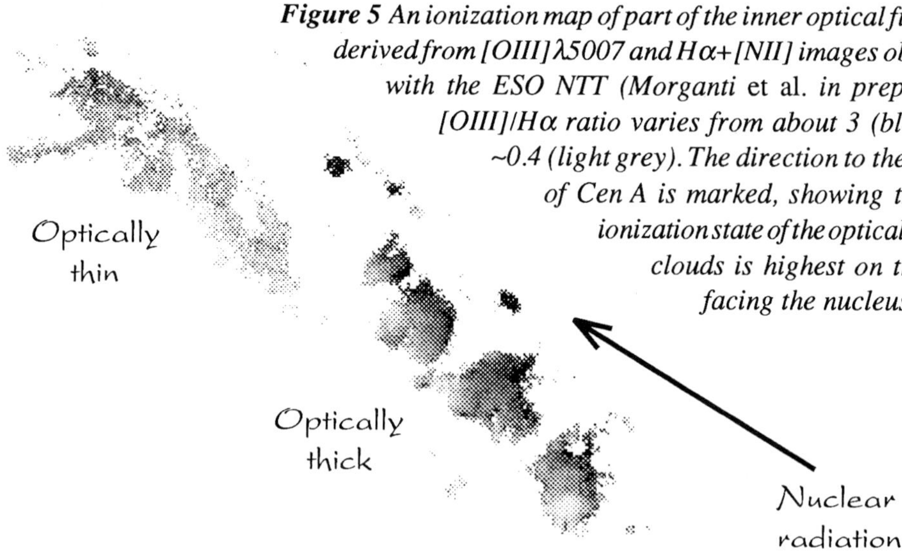

***Figure 5** An ionization map of part of the inner optical filament derived from [OIII]λ5007 and Hα+[NII] images obtained with the ESO NTT (Morganti* et al. *in prep.). The [OIII]/Hα ratio varies from about 3 (black) to ~0.4 (light grey). The direction to the centre of Cen A is marked, showing that the ionization state of the optically thick clouds is highest on the side facing the nucleus.*

rower towards longer wavelengths. This would be consistent with the conclusions of Padovani & Urry (1990, 1991), based on radio and X-ray selected samples of BL Lac objects, that the Lorentz factor γ of the beam-emitting jet is about 3 for the X-ray and about 8 for the radio-emitting component (see figure 3).

A rather direct observational verification of the basic idea that ionizing photons are coming from the direction of the nucleus is provided by the ionization map of the inner filament constructed from the ratio of [OIII]λ5007 and Hα+[NII] continuum-free line images (Morganti *et al.* in prep.) shown in figure 5. Here the [OIII]/Hα ranges from about 3 on the edges of the optically thick clouds facing the nucleus down to ~0.4 on the far side. This gradient is not seen so clearly in the fainter filaments further to the north. This is probably because these features are predominantly optically thin — as indicated by the spectroscopy in Morganti *et al.* (1991) — rather than the structure being unresolved.

In the optical band, the Cen A beam is so far only seen by virtue of its line fluorescence. In the infrared and closer to the nucleus, however, Joy *et al.* (1991) find evidence for a remarkably blue ($f_\nu \sim \nu^{+2.9}$ after correction for the extinction due to the dust band) jet extending some 10arcsec in the direction of the X-ray and radio jets. This is remarkably reminiscent of the blue ($f_\nu \sim \nu^{+3}$), polarized, scattering source in PKS2152–69 and suggests that dust (Rayleigh) scattering of the BL Lac beam could be responsible for this feature. The jet spectum is shown in Figure 4. This may also explain the apparent excess at J in the de-reddened nuclear flux measurements, if there is a scattered, extended component contributing to the Bailey *et al.* (1986) measurements — which do indeed show polarization perpendicular to the jet axis.

It is clear in the case of Cen A and in other objects that the photon-gas interaction is not the whole story. The origin and kinematic evolution of the excited gas very probably owes much to interactions with the particle jet. The chaotic velocity field and linear streamers seen in the Cen A filaments attest to that (see the contribution by Taylor, Morganti & Fosbury in these proceedings).

3 IMPLICATIONS FOR HIGH Z RADIO GALAXIES

The efficiency of dust scattering in the ultraviolet has important implications for the observations of the structure of radio galaxies at redshifts ≥ 1 where the rest-frame ultraviolet is shifted into the optical spectrum. The contrast between a Rayleigh-scattered nuclear continuum against the evolved stellar population of a galaxy is a maximum somewhere shortward of the H & K break and so such sources are the natural place to look for scattering effects. The discovery that high z radio sources are often elongated along their radio axes (McCarthy *et al.* 1987, Chambers *et al.* 1987) begged the question of a beaming/scattering interpretation (Tadhunter *et al.* 1989). Alternative explanations based on the concept of 'jet-induced star formation' have frequently been proposed (e.g., Rees 1989) although these find little support in the more detailed observations of low redshift powerful radio galaxies and quasars. The discovery of perpendicular linear polarization in a few of the high redshift sources (di Serego Alighieri *et al.* 1989, Scarrott *et al.* 1990, Jannuzi & Elston 1991) lends considerable weight to the scattering hypothesis although dust processes alone do not readily explain the extensions which are seen also in the infrared. The critical observations in this field of study are extremely difficult with current 4m class telescopes and we look forward to the capabilities to be offered by the new, much larger photon collecting areas.

4 CONCLUSIONS

Although direct measurements of the small-scale geometric structure of the central engines of active galaxies is severely limited by angular resolution limitations, studies of their effect on the surrounding medium — up to scales of megaparsecs in giant sources — allow us to see particle jets and detect both extrinsic and intrinsic photon anisotropies. The blazar/BL Lac beams in particular carry information in their spectra, polarization and variability about regions very close to the site of jet formation. In the case of Centaurus A we have made a first attempt to measure in some detail the properties of a beam which is not pointed towards us.

References

Antonucci, R.R.J. & Miller, J.S., 1985. *Astrophys. J.*, **297**, 621

Antonucci, R.R.J., 1984. *Astrophys. J.*, **278**, 299

Bailey, J., Sparks, W.B., Hough, J.H. & Axon, D.J., 1986. *Nature*, **322**, 150

Barthel, P., 1989. *Astrophys. J.*, **336**, 606

Blanco, V.M., Graham, J.A., Lasker, B.M. & Osmer, P.S., 1975. *Astrophys. J.*, **198**, L63

Burns, J.O., Feigelson, E.D. & Schreier, E.J., 1983. *Astrophys. J.*, **273**, 128

Chambers, K.C., Miley, G.K. & van Breugel, W., 1987, *Nature, 329*, 604

di Serego Alighieri, S., Binette, L., Courvoisier, T. J-L., Fosbury, R.A.E. & Tadhunter, C.N., 1988. *Nature*, **334**, 591

di Serego Alighieri, S., Fosbury, R.A.E., Quinn, P.J. & Tadhunter, C.N., 1989. *Nature*, **341**, 307

Feigelson, E.D. & Schreier, E.J., Delvaille, J.P., Giacconi, R., Grindlay, J.E.& Lightman, A.P., 1981. *Astrophys. J.*, **251**, 31

Fosbury, R.A.E., di Serego Alighieri, S., Courvoisier, T. J.-L., Snijders, M.A.J., Walsh, J. & Wilson, W.,1990. In: *"Evolution in Astrophysics", IUE astronomy in the era of new space missions,* Toulouse, France, 29 May – 1 June 1990. ESA SP-**310**

Ghisellini, G., Maraschi, L., Tanzi, E.G. & Treves, A., 1986. *Astrophys. J.*, **336**, 681

Israel, F.P., van Dishoek, E.F., Baas, F., Koornneef, J., Black, J.H. & de Graauw, T., 1990. *Astron. Astrophys.*, **227**, 342

Jannuzi, B.T. & Elston, R. 1991. *Astrophys. J.*, **366**, L69

Joy, M., Harvey, P.M., Tollestrup, E.V., Sellgren, K., Mc Gregor, P.J. & Hyland, A.R., 1991. *Astrophys. J.*, **366**, 82

Joy, M., Lester, D.F.,Harvey, P.M. & Ellis, H.B., 1988. *Astrophys. J.*, **326**, 662

McCarthy, P.J., van Breugel, W., Spinrad, H. & Djorgovski, S., 1987. *Astrophys. J.*, **321**, L29

Morganti, R., Robinson, A., Fosbury, R.A.E., di Serego Alighieri, S., Tadhunter, C.N. & Malin, D.F., 1991. *Mon. Not. R. astr. Soc.*, **249**, 91

Morini, M., Anselmo, F., & Molenti, D., 1989. *Astrophys. J.*, **347**, 750

Orr, J.J.L. & Browne, I.W.A., 1982. *Mon. Not. R. astr. Soc.*, **200**, 1067

Padovani, P. & Urry, C.M., 1990. *Astrophys. J.*, **356**, 75

Padovani, P. & Urry, C.M., 1991. *Astrophys. J.*, **368**, 373

Pogge, R.W., 1989. In: *ESO workshop No. 32, "Extranuclear activity in galaxies"*, Garching, May 1989. eds. Meurs, E.J.A. & Fosbury, R.A.E., p411

Rees, M.J., 1989. *Mon. Not. R. astr. Soc.*, **239**, 1p

Robinson, A., Binette, L., Fosbury, R.A.E., & Tadhunter, C.N., 1987. *Mon. Not. R. astr. Soc.*, **227**, 97

Scarrott, S.M., Rolph, C.D. & Tadhunter, C.N., 1990, *Mon. Not. R. astr. Soc.*, **243**, 5p

Tadhunter, C.N., Fosbury, R.A.E. & di Serego Alighieri S., 1989. In: *"BL Lac objects, lecture notes in physics"*. eds. Maraschi, L., Maccacaro, T. & Ulrich, M.-H., p321

Tadhunter, C.N., Fosbury, R.A.E., Binette, L.A., Danziger, I.J. & Robinson, A. 1987. *Nature*, **325**, 504

Tadhunter, C.N., Fosbury, R.A.E., di Serego Alighieri, S., Bland, J., Danziger, I.J., Goss, W.M., McAdam, B. & Snijders, M.A.J., 1988. *Mon. Not. R. astr. Soc.*, **235**, 403

Tsvetanov, Z., Tadhunter, C., Pérez, E. & González-Delgado, R., 1989. In: *"ESO workshop No. 32, "Extranuclear activity in galaxies"*, Garching, May 1989. eds. Meurs, E.J.A. & Fosbury, R.A.E., p19

Urry, C.M. & Padovani, P., 1990. In: *Proc. Georgia State University Conference, "Variability of active galactic nuclei"*, May 1990

Spectropolarimetry of AGN

R. W. Goodrich

California Institute of Technology/Palomar Observatory

1 THE USE OF SPECTROPOLARIMETRY IN AGN RESEARCH

Spectropolarimetry has already shown itself to be a powerful tool in several branches of astrophysics, including research on active galaxies. Recent work on reorganizing the classification scheme of Seyfert galaxies has been motivated to a large extent by spectropolarimetric observations that some Seyfert 2s harbor Seyfert 1 nuclei which are obscured from our direct view, but are visible through scattered (hence polarized) light (Miller and Goodrich 1990). This work has led to so-called "unification schemes," in which it is postulated that there is really only one type of Seyfert galaxy, which can look different depending on our viewing angle of its nucleus.

Another type of unification scenario has been proposed for radio galaxies and quasars. In this hypothesis relativistic Doppler boosting is the anisotropy mechanism, and our classification of what we see depends on how close to the relativistic outflow axis our line of sight is. It is clear that a number of anisotropy mechanisms, such as obscuration and Doppler boosting, are operating, and one of the current applications of spectropolarimetry is to try to disentangle the effects so that we may better understand the truly interesting physics within the central engines.

Another application of spectropolarimetry is to probe the geometry and the kinematics within the central regions of AGN. In a number of Seyfert galaxies "moving mirrors" have been discovered as high-polarization features within the hydrogen (or sometimes [O III]) line profiles (e.g. Goodrich 1989). These mirrors not only allow us a different viewing angle into the nucleus, but can also imprint their own signature onto the line profile. For example, a dust mirror will be at too low a temperature to significantly broaden the reflected line profile, while an electron mirror will usually produce a significant broadening. If the intrinsic profile is known, an estimate of the temperature of the electrons can be made (Miller, Goodrich, and Mathews 1991).

2 OBSERVATIONS

At Palomar Observatory we have built a new spectropolarimeter for the 5-m telescope, which makes it one of the most powerful instruments ever used in this type of work. Additions to the Cassegrain double spectrograph include a superachromatic halfwave plate and a calcite polarizing beamsplitter of novel design. We have initiated a study of a sample of radio galaxies including both narrow- and broad-line radio galaxies (NLRGs and BLRGs). The observational program is in two parts: (1) to search for highly polarized radio galaxies, and (2) to study in detail those high-polarization objects already known or newly found. A third goal for the long term is to study the polarization variability in radio galaxies—this gives us a much better idea of what is causing both spectral variability and polarization in these objects.

3 PRELIMINARY RESULTS

In the highly polarized blazar 3C 371 we find that the narrow emission lines such as Hα, [O I] λλ6300,6363, and [S II] λλ6716,6731 are unpolarized. Figure 1 shows the polarization and integrated-light spectra of 3C 371, demonstrating the drop in *P* that occurs across the emission features. This behaviour is consistent with the expectation that the featureless continuum in 3C 371 is predominantly highly-polarized synchrotron radiation. 3C 371 is of particular interest because it is a relatively low-luminosity and low-ionization example of the blazar class. In fact Miller (1975) has shown that the emission spectrum of 3C 371 closely resembles that of the nearby LINER NGC 1052, indicating that even the lowest luminosity AGN can have highly relativistic particles in their central regions.

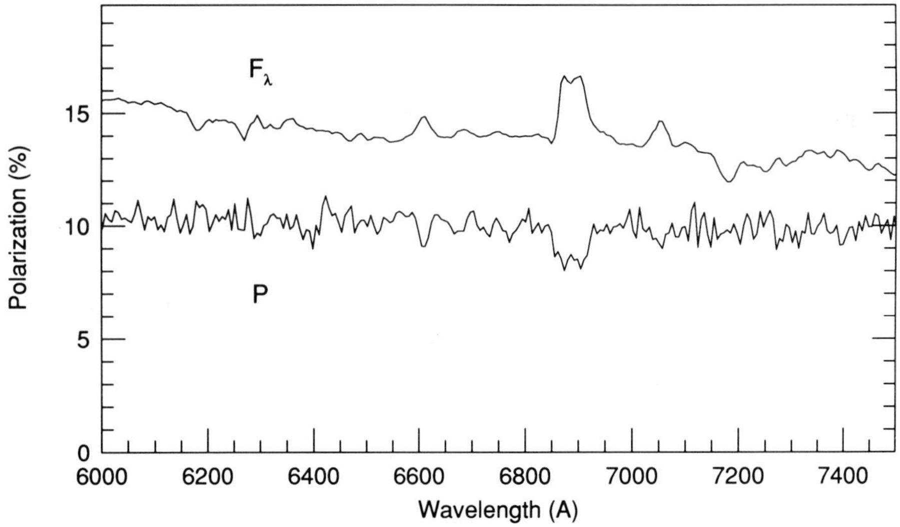

Figure 1. Polarization and total flux of the blazar 3C 371. Note the drop in *P* across the narrow Hα, [O I], and [S II] emission lines.

Another interesting aspect of blazars like 3C 371 is the lack of a visible broad Hα line. According to unification scenarios blazars are seen along a line-of-sight close to the radio jet axis. Presumably this viewing angle also gives us the best view of the broad-line region. Yet in some blazars like 3C 371 broad lines are not seen, even though narrow lines are readily visible. In some AGN, then, there must be little if any gas in the BLR, a situation which was previously thought to characterise Seyfert 2 galaxies and NLRGs. However, some unification scenarios now predict that these latter objects *do* actually contain BLRs, but they are hidden from our direct view.

The broad-line radio galaxy 3C 109 was studied polarimetrically by Rudy et al. (1984). They suggested that both the polarization and the high Balmer decrement of the broad lines were consistent with transmission through dust. Our spectropolarimetry, shown in Figures 2 and 3, confirms this picture. The shape of $P(\lambda)$ is consistent with the shape created by transmission through aligned dust grains in our own Galaxy's interstellar medium. The interstellar polarization along the line-of-sight towards 3C 109 is only ~1% from Galactic dust, so the rest of the polarization must be due to dust within the host galaxy of the AGN.

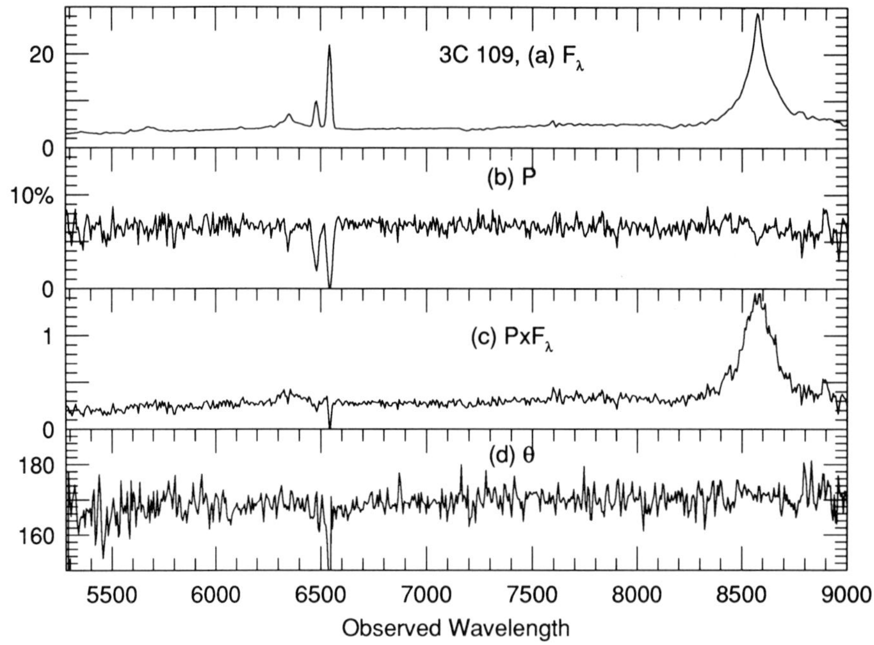

Figure 2. Polarization structure in the BLRG 3C 109. In panel *(b)* note the drop in *P* across the narrow lines of [O III] and Hβ.

After correction for Galactic polarization and reddening, we can calculate the minimum amount of reddening to be expected from the polarizing dust. In our own Galaxy the maximum value of $P/E(B-V)$ is 9%. Since the polarization in 3C 109, corrected for the Galactic component, is ~7%, the implied $E(B-V)$ is 0.78. Dereddening the observed spectrum by this amount, however, produces a very blue continuum; α=1.1±0.3 where $f_\nu \propto \nu^\alpha$. In general, the bluest continua seen in AGN have ν=0.5 (e.g. Barvainis 1990). Goodrich and Cohen (1991) explore in more detail the possible explanations for the discrepancy between the observed polarization and the continuum reddening.

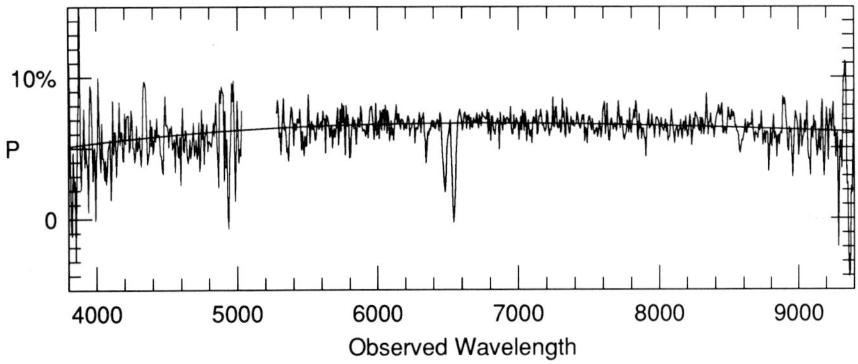

Figure 3. The continuum polarization of 3C 109 is here fit to the "Serkowski law," the $P(\lambda)$ shape caused by transmission through dichroic dust grains in our own Galaxy. Two components have been used: the first due to Galactic polarization from our own interstellar medium, and the second from dust within the host galaxy of 3C 109 itself. The drop in P towards the blue is diagnostic of dust transmission; dust reflection often shows a rise to the blue.

Another very interesting object is 3C 227. This BLRG, like the better-known 3C 390.3, has a very peculiar broad emission profile, which peaks well to the blue of the narrow emission lines. The polarization structure across the Hα line is equally interesting, showing both an increase over the continuum polarization and a rotation of the position angle. Figure 4 displays the Hα spectropolarimetry data.

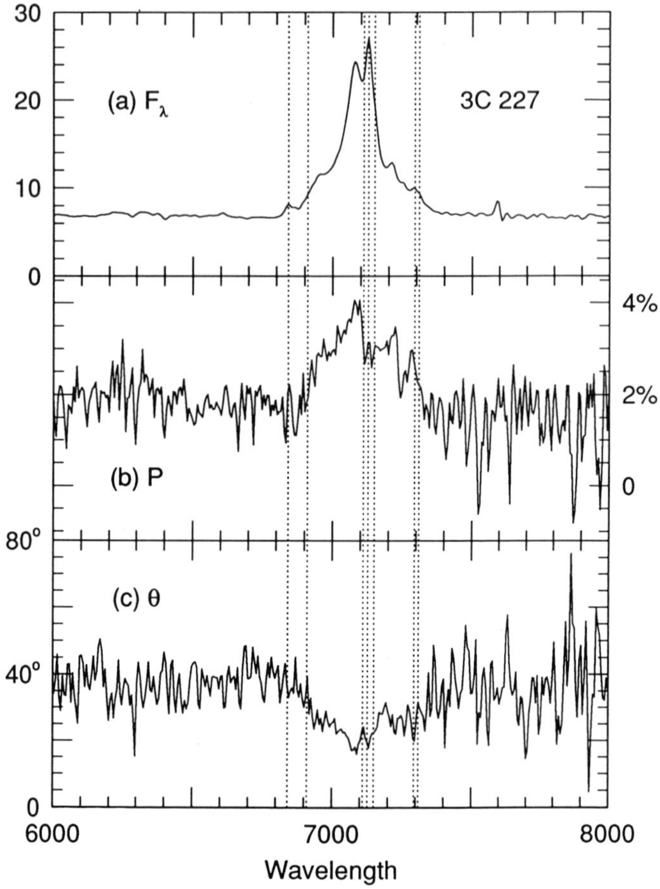

Figure 4. The polarization and p.a. across the broad Hα line in 3C 227. The vertical dotted lines show the positions of narrow lines of [O I] λλ6300,6363, [N II] λ6548, Hα, [N II] λ6583 and [S II] λλ6716, 6731 (left to right). Broad Hα peaks well to the blue of narrow Hα. Note that P increases across the broad line, and θ rotates by ~20°.

The BLRG 3C 445, like 3C 227, has an asymmetric broad Hα emission line, although it is not so dramatic as in 3C 227. Our spectropolarimetry of this object, presented in Figure 5, shows that the polarization rises across the blue wing of Hα. The position angle also rotates through this wing. Such a polarization feature could be caused by an inflowing cloud of electrons or dust. Alternatively, the polarization might be caused by transmission through aligned dust grains. If the BLR gas is outflowing, the dust must be located within the BLR clouds themselves, whereas inflowing gas requires that the dust be situated between the BLR clouds.

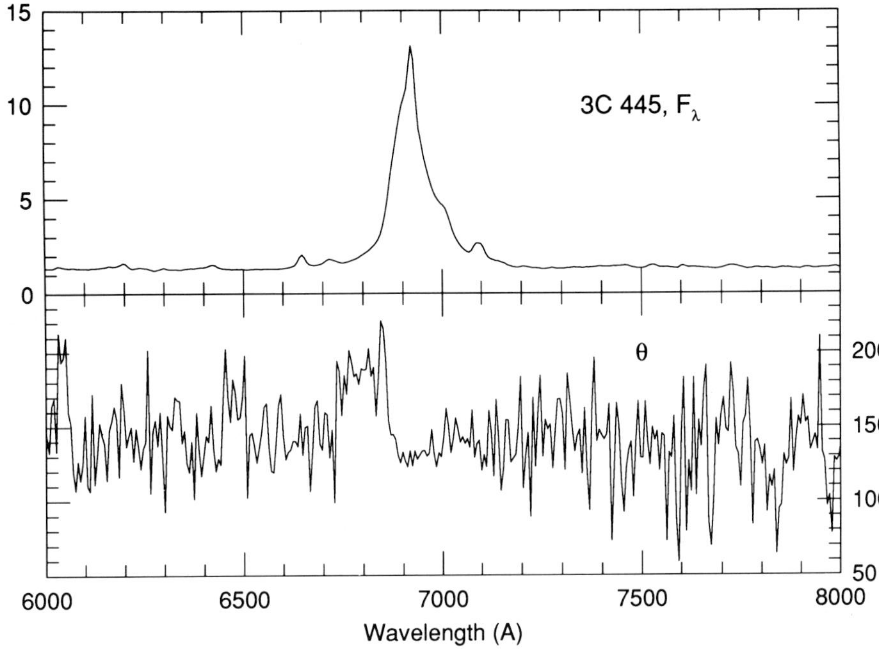

Figure 5. In 3C 445, θ shows a large (~50°) rotation in the far blue wing of Hα.

4 CONCLUSIONS

Spectropolarimetry has proven to be a powerful tool in research on Seyfert galaxies, and we are finding it equally effective in studying radio galaxies. Radio galaxies, perhaps because of the method of their discovery, seem to show more evidence of dust than do the Seyferts, but "complete" samples of the latter are difficult to construct. In several BLRGs polarization structure across the broad emission line profiles should provide a useful diagnostic for the internal kinematics and geometry of the active nucleus. Further study will hopefully allow us to use polarization and spectral variability as yet another diagnostic tool.

5 REFERENCES

Barvainis, R. 1990, ApJ, 353, 419.

Goodrich, R. W. 1989, ApJ, 340, 190.

Goodrich, R. W., and Cohen, M. H. 1991, ApJ, submitted.

Miller, J. S. 1975, ApJ, 200, L55.

Miller, J. S., and Goodrich, R. W. 1990, ApJ, 355, 456.

Miller, J. S., Goodrich, R. W., and Mathews, W. G. 1991, ApJ, 378, 47.

Rudy, R. J., Schmidt, G. D., Stockman, H. S., and Tokunaga, A. T. 1984, ApJ, 278, 530.

AGN Classification and Evolutionary Scheme

B.V. KOMBERG

Astro Space Center of P.N. Lebedev Physical Institute, 53 Leninsky pr., Moscow, 117924, Russia

1 INTRODUCTION

For the past almost 50 years, since the first evidence had been obtained on the unusual properties of nucleus region of some galaxies (Seyfert, Ambarzumian, Shklovsky, Zwicky, Burbidge, Arp, et al.), the variety of the AGN observed has grown up to such an extent that it becomes an actual point to review the AGN classification. Certainly, one should keep in mind that this pile of AGN observational properties could well be ascribed not only to existence of the various AGN - types but be attributed to some other aspects, say as :

- cosmological evolution phase,
- luminosity burst phase,
- radio jet orientation with respect to line of sight,
- radiation diagramm width,
- host galaxy properties,
- host galactic cluster properties.

When considering AGN classification the effects of cosmological evolution should be taken into account. Otherwise, it would be hard to account for the fact that :

- luminosity functions AGN of various types (in optical and radio) can match each other,
- AGN of many types have in average the same slope $L_r \sim L_{opt}^2$ on the $L_r - M_{pg}$ plane,
- radio sources (RS) of various types form the same sequences ("main" and "giant" according to Shklovsky) on the $L_r - l_2$ plane. [1]

2. EVOLUTION OF QSO-TYPE OBJECTS TO THE LESS LUMINOUS TYPES OF AGN

The first (naive) evolutionary schemes (for example [2]) deals mostly with the following questions :

- what is the role of recurrence in the QSO evolution?
- What is the place of QSO in the evolutionary sequence?

The main objection to the "pure recurrence", when AGN are considered as an excited state of a normal nucleus, is the peculiarity of their "host galaxies". For instance, some specific properties of the central parts of the host galaxies can be:
- high luminosity,
- high stellar concentration to the centre,
- low specific angular momentum,
- high dust and gas abundance in central regions of the host galaxies,
- evidence for interaction and/or merging.

QSO may be an initial bright and short ($\leq 10^7$ years) stage of the AGN evolution :
- this well accounts for the low spatial density of the nearby (Z < 0.5) QSO with respect to less active nuclei,
- the sharp growth of bright QSO comoving spatial density with Z - because to Z = 0 the most QSOs became less luminous AGN (LAGN),
- the full AGN life-time (from QSO to LAGN) may be up to billions of years. This agrees with the synchrotron loss time for relativistic particles in weak magnetic fields of extended RG-components (radio galaxy components).

Some facts argue in favour of the short initial AGN stage as QSO :
- At $M_{\rm QSO} < 10^9\ M_\odot$ - energy restrictions.
- Real close QSO - pairs are relatively rare (10^{-3}) with respect to those of bright galaxies ($\tau_{\rm QSO}/\tau_{\rm gal} \approx 10^{-3}$) [3].
- The number of QSO existing within the interval $0 \leq Z < 3$ should not exceed that of bright galaxies [4] :

$$N_{\rm QSO}^{\rm tot} = \int_0^{Z^X} \frac{n_o}{\tau_{\rm QSO}} f(Z) \frac{dt}{dZ} dZ < N_{g\,\rm EG}^{\rm tot}$$

This condition is fulfilled at $n_\circ = 10^{-6}\ Mpc^{-3}$,

$$Z^X = 3,\ \ f(Z) \sim (1+Z)^6,\ \ \tau_{\rm QSO} \leq 10^7 \text{ years}$$

3. NEW ASPECTS

There are many arguments in favour of an AGN luminosity anisotropy (activity cones):
- the existence of radio and optical jets and the linear symmetry of RG structures,
- the alignment of the radiostructure and the extended line regions (ELR) in strong RG,
- the deficit of ionizing radiation with respect to the observed flux in recombination lines,
- the correlation between the line widths in AGN spectra and the dimensions of the

radio structure,
- the wide wings in polarized light of the permitted lines of Sy II (Seyfert II),
- the dependence between the RG nucleus optical variability amplitude and the asymmetry degree of extended RG-components [5].

There is a possible dependence of the radiation anisotropy on the radiation band (e.g. the X-ray width is greater than the radio one). For BL Lac type objects this accounts for the greater spatial density in X-ray than in radio [6].

The radiation anisotropy possibly depends on the thickness (determined by the accretion rate) near the nucleus disk : the greater it is, the higher is the anisotropy. This may well account for the small cosmological evolution of BL Lac type objects (it is large for young distant objects) [6].

4 SOME CONSEQUENCES OF THE HYPOTHESIS ON THE EXISTENCE OF TWO QSO GENERATIONS

The existence of two QSO generations was suggested in [7] :

QSOs II are very active nuclei in giant galaxies, a final product of smaller members merging in clusters and groups at $Z < 3$. Most of the bulk of the presently observable QSOs evidently belong to this type.

QSOs I form an hypothetic population of less active nuclei of small primary galaxies which have not passed the merging stage. These objects may provide a major contribution to the QSO counting at $Z > 4$ and $m \geq 24 - 26$.

QSOs may evolve either into radio galaxies identified with gE-galaxies or into cD-galaxies with weak radio emission. This obviously depends on the fraction of S-galaxies in the overall number of merging components (These are S-galaxies that provide the magnetic fields in "mergers" and the gas, which are necessary to trigger the radio emission). Thus, nuclei of the radio loud QSSII-type may be formed in the case of multiple merging S-galaxies, and nuclei of the radio quiet QSG II-type may be formed in the case of multiple merging E or SO-galaxies. Nuclei of Sy G, within these scheme, may be formed only when not too many SG are merging, namely, when their number is insufficient either for the global isotropization of the resulting system or for a RG to appear.

QSSII formation associated with the merging of galaxies, is accompanied by a growing rate of star formation in the system and by an intense emission between 30 and 100 microns of the dust heated to T = 90 - 30 K (sometimes the intensity of the emission of QSS II in the IR range reaches several times 10^{12} $L_{\odot}$). Thus, it is not impossible that the strong background emission at few hundreds of microns (region of the cosmic background radiation) may be due to the epoch of maximal rate of merging process and/or QSO II formation. In that case, it may reflect processes that occur at Z = 2.5 - 3.5 [8].

Hypothetic QSO I (which are less active nuclei in single small primary galaxies, by definition) are probably not strong sources either in radio or in IR and are not frequent in regions of higher spatial density of galaxies. Evidently, their descendants, far advanced in evolution, should be looked for in large-scale voids where the density of galaxies is minimal. Owing to the star formation rate decrease during the collisionless evolution, these assumed to be descendants of QSO I should preserve gas enough for active star formation in the present epoch, consequently, they should be blue. Search for such objects in voids may give better insight into the properties of primary galaxies and their evolution advantages.

REFERENCES

[1] I.S. Shklovsky, (1962), Astron. J., (USSR), **39**, 591

[2] B.V. Komberg, L.M. Ozernoy, (1970), Astrophys. and Sp. Sc. **6**, 450
B.V. Komberg, (1983), Sov. Astron., **26**(6), 643

[3] B.V. Komberg, (1989), Astrofizika (USSR) **30**(2), 399

[4] B.V. Komberg, R.A. Syunyaev, (1971), Sov. Astron. **15**, 183

[5] B.V. Komberg, (1975), Preprint N°252, Space Res. Inst. (Moscow)

[6] B.V. Komberg, (1991), Proceedings of Symposium on *Variability of Blazars*, Turku, Finland

[7] B.V. Komberg, (1984), Astrofizika (USSR), **20**, 73
B.V. Komberg, (1989), Proceedings Special Astrophys. Obs. **61**, 134

[8] V.V. Burdyuzha, B.V. Komberg, (1990), Astron. Astrophys. **234**, 40

What Does it Take to Make a Radio Galaxy ?

P.A.G. SCHEUER

Mullard Radio Astronomy Observatory, Cavendish Laboratory, U.K.

1 THE QUESTION

The question I want to discuss is: *What is the difference between elliptical and spiral galaxies that is relevant to making active nuclei produce radio jets always in one, never (at any rate, never with much power) in the other?* For the moment I shall include the host galaxies of radio quasars among the ellipticals and of radio-quiet QSOs among spirals, though that is likely to be a very rough description at best.

I would begin with three assertions (which are largely implied in the statement of the question)

1. AGN occur in both E and S galaxies.
2. AGN in S galaxies do not produce powerful radio galaxies.
3. AGN in E galaxies always produce powerful radio galaxies.

Everyone in the field is conscious of "facts" 1 and 2, and also knows that "fact" 2 is not quite true; certainly there are Seyferts with significant radio emission. I am not sure that "fact" 3 is as widely appreciated, but it may well be just as significant that an E galaxy is a sufficient condition for an AGN to make a radio galaxy as that it is a necessary condition.

2 THE CONSTRAINTS

Let us begin with a list of the principal differences between spiral and elliptical galaxies:

(i) Shape. It is plausible to interpret this as a difference in

(ii) specific angular momentum. It is harder to decide by observation whether individual stellar orbits in elliptical galaxies have low angular momentum (and therefore dive closer to the nucleus) or whether stars are in rather round orbits, i.e. orbits in very varied orientations.

(iii) Much less interstellar gas and

(iv) much less recent star formation in ellipticals than in spirals

(v) Typically, ellipticals are in dense clusters or groups, while, typically, spirals are associated with much less densely populated regions or the outskirts of dense clusters.

A difference between E and S galaxies must fulfil two important conditions if it is to be relevant to our question:

(i) It must not interfere with the formation of AGN in spirals. That seems to indicate a process acting further out than the AGN. Thus, for example, the fuel supply rate seems unlikely to be the key, since AGN of the whole range of total luminosities occur in both radio-loud and radio-quiet objects. If we believe that the luminosity is even very roughly related to the Eddington luminosity, the same comment applies to the mass of the black hole.

(ii) On the other hand, it is quite clear that the crucial difference must occur or act on scales much smaller than 1 parsec, since radio jets are observed down to $\ll 1$ parsec. Thus, for example, one cannot directly blame the cluster environment of most ellipticals for the difference. Simple theories of "stifling the radio source with the denser interstellar medium in spirals" also fail this condition. One would need to postulate a very much denser environment on the sub-parsec scale in spirals (and then the broadly similar spectroscopy of AGN in spirals and many ellipticals might become puzzling).

The two conditions seem to contradict each other. I conclude that we have to look for something that changes the nature of the AGN in some qualitative but fairly subtle way, without grossly changing its luminosity. There is a very limited range of inputs that determine the nature of an AGN: the fuel supply, the mass of the black hole, and its rotation. We have already noted that in some sense the fuel supply rate is similar for AGN in radio-loud and radio-quiet objects since both cover the whole range of luminosities; however, its composition might affect the way accretion works (e.g. through cooling rates), and the supply might be continuous or episodic. In an extreme form of the latter, the central black hole could be spun up by the arrival of a very large amount of matter, after which the power of the AGN is supplied largely by the rotational energy of the black hole, as in the model introduced by Rees, Begelman, Blandford and Phinney (1982). Regarding the mass of the black hole, the Eddington limit indicates that masses up to 10^8 $M_\odot$ occur in some radio-quiet AGN; somewhat more massive black holes could occur in radio galaxies, but masses in excess of 10^{10} $M_\odot$ seem to be ruled out on dynamical grounds, at any rate in the nearest giant ellipticals. Here I want to draw attention to another distinction between AGN in spirals and ellipticals, largely because it fulfils both conditions (i) and (ii) above and it looks like an inevitable consequence of mergers.

3 THE CONSEQUENCES OF CANNIBALISM

It is now widely believed that elliptical galaxies are the results of mergers, and massive ellipticals may be the result of a whole sequence of mergers. One of the surprising things about mergers is how quickly the stars seem to relax to a rounded configuration and how quickly the gas and also any supermassive object can settle into the centre. Begelman, Blandford and Rees (1980) discussed the fate of central black holes in mergers; they were primarily interested in the formation of binary black holes, one from each of the merging galaxies, but many of their arguments are relevant here. More recently, numerical simulations of mergers have shown that much of the gas can accumulate near the centre; thus Barnes and Hernquist (1991) find that in a merger of two equal disk galaxies, all the gas is within 200 pc of the centre after just a couple of rotation periods of the merged galaxy. Nature seems to be less tidy, as many ellipticals (and radio galaxies in particular) show dust features and/or extended narrow-line emission; nevertheless, it seems quite likely that a new fuel supply into the AGN can be established soon after a merger.

The point which now arises is that the newly installed fuel supply (accretion disk) has angular momentum in a direction unrelated to the spin of the black hole that it feeds. The black hole will retain its original spin rate and direction, since its spin interacts negligibly with the large-scale matter distribution around it. (In a well-behaved disk galaxy, on the other hand, we expect the black hole to have the same spin direction as the accretion disk that has built it up over the ages.)

Although the spin of the black hole does not interact with the surrounding galaxy as a whole, it will interact significantly with the accretion disk in its immediate neighbourhood. There are two forms of interaction: classical precession due to the gravitational quadrupole moment of the spinning black hole, and the Lense-Thirring precession. The two are comparable one Schwarzschild radius r_S from the black hole, but at larger radii r the Lense-Thirring precession wins by a factor $\sqrt{(r/r_S)}$; therefore we need only consider the Lense-Thirring precession.

The action of Lense-Thirring precession in accretion disks was discussed in a short and famous paper by Bardeen and Petterson (1975). A ring of matter at radius r has a precession period of the order of $(r_S/c)(r/r_S)^3$. Differential precession at small r will be damped out by viscous processes, so that, very close to the black hole, the accretion disk becomes flat and normal to the spin axis of the black hole. Meanwhile, the viscous processes (described by the parameter α in standard models) feed material inwards towards the black hole, and significant precession will only occur out to the radius where the inflow time equals the precession period. To guide further thought, Table 1 lists some relevant time scales

TABLE 1

r	r/r_S	τ_{circ}	τ_{LT}
3×10^{12} m	10	3 days	2 months
3×10^{13} m	100	4 months	200 years
3×10^{14} m	1000	9 years	2×10^5 years
3×10^{15} m	10000	300 years	2×10^8 years

Keplerian orbital periods τ_{circ} and Lense-Thirring precession times τ_{LT} for a maximally rotating 10^8 $M_\odot$ black hole.

It is quite clear that the warped transition region must be within 10^{15} m; conventional models would put it at 10^{13} to 10^{14} m. In any event, it is well inside the smallest radius at which radio jets are observed using VLBI. (In fact, that conclusion is so robust that it would be true even if there were only classical precession).

Before going on, we should pause to consider the precession of the black hole due to the presence of the accretion disk and the operation of Newton's third law. The angular momentum of a maximally rotating black hole is $(1/2)Mr_Sc$, as compared with the angular momentum $mc\sqrt{(rr_S)}$ of mass m in circular orbit at radius r, and as the mass m of the disk up to the radius of the warped region will be several orders of magnitude smaller than the mass M of the black hole, the reaction on the black hole causes negligible precession. Furthermore, the mass accreted onto the black hole will take 10^8 years to align the black hole more or less well with the new accretion disk

Thus we have a difference between AGN in elliptical and spiral galaxies which is qualitative, which should appear in nearly all cases, which does not interfere with the production of an AGN of high luminosity, and which acts on a scale $\ll 1$ parsec for any plausible set of parameters.

That is all I would really claim. If you ask me for a *mechanism* by which this difference acts to favour the diversion of energy into the jets characteristic of radio sources, then I have no good answer. It is true (as Bardeen and Petterson already pointed out) that the warped region of the disk intercepts a substantial fraction of the total luminosity of the quasar, a power much greater than that generated by dissipation of gravitational energy within that region. However, the warped region is either transparent and so absorbs little, or is opaque and easily re-radiates the incident radiation without any spectacular rise in temperature; unless someone can think of a really powerful greenhouse effect for accretion disks we cannot expect large effects on the disk. We would perhaps like the crucial difference to encourage the growth of a magnetised corona over the disk, for the twisting of magnetic fields is the best suggestion so far for making the jets, but here we enter a far more complex situation which is perhaps not well understood and which I certainly do not understand.

Finally, I want to note a rather different consequence of the picture I have suggested. If my story about the consequences of mergers is broadly true, then radio galaxies contain black holes grossly misaligned with the outer parts of the accretion disks that feed them - regardless of whether that is crucial for making them into radio galaxies. Some of the nearer radio galaxies have quite well-defined dust lanes, and it is well known that the radio axes are generally within 30° of the normals to these dust lanes (Kotanyi and Ekers, 1979). Thus we are faced with one or both of two conclusions:

(i) The direction of the radio jets is not defined by the innermost parts of the AGN, but by the material and/or magnetic structure at radii around 10^{14} to 10^{15} m;

(ii) The observations of dust lanes are restricted to nearby ellipticals containing ancient and feeble radio nuclei in which the black hole has been partially re-aligned with the angular momentum of the fuel supply.

REFERENCES

Bardeen, J.M. and Petterson, J.A., 1975. Astrophys. J. **195**, L65
Barnes, J.E. and Hernquist, L.E., 1991. Astrophys. J. **370**, L65
Begelman, M.C., Blandford, R.D. and Rees, M.J., 1980. Nature **287**, 307
Kotanyi, C.G. and Ekers, R.D., 1979. Astron. Astrophys. **73**, L1
Rees, M.J., Begelman, M.C., Blandford, R.D. and Phinney, E.S., 1982. Nature **295**, 17